Disaster Risk Reduction

Methods, Approaches and Practices

Series Editor

Rajib Shaw, Keio University, Shonan Fujisawa Campus, Fujisawa, Japan

Disaster risk reduction is a process that leads to the safety of communities and nations. After the 2005 World Conference on Disaster Reduction, held in Kobe, Japan, the Hyogo Framework for Action (HFA) was adopted as a framework for risk reduction. The academic research and higher education in disaster risk reduction has made, and continues to make, a gradual shift from pure basic research to applied, implementation-oriented research. More emphasis is being given to multi-stakeholder collaboration and multi-disciplinary research. Emerging university networks in Asia, Europe, Africa, and the Americas have urged process-oriented research in the disaster risk reduction field. With this in mind, this new series will promote the output of action research on disaster risk reduction, which will be useful for a wide range of stakeholders including academicians, professionals, practitioners, and students and researchers in related fields. The series will focus on emerging needs in the risk reduction field, starting from climate change adaptation, urban ecosystem, coastal risk reduction, education for sustainable development, community-based practices, risk communication, and human security, among other areas. Through academic review, this series will encourage young researchers and practitioners to analyze field practices and link them to theory and policies with logic, data, and evidence. In this way, the series will emphasize evidence-based risk reduction methods, approaches, and practices.

All submitted proposals are carefully reviewed by the series editor.

Publication Ethics and Publication Malpractice Statement: https://www.springernature.com/gp/policies/book-publishing-policies

Ethical, Conflict-of-Interest, and Editorial Policy Statement: https://www.springer.com/us/authors-editors/journal-author/journal-author-helpdesk/before-you-start/before-you-start/1330#c14214

Mohammed Firoz C · Lalit Kumar Dashora ·
Rajib Shaw
Editors

Nature-Based Solutions for Urban and Peri-Urban Areas

For Resilient and Sustainable Urbanization

Springer

Editors
Mohammed Firoz C
Department of Architecture and Planning, Centre for Climate Resilience and Disaster Management
National Institute of Technology Calicut
Kozhikode, Kerala, India

Lalit Kumar Dashora
School of Global Studies
Thammasat University
Bangkok, Thailand

Rajib Shaw
Graduate School of Media and Governance
Keio University
Fujisawa, Kanagawa, Japan

ISSN 2196-4106 ISSN 2196-4114 (electronic)
Disaster Risk Reduction
ISBN 978-981-96-5932-6 ISBN 978-981-96-5933-3 (eBook)
https://doi.org/10.1007/978-981-96-5933-3

This Springer imprint is published by the registered company Springer Nature Singapore Pte Ltd.
The registered company address is: 152 Beach Road, #21-01/04 Gateway East, Singapore 189721, Singapore

Preface

Increase in urbanization, often unaddressed, especially in the developing world leads to multifaceted problems leading to climate change and disaster risks. Such urbanizations contribute to sprawl imparting extreme pressure on the peri-urban areas which results in environmental loss and degradation. Unfortunately, the urban planning process and procedures in practice has not been fruitful in most of the world in addressing the aforementioned climate risks and disasters. There exists the need for proper building codes and bylaws and zonal regulation from a regional to a local and even to building level. Most of the existing infrastructure, which are built or yet to be built need to be resilient to address these risks. The impact of such risks will be severe on the vulnerable populations, thus widening the issues of social inequality leading to exclusions of selected communities.

In order to address such issues, Nature-based solutions (NbS) are deemed to be the most appropriate and sustainable approach models using natural ecosystems. The key features of the NbS include restoration of the ecosystems (wetlands, mangroves, greenspaces, etc.); improving the resilience of the urban areas to address biodiversity loss, problems related to air and water quality, sea-level rise, urban heat islands, flood and droughts etc. There exist multiple levels of solutions, namely restoration of depleted wetlands, mangroves, improving urban green cover, afforestation, controlling soil erosion, etc.

However, the implementation of the NbS in urban areas is always affected by a myriad of issues with respect to spatial planning and zoning, impact on the existing populations and neighborhoods, reworking on the existing urban infrastructures and the associated issues with respect to finance, governance, etc. thus necessitating the need for policy changes.

Through an interdisciplinary lens, this book therefore intends to address the environmental issues of urbanization using nature-based solutions as a saviour. Accordingly, empirical, theoretical and case-study-based research across the globe are structured as the contents of the book. Apart from the introduction, the book has four sections: (1) Spatial Planning perspectives for Urban and Peri-Urban Areas, (2)

Urban Infrastructure Solutions and Development Strategies, (3) Traditional Methods and Community-Driven Approaches and 4) Governance and Policies. It can be assumed that the book shall address the needs of a wide range of policymakers, researchers, and practitioners, students and academia from diverse disciplines.

Kozhikode, India — Mohammed Firoz C
Bangkok, Thailand — Lalit Kumar Dashora
Fujisawa, Japan — Rajib Shaw

About This Book

Increased urbanization and unplanned development lead to sprawl imparting extreme pressure on the urban and peri-urban areas, which results in the loss of ecosystem, biodiversity, green spaces, forest cover, wetlands, rivers, etc. The resultant effect is the disaster and climate risks. In order to address this, Nature-based solutions (NbS) are deemed to be the most appropriate and sustainable approach. Implementation of the NbS in urban areas is always affected by a myriad of issues with respect to spatial planning and zoning, impact on the existing populations and neighborhoods, reworking on the existing urban infrastructures, the associated issues with respect to finance, governance, etc. thus necessitating the need for policy changes.

Through an interdisciplinary lens, this book intends to address the environmental issues of Urbanization using NbS as a savior. Accordingly, empirical, theoretical and case study-based research across the globe are structured as the contents of the book. Apart from the introduction, the book has four parts: (1) Spatial Planning Perspectives for Urban and Peri-Urban Areas, (2) Urban Infrastructure Solutions and Development Strategies, (3) Traditional Methods and Community-Driven Approaches and (4) Governance and Policies. It can be assumed that the book shall address the needs of a wide range of policymakers, researchers, practitioners, students and academia from diverse disciplines.

Contents

Editors and Contributors

About the Editors

Mohammed Firoz C is currently the Chairperson of the Centre for Climate Resiliance and Disaster Management and former Head of the Department of Architecture and Planning, at the National Institute of Technology (NIT) Calicut, India. Besides NIT Calicut, he has teaching experience as a visiting faculty at the Keio University, Japan, AIT Bangkok and at the Architectural Association London. Dr. Firoz graduated from NIT Calicut as a university rank holder, completed his post-graduation in Urban and Regional Planning from CEPT University, a Postgraduate Diploma in Urban Environment Management and Law from the National Law University, Delhi, India, and Ph.D. from the Indian Institute of Technology (IIT) Kharagpur. He is actively involved in funded research projects, publishing as well as consultancy and university services. He has written 50 papers in indexed journals and book chapters. He has also guided 5 PhDs and is currently guiding 11 more. He was also on the advisory panel of more than 21 Ph.D. theses as its chairperson/ member, principal guide for 30 master's theses and 70 undergraduate theses. He is associated with the editorial board of three reputed journals besides being a reviewer for several journal papers.

Lalit Kumar Dashora is currently associated with the School of Global Studies, Research Center for Sustainable Development and Innovation, Thammasat University, Bangkok, Thailand. He is a specialist in Urban Risk and Resilience. Formerly, he worked as a Senior Technical Specialist at the Asian Disaster Preparedness Center (ADPC) in Bangkok. Prior to joining ADPC, he worked with the Asian Development Bank (ADB) in Dushanbe (Tajikistan), and the Government of Gujarat (India). He holds a Master's degree in Geo-Information Science for Earth Observation, Environmental Modelling and Management (ITC, The Netherlands), and Urban and Regional Planning (School of Planning, CEPT University, Ahmedabad). Risk-informed, climate-compatible planning and development is at the forefront of his experience. Over the past 18+ years, he has been actively engaged in various high-impact and strategic programmes and projects, such as for the WMO-supported Climate Risk and Early Warning Systems (CREWS) in Southeast Asia and Pacific; for DFID (with UK Met Office) supported Evidence Review: Options Analysis for a Regional Climate Change Programme to Deliver More Effective Climate Services, Early Warning and Disaster Risk Reduction; for Norad supported Urban Resilience to Climate Extreme in Southeast Asia quote a few notable ones.

Rajib Shaw is a Professor at the Graduate School of Media and Governance of Keio University, Japan. He is the Co-founder of a Delhi (India) based social entrepreneur startup Resilience Innovation Knowledge Academy (RIKA), and RIKA Institute. He is the Co-chair of the United Nations Asia Pacific Science Technology Advisory Group (AP-STAG) and was CLA for IPCC's 6th Assessment Report. Professor Shaw is the recipient of the prestigious "Pravasi Bharatiya Samman Award (PBSA)" in 2021 for his contribution to the education sector. He also received the United Nations Sasakawa Award in 2022 for his lifetime contributions to the field of disaster risk reduction. He published 79 books and over 450 research papers. More about his work can be found at: www.rajibshaw.org.

Contributors

Nazish Abid University of Bahrain, Zallaq, Bahrain

Vinayak S. Adane Department of Architecture and Planning, Visvesvaraya National Institute of Technology, Nagpur, India

Divya S. Agarwal Assistant Professor (temporary), Department of Architecture and Planning, MANIT, Bhopal, M.P., India

Casper Boongaling Agaton Department of Community and Environmental Resource Planning, College of Human Ecology, University of the Philippines Los Baños, Laguna, Philippines

Sudha Arlikatti Amrita School for Sustainable Futures, Amrita Vishwa Vidyapeetham, Amritapuri, India

Sunny Bansal Anant National University, Ahmedabad, India

Vidhu Bansal Anant National University, Ahmedabad, India

Dipanksha S. Bawankar School of Planning and Architecture, New Delhi, India

Alka Bharat Professor, Department of Architecture and Planning, MANIT, Bhopal, M.P., India

Antonija Bogadi SYNYO GmbH, Vienna, Austria

Dominic Boyer Center for Coastal Futures and Adaptive Resilience, Rice University, Houston, USA

Marta Brković Dodig Faculty for Media and Communications, Singidunum University, Belgrade, Serbia

David Brown School of Engineering & Aviation, Central Queensland University, Rockhampton, Australia

Ombretta Caldarice Politecnico di Torino, Turin, Italy

Ranit Chatterjee Resilience Innovation Knowledge Academy (RIKA) India, New Delhi, India;
Keio University, Tokyo, Japan

Bidisha Chattopadhyay School of Planning and Architecture, New Delhi, India

Souresh Cornet Amrita School for Sustainable Futures, Amrita Vishwa Vidyapeetham, Amritapuri, India

Ambika Dabral Resilience Innovation Knowledge Academy (RIKA) India, New Delhi, India

Lalit Kumar Dashora Thammasat University, Bangkok, Thailand

Meenakshi Dhote School of Planning and Architecture, New Delhi, India

Aishwarya Dwivedi Maulana Azad National Institute of Technology, Bhopal, India

Jessica Eisma Department of Civil Engineering, The University of Texas at Arlington, Arlington, USA

Hanliang Fu School of Management, Xi'an University of Architecture and Technology, Xi'an, China

Sheikh Omar Fye Amrita School for Sustainable Futures, Amrita Vishwa Vidyapeetham, Amritapuri, India

Namrata T. Gaurkhede Department of Architecture and Planning, Visvesvaraya National Institute of Technology, Nagpur, India

Krishnakali Ghosh Resilience Innovation Knowledge Academy (RIKA) India, New Delhi, India

Shufen Guo School of Public Administration, Shanxi University of Finance and Economics, Taiyuan, China

Mazharul Haque National Institute of Technology, Patna, India

Neetu Kapoor Department of Architecture, National Institute of Technology, Hamirpur, Himachal Pradesh, India

Ganesh Keremane UniSA Business, University of South Australia, Adelaide, SA, Australia

Claire Connolly Knox School of Public Administration, University of Central Florida, Orlando, USA

Charleton Dunn L. Llosa Department of Community and Environmental Resource Planning, College of Human Ecology, University of the Philippines Los Baños, Laguna, Philippines

Angela Million Technische Universität Berlin, Berlin, Germany

Mohammed Firoz C Department of Architecture and Planning, National Institute of Technology Calicut, Kozhikode, India

Krishnadas Sasidharan Nair Department of Architecture and Planning, National Institute of Technology Calicut, Kozhikode, India

Bruna Pincegher Politecnico di Torino, Turin, Italy

Maria Pizzorni University of Southern Denmark, Odense, Denmark

Mina Radić Sibinović Technische Universität Berlin, Berlin, Germany

Edgar M. Reyes Jr. Department of Community and Environmental Resource Planning, College of Human Ecology, University of the Philippines Los Baños, Laguna, Philippines

Nabanita Saha Department of Planning and Architecture, National Institute of Technology Rourkela, Rourkela, India

Somi Sareen National Institute of Technology, Patna, India

Basudatta Sarkar Department of Planning and Architecture, National Institute of Technology Rourkela, Rourkela, India

Mohammad Newaz Sharif School of Public Administration, University of Central Florida, Orlando, USA

Rajib Shaw Resilience Innovation Knowledge Academy (RIKA) India, New Delhi, India;
Graduate School of Media and Governance, Keio University, Tokyo, Japan

Rubika Singh Department of Architecture, National Institute of Technology, Hamirpur, Himachal Pradesh, India

Rajat Soni Maulana Azad National Institute of Technology, Bhopal, India

Nicola Tollin University of Southern Denmark, Odense, Denmark

Zhifang Wu UniSA Business, University of South Australia, Adelaide, SA, Australia

Chapter 1
Nature-Based Solutions for Urban and Peri-Urban Areas for a Sustainable and Resilient Future: An Introduction

Krishnadas Sasidharan Nair, Mohammed Firoz C, Lalit Kumar Dashora, and Rajib Shaw

Abstract Accelerating urbanization presents considerable issues, including urban sprawl, loss of habitat, and increased vulnerability to climate change and disasters, in the urban and peri-urban regions. The aforementioned challenges disproportionately impact the marginalized populace, exposing social inequalities. Strategic urban planning, integrated with Nature-based Solutions (NbS), can become essential to mitigate such risks. NbS, generally utilize natural ecosystems to enhance climate resilience, biodiversity, and human well-being, and thereby have emerged as a viable strategy for addressing urban environmental issues. These solutions can alleviate climate challenges, facilitate ecosystem restoration, and improve urban livability for the inhabitants by offering multifunctional benefits. Nonetheless, the execution of strategies encounters challenges like inadequate funds, governance issues, spatial limitations, and the potential for gentrification. Overcoming these obstacles necessitates participatory governance, creative and efficient financial mechanisms, and a transition to adaptable urban planning that incorporates both green and the existing grey infrastructure in a given region. NbS can markedly improve urban resilience and sustainability when effectively integrated into urban planning frameworks, guaranteeing equitable benefit distribution and fostering social equity while meeting the requirements of vulnerable populations.

Keywords Nature-based Solution (NbS) · Urban resilience · Urban planning

K. S. Nair (✉) · M. Firoz C
Department of Architecture and Planning, Centre for Climate Resilience and Disaster Management, National Institute of Technology Calicut, Kozhikode, India
e-mail: ar.krishnadas.sn@gmail.com

L. K. Dashora
Thammasat University, Bangkok, Thailand

R. Shaw
Graduate School of Media and Governance, Keio University, Fujisawa, Japan

M. Firoz C et al. (eds.), *Nature-Based Solutions for Urban and Peri-Urban Areas*, Disaster Risk Reduction, https://doi.org/10.1007/978-981-96-5933-3_1

1.1 Introduction

With an estimated 68% of the world's population residing in urban regions by 2050, the fast-paced urbanization over the past few decades has drastically altered the global landscape (World Urbanization Prospects 2019). According to Seto et al. (2012) this unprecedented urban growth is accompanied by major obstacles. These challenges include urban sprawl, increment in population, and drastic changes in land-use and land-cover (LULC), all of which expose the risks associated with climate change and natural disasters. Urban sprawl regularly impacts the peri-urban areas, which disrupts the natural ecosystems of the region due to unplanned growth and intensifies vulnerabilities to flooding, heatwaves, and other extreme weather events, furthermore, resulting in habitat loss and fragmentation. This not only threatens biodiversity but also diminishes ecosystem services essential for human well-being, such as clean air and water (Das et al. 2024). The impact of urbanization on ecosystem services is a field of continuing study, and the association with NbS has garnered increasing significance over time. According to Pelling et al. (2015), these kinds of obstacles also have a disproportionate impact on disadvantaged populations, which not only exacerbates existing social inequities but also prevents marginalized people from accessing crucial resources and opportunities.

Henceforth, it is imperative that strategic urban planning be implemented in order to achieve sustainable and equitable urban expansion since the interaction between urbanization and climate hazards highlights the urgency of the situation. It is possible for effective urban planning to address the underlying causes of these problems by incorporating land-use management, the creation of resilient infrastructure, and policy frameworks that are inclusive (Gencer 2013). Urban planning needs to incorporate innovative techniques to create resilience in the context of disaster and climate threats. Some examples of these strategies include risk-sensitive land-use planning, green infrastructure, and community-based disaster preparedness programs (Davoudi et al. 2013). NbS are particularly promising because they offer multifunctional benefits by harnessing natural ecosystems to mitigate climate threats, promote biodiversity, and improve human well-being. These benefits can be achieved by leveraging natural infrastructure. According to Raymond et al. (2017) and Bush and Doyon (2019), NbS have the potential to alleviate socioeconomic inequities and promote inclusive urban development by focusing on the needs of disadvantaged populations. Henceforth, for the purpose of building a sustainable and resilient urban future, the incorporation of NbS into urban planning methods can be a crucial step that shall be taken as urban areas continue to expand. Thus, integrating NbS not only mitigates immediate climate and disaster threats but also enhances long-term resilience by strengthening natural and social systems. As discussed, NbS can bolster urban resilience while promoting biodiversity, environmental services, and community welfare. Further sections will explore NbS, including its defining attributes, multifunctional advantages, and compatibility with urban resilience and sustainability objectives. The discussion will also address the challenges and obstacles to implementing NbS in urban and peri-urban areas, highlighting the necessity of inclusive governance, strong regulatory frameworks, and innovative finance mechanisms to surmount these difficulties.

1.2 Nature-Based Solutions: Concept, Evolution, and Applications

NbS constitute a transformative approach to addressing global environmental issues by leveraging the capabilities of natural systems. According to the International Union for Conservation of Nature (IUCN), NbS are initiatives that safeguard, sustainably manage, and rehabilitate ecosystems to tackle societal difficulties while delivering advantages for human well-being and biodiversity (Cohen-Shacham et al. 2016). In the last twenty years, NbS have developed into a multidisciplinary framework, gaining significance in relation to climate change, biodiversity decline, and urban sustainability (Eggermont et al. 2015). Their growing adoption in policy and practice across various geographical and socio-political contexts demonstrates their efficacy in tackling numerous urban environmental issues, such as air and water quality, heat islands, and ecosystem degradation (Cohen-Shacham et al. 2016; Seddon et al. 2020).

Urban diversity encompasses multiculturalism, racialized neighborhoods, and under-represented cultural groups. Reflecting systematic disparities in the access and distribution of urban natural resources, these elements influence the interactions and priorities of different groups in urban environments. In such a setting, NbS can be regarded as a way to embrace meanings and societal values. Considering that a primary objective of NbS is to foster effective human–nature interactions in multicultural urban environments, incorporating multiculturalism into these solutions can further acknowledge the diverse social and cultural values linked to nature and enhance the relevance of projects to a broader cultural spectrum (Frantzeskaki et al. 2019).

Moreover, NbS possess significant potential for the restoration of ecosystems and improving ecosystem services, such as wetlands, mangroves, forests, and urban green spaces, which would also facilitate effective human-nature interactions in multicultural urban settings. These ecosystems serve as natural buffers, mitigating the effects of natural calamities such as floods, droughts, and sea-level rise, thereby ensuring a secure environment for the populace. Adopting a nature-based solution can also raise living standards in various aspects, since green urban infrastructure, such as parks and tree-lined streets, improves air quality and temperature regulation, thus alleviating urban heat stress (Gill et al. 2007).

NbS also provide scalable strategies for climate adaptation, such as afforestation and the conservation of natural floodplains to manage water flows and alleviate the effects of extreme weather events (Jato-Espino et al. 2018). These solutions offer co-benefits, including carbon sequestration and enhanced urban livability, which are essential for alleviating the impacts of climate change (Seddon et al. 2020). Notable instances of NbS integration in urban design are evident in cities such as Singapore and Copenhagen, which have illustrated the efficacy of these strategies in improving climate resilience and urban sustainability (Kabisch et al. 2017).

NbS are inherently connected to the notion of ecosystem services, which denote the advantages humans obtain from natural surroundings. Moreover, locally provided ecosystem services significantly influence urban quality of life and should be explicitly included in the formulation of plans for sustainable development, livability, and resilience in urban environments (Elmqvist et al. 2013). Furthermore, NbS tackle problems like soil erosion and land degradation via reforestation and sustainable land management techniques, as demonstrated by China's effective restoration of the Loess Plateau, which mitigated soil erosion, enhanced agricultural productivity, and augmented carbon sequestration (Buckingham 2016).

Despite the numerous advantages of NbS, several problems endure in its execution. These encompass restricted funding, governance intricacies, inadequate technical proficiency, and insufficient stakeholder knowledge. Overcoming institutional and policy obstacles is essential for the expansion of NbS and their incorporation into conventional urban planning and climate adaptation frameworks.

1.3 The Implementation Challenges of Nature-Based Solutions

Implementing NbS encounters numerous problems that affect its practicality, scalability, and efficacy. Spatial planning and zoning typically clash with the prerequisites for NbS integration, as urban areas often have inadequate open spaces due to high-density growth patterns. This spatial constraint requires substantial modifications to existing land-use plans, which may be resource-demanding and politically delicate. Urban development intensifies land competition, complicating the allocation of areas for NbS without interfering with current land uses (Nesshöver et al. 2017; Pauleit et al. 2017).

NbS projects may also affect current communities and neighborhoods, especially through unexpected outcomes like gentrification and displacement. Initiatives to establish green spaces and infrastructure may elevate property values, thus exacerbating socioeconomic imbalances. These disruptions underscore the necessity of inclusive planning approaches to guarantee equitable distribution of NbS benefits and to prevent unfavorable impacts on vulnerable groups (Anguelovski et al. 2018; Kabisch et al. 2017). Such disparities emphasize the necessity for urban planning frameworks that integrate social fairness as a fundamental premise.

The incorporation of NbS with current urban infrastructure is a major challenge. Modifying grey infrastructure to integrate green solutions frequently necessitates specialized technical knowledge and interdisciplinary cooperation. Conventional stormwater drainage systems may be incompatible with NbS, resulting in delays and elevated costs. This incompatibility may diminish the viability and attractiveness of NbS projects for stakeholders (Frantzeskaki et al. 2017; Meerow et al. 2016).

Financial limitations are again a major barrier to the implementation of NbS. Initial investments for ecological restoration, green infrastructure construction, and continuous upkeep are frequently considerable. In contrast to constructed solutions, NbS generally lack robust finance structures, complicating the acquisition of essential financial resources. The monetization of ecosystem services, including carbon sequestration and flood mitigation, continues to pose challenges, hampering the demonstration of explicit economic advantages to policymakers and investors (Seddon et al. 2020).

Governance and policy frameworks are essential for surmounting the above-listed obstacles. Disjointed governance frameworks and redundant responsibilities among stakeholders impede coordinated efforts, diminishing the efficacy of NbS initiatives. The lack of standardized norms and regulatory frameworks intensifies discrepancies in the application of NbS across various locations. To tackle these governance problems, strong policy measures are necessary to promote interdisciplinary collaboration, clarify responsibilities, and define explicit objectives for NbS efforts (Albert et al. 2021; Van Der Jagt et al. 2019). Hence, NbS-related studies still need to be prompted more in the research and practice domains through several good practices from different contexts across the globe.

The book therefore intends to several case based empirical and theoretical applications of NbS from different contexts of the world. They are structured into subthemes and are chaptalized accordingly. It can be assumed that the chapters presented shall address the needs of a wide range of policymakers, researchers, practitioners, students, and academia from diverse disciplines.

1.4 Purpose and Structure of the Book

The book titled "Nature-based Solutions for Urban and Peri-Urban Areas: For a Sustainable and Resilient Urbanization" serves as a comprehensive guide to understanding and applying NbS in urban and peri-urban settings. It explores the theoretical underpinnings, practical applications, and policy implications of NbS. Through an interdisciplinary lens, this book hereby intends to address the environmental issues of urbanization using NbS as a savior. Accordingly, empirical, theoretical, and case study-based research across the globe is structured as the contents of the book. Apart from the introduction, the book has four sections:

(1) Spatial planning perspectives for urban and peri-urban areas,
(2) Urban infrastructure solutions and development strategies,
(3) Traditional methods and community-driven approaches, and
(4) Governance and policies.

It can be presumed that the book shall address the needs of a wide range of policymakers, researchers, practitioners, students, and academia from diverse disciplines.

The book covers various aspects of NbS through these four thematic approaches. The dialogues within Spatial Planning Perspectives concentrate on NbS as a way to tackle urban challenges, including heat mitigation and the efficient utilization of peri-urban areas, while integrating planning approaches to optimize open spaces and promote climate-responsive urban growth. The theme of Urban Infrastructure Solutions and Development Strategies explores innovative applications of NbS in urban settings, including climate adaptation strategies such as green infrastructure, vertical gardens, and water management systems, highlighting their significance in enhancing urban resilience. The Traditional Methods and Community-Driven Approaches section illustrates the interplay between traditional practices and NbS, demonstrating their efficacy in tackling modern challenges such as coastal erosion, drought, and flooding while underscoring the significance of community involvement in the execution of green infrastructure. The Governance and Policies subject emphasizes the necessity for robust policy frameworks and governance models to promote the adoption of NbS, addressing instruments for risk-informed decision-making and ways to surmount implementation barriers via adaptive governance and stakeholder engagement. Collectively, these themes provide an extensive perspective on the diverse functions NbS can fulfil in developing sustainable and resilient urban settings.

The book consists of 16 chapters, with the first being the introduction. The second chapter examines the reduction of the Urban Heat Island (UHI) effect by analyzing geographical distributions of urban blue-green infrastructure, highlighting the substantial impact of urban greenery and water bodies on ambient temperatures and microclimates, with Nagpur City in India being chosen as the study area. The third chapter elaborates on the UHI problem by examining its many causes, measurement techniques, and prospective NbS-based mitigation measures, while also presenting a case study of Jhansi City that utilizes remote sensing and GIS capabilities for a comprehensive knowledge of UHI impacts. The fourth chapter highlights peri-urban areas as essential buffers within urban ecosystems and explores adaptive planning strategies, including green infrastructure and zoning regulations, to align urban growth with ecosystem conservation, illustrated by a study of Varanasi, assessing peri-urban open spaces. The NbS with agile governance ideas, emphasizing smaller tier-II industrial cities are expressed in the fifth chapter and it connects and examines citizen participation, administrative flexibility, and procedural structures to effectively incorporate NbS into swiftly evolving urban environments, proposing a convergence model to improve resilience and inclusivity in urban design.

Chapters 6 to 9 examine various approaches for addressing urban challenges using NbS, prioritizing the enhancement of urban resilience and sustainability. Chapter 6 analyses the notion of Urban Climate Shelters (UCS), especially within schoolyards, as a nature-based strategy to safeguard at-risk populations from heat, flooding, and drought. It examines case studies from European cities, emphasizing the co-benefits of NbS for urban resilience and the feasibility of replicating similar techniques in other regions. Chapter 7 examines the significance of vertical greenery systems (VGS) as a vital element of green infrastructure, emphasizing their development in Belgrade, Serbia. This study examines the planning and design of Vertical Green Structures (VGS) in public areas, identifies deficiencies, and offers solutions for further integration of VGS in urban planning. The application of NbS for stormwater management in slums, specifically in flood-prone regions of India is explained in Chap. 8. This chapter also examines the problems and opportunities associated with the implementation of NbS in urban slums, utilizing Nagpur as a case study. The chapter emphasizes the significance of context-specific solutions and the capacity of NbS to enhance urban resilience via sustainable water management. Chapter 9 examines the cooling capacity of Urban Green Areas (UGA) in mitigating the Urban Heat Island (UHI) phenomenon in Bhopal as a case study. The research measures the cooling impact of UGAs, examining variables such as sizes, shape, and vegetation characteristics, and offers quantitative data for implementing NbS to mitigate UHIs and enhance urban sustainability.

Chapter 10 examines the drought conditions in Banaskantha, Gujarat, employing indicators such as the Standardized Precipitation Index (SPI) and the Vegetation Condition Index (VCI) to evaluate arid intervals and plant-soil moisture levels. It underscores the integration of NbS with conventional and technological methods to improve soil moisture, mitigate runoff, and facilitate agricultural and ecological restoration. Chapter 11 examines the effects of urban floods on at-risk populations in Texas, USA, emphasizing green stormwater infrastructure (GSI) as a flood mitigation strategy. By engaging stakeholders and drawing lessons from Hurricane Harvey, it underscores GSI's significance in resilient land-use planning. Chapter 12 analyses Alappad, India, where coastal erosion intensifies risks following the tsunami. It advocates a hybrid mitigation plan that integrates structural and NbS, such as TetraPot-Mangroves and bio-shields, to protect the coastline and livelihoods. The water security in Taiyuan City, China, which is Managed Aquifer Recharge (MAR) for the purpose of stormwater recycling is addressed in Chap. 13. It underscores the importance of community involvement and place attachment, connecting social norms and emotional perceptions to the effective implementation of NbS.

Chapters 14, 15, and 16 within the Governance and Policies topic examine critical problems and strategies for the implementation of NbS. Chapter 14 explores inadequacies in municipal policies, frameworks, and instruments, highlighting the challenges of incorporating NbS into disaster risk management and urban planning while addressing obstacles to their effective implementation. Adaptive governance models, offering a technique to evaluate and improve governance capacity for NbS, based on empirical research conducted in Vienna and Tucson are expressed in Chap. 15. It emphasizes the significance of stakeholder networks and knowledge exchange in addressing governance difficulties. Chapter 16 examines the factors influencing the successful implementation of NbS, emphasizing obstacles such as insufficient stakeholder engagement, restricted resources, and unsuitable policies. It classifies these difficulties according to the International Union for Conservation of Nature (IUCN) Global Standards and provides solutions for participatory planning, adaptive management, and enhanced transparency. Together, these chapters collectively offer an extensive perspective on overcoming governance and policy challenges to effectively implement NbS, with the objective of fostering more resilient and sustainable urban settings.

References

Albert C, Brillinger M, Guerrero P, Gottwald S, Henze J, Schmidt S, Ott E, Schröter B (2021) Planning nature-based solutions: principles, steps, and insights. Ambio 50(8):1446–1461. https://doi.org/10.1007/s13280-020-01365-1

Anguelovski I, Connolly JJT, Masip L, Pearsall H (2018) Assessing green gentrification in historically disenfranchised neighborhoods: a longitudinal and spatial analysis of Barcelona. Urban Geogr 39(3):458–491. https://doi.org/10.1080/02723638.2017.1349987

Buckingham K (2016) Beyond trees: restoration lessons from China's Loess plateau. In: Song L, Garnaut R, Fang C, Johnston L (eds) China's new sources of economic growth, ANU Press, vol 1, pp 379–396. https://www.jstor.org/stable/j.ctt1rrd7n9.23

Bush J, Doyon A (2019) Building urban resilience with nature-based solutions: how can urban planning contribute? Cities 95:102483. https://doi.org/10.1016/j.cities.2019.102483

Cohen-Shacham E, Walters G, Janzen C, Maginnis S (eds) (2016) Nature-based solutions to address global societal challenges. In: IUCN international union for conservation of nature. https://doi.org/10.2305/IUCN.CH.2016.13.en

Das S, Choudhury MR, Chatterjee B, Das P, Bagri S, Paul D, Bera M, Dutta S (2024) Unraveling the urban climate crisis: exploring the nexus of urbanization, climate change, and their impacts on the environment and human well-being—a global perspective. AIMS Public Health 11(3):963–1001. https://doi.org/10.3934/publichealth.2024050

Davoudi S, Brooks E, Mehmood A (2013) Evolutionary resilience and strategies for climate adaptation. Plan Pract Res 28(3):307–322. https://doi.org/10.1080/02697459.2013.787695

Eggermont H, Balian E, Azevedo JMN, Beumer V, Brodin T, Claudet J, Fady B, Grube M, Keune H, Lamarque P, Reuter K, Smith M, Van Ham C, Weisser WW, Le Roux X (2015) Nature-based solutions: new influence for environmental management and research in Europe. GAIA-Ecol Perspect Sci Soc 24(4):243–248. https://doi.org/10.14512/gaia.24.4.9

Elmqvist T, Fragkias M, Goodness J, Güneralp B, Marcotullio PJ, McDonald RI, Parnell S, Schewenius M, Sendstad M, Seto KC, Wilkinson C (eds) (2013) Urbanization, biodiversity and ecosystem services: challenges and opportunities: a global assessment. Springer Netherlands. https://doi.org/10.1007/978-94-007-7088-1

Frantzeskaki N, McPhearson T, Collier MJ, Kendal D, Bulkeley H, Dumitru A, Walsh C, Noble K, Van Wyk E, Ordóñez C, Oke C, Pintér L (2019) Nature-based solutions for urban climate change adaptation: linking science, policy, and practice communities for evidence-based decision-making. Bioscience 69(6):455–466. https://doi.org/10.1093/biosci/biz042

Frantzeskaki N, Borgström S, Gorissen L, Egermann M, Ehnert F (2017) Nature-based solutions accelerating urban sustainability transitions in cities: lessons from Dresden, Genk and Stockholm cities. In: Kabisch N, Korn H, Stadler J, Bonn A (eds) Nature-based solutions to climate change adaptation in urban areas, Springer International Publishing, pp 65–88. https://doi.org/10.1007/978-3-319-56091-5_5

Gencer EA (2013) The interplay between urban development, vulnerability, and risk management: a case study of the Istanbul metropolitan area. Springer Berlin Heidelberg 7. https://doi.org/10.1007/978-3-642-29470-9

Gill SE, Handley JF, Ennos AR, Pauleit S (2007) Adapting cities for climate change: the role of the green infrastructure. Built Environ 33(1):115–133. https://doi.org/10.2148/benv.33.1.115

Jato-Espino D, Sañudo-Fontaneda LA, Andrés-Valeri VC (2018) Green infrastructure: cost-effective nature-based solutions for safeguarding the environment and protecting human health and well-being. In: Hussain CM (ed) Handbook of environmental materials management, Springer International Publishing, pp 1–27. https://doi.org/10.1007/978-3-319-58538-3_46-1

Kabisch N, Korn H, Stadler J, Bonn A (eds) (2017) Nature-based solutions to climate change adaptation in urban areas: linkages between science, policy and practice. Springer International Publishing. https://doi.org/10.1007/978-3-319-56091-5

Meerow S, Newell JP, Stults M (2016) Defining urban resilience: a review. Landsc Urban Plan 147:38–49. https://doi.org/10.1016/j.landurbplan.2015.11.011

Nesshöver C, Assmuth T, Irvine KN, Rusch GM, Waylen KA, Delbaere B, Haase D, Jones-Walters L, Keune H, Kovacs E, Krauze K, Külvik M, Rey F, Van Dijk J, Vistad OI, Wilkinson ME, Wittmer H (2017) The science, policy and practice of nature-based solutions: an interdisciplinary perspective. Sci Total Environ 579:1215–1227. https://doi.org/10.1016/j.scitotenv.2016.11.106
Pauleit S, Zölch T, Hansen R, Randrup TB, Konijnendijk Van Den Bosch C (2017) Nature-based solutions and climate change—four shades of green. In: Kabisch N, Korn H, Stadler J, Bonn A (eds) Nature-based solutions to climate change adaptation in urban areas. Springer International Publishing, pp 29–49. https://doi.org/10.1007/978-3-319-56091-5_3
Pelling M, O'Brien K, Matyas D (2015) Adaptation and transformation. Clim Change 133(1):113–127. https://doi.org/10.1007/s10584-014-1303-0
Raymond CM, Frantzeskaki N, Kabisch N, Berry P, Breil M, Nita MR, Geneletti D, Calfapietra C (2017) A framework for assessing and implementing the co-benefits of nature-based solutions in urban areas. Environ Sci Policy 77:15–24. https://doi.org/10.1016/j.envsci.2017.07.008
Seddon N, Chausson A, Berry P, Girardin CAJ, Smith A, Turner B (2020) Understanding the value and limits of nature-based solutions to climate change and other global challenges. Philos Trans R Soc B Biol Sci 375(1794):20190120. https://doi.org/10.1098/rstb.2019.0120
Seto KC, Güneralp B, Hutyra LR (2012) Global forecasts of urban expansion to 2030 and direct impacts on biodiversity and carbon pools. Proc Natl Acad Sci 109(40):16083–16088. https://doi.org/10.1073/pnas.1211658109
Van Der Jagt APN, Smith M, Ambrose-Oji B, Konijnendijk CC, Giannico V, Haase D, Lafortezza R, Nastran M, Pintar M, Železnikar Š, Cvejić R (2019) Co-creating urban green infrastructure connecting people and nature: a guiding framework and approach. J Environ Manag 233:757–767. https://doi.org/10.1016/j.jenvman.2018.09.083
World Urbanization Prospects: The 2018 Revision (2019) United Nations

Part I
NBS—Spatial Planning Perspectives for Urban and Peri-Urban Areas

Chapter 2
Nature-Based Solutions (NbSs) as Spatial Planning Strategy to Mitigate Urban Heat Island (UHI) Effect: A Case of Nagpur City, India.

Dipanksha S. Bawankar, Bidisha Chattopadhyay, and Meenakshi Dhote

Abstract Increasing urbanization, concretization, and reducing urban green space (UGS) are seen as leading causes for significantly warmer cities (IPCC 2022). There is a growing realization that solutions based on natural processes and ecosystems can help to tackle the pressing challenge of the Urban Heat Island (UHI) effect. However, in the context of India, there are significantly fewer studies focused on the application of nature-based solutions (NbSs) for the mitigation of UHI in spatial planning. The present study was undertaken for Nagpur, a city in the drought-prone Vidarbha region in Maharashtra, a western state of India. The city lies in the Tropical Savanna climate (Aw) characterized by extremely hot summers. The study was undertaken at two spatial scales. The city-level assessment involved spatiotemporal analysis, and of a detailed scale of urban blue-green spaces through spatial temperature measurements with a handheld temperature measuring device. The major findings of the study suggest that there has been a drastic reduction in Nagpur's green cover in the selected time period and areas with higher temperatures are concentrated in the old city area with higher built density and in areas with bare lands. Areas with blue-green spaces exhibited lower temperatures compared to the surrounding areas. In addition, the result showed that the size, shape of a UGS and canopy cover within the UGS play a very important role in lowering the ambient temperature. Similarly, the size of the waterbody emerged as a significant influencing factor for mitigating heat in the area. The understanding from the study could play a critical role in the informed siting of urban green and blue spaces particularly in the climatic regime of the study area.

Keywords Nature-based solutions (NBSs) · Urban heat island (UHI) · Urban blue-green infrastructure (BGI) · Land surface temperature (LST) · Spatial UHI mitigation · Microclimate

---

D. S. Bawankar (✉) · B. Chattopadhyay · M. Dhote
School of Planning and Architecture, New Delhi, India
e-mail: dipankshabawankar09@gmail.com

M. Firoz C et al. (eds.), *Nature-Based Solutions for Urban and Peri-Urban Areas*, Disaster Risk Reduction, https://doi.org/10.1007/978-981-96-5933-3_2

2.1 Introduction

Urbanization and its associated anthropogenic activities lead to a significant modification in the surface characteristics due to a reduction in vegetation, soil sealing and albedo variation, affecting the local energy balance (Oke 1982). Moreover, factors such as decreased sky view factors, high heat capacity in materials used, human-generated heat sources and urban morphology all play crucial roles in determining UHI intensity (Sharma and Bharat 2009; Blocken 2015; Nakata-osaki et al. 2015). In addition to the UHI being experienced by cities, rising temperatures due to climate change pose one of the notable anthropogenic threats to urban livelihoods, as well as to human and environmental well-being (NDMA 2019; WHO 2014). There is a likelihood of an increase in frequency, duration and intensity of heatwaves under climate change, causing a significant increase in heat-related deaths and further increasing vulnerability, particularly in cities of the Global South (IPCC 2022; Perkins et al. 2012; Stewart and Oke 2012).

Heat mitigation strategies through the planning of Blue-Green Infrastructure (BGI), under the umbrella of NbS, have proven the potential to enhance urban cooling results along with varied ecosystem services as co-benefits (Santamouris et al. 2018; Yu et al. 2020). It is also clear from the literature that it is not a case of "one size fits all" as the effectiveness of BGI is dependent on climatic and geographical conditions (Almaaitah et al. 2021). Further, it is critical to understand the functioning of BGI as a poorly designed system may exacerbate heat stress (Gunawardena et al. 2017). Factors such as shading conditions, size and shape of BGI, and vegetation density play a significant role in heat mitigation (Santamouris et al. 2018; Yu et al. 2020). However, the factors of BGI that influence heat mitigation have seldom been given thought while preparing Master Plans (Sánchez and Govindarajulu 2023). While the importance of BGI in mitigating UHI effects and increasing urban resilience is widely recognized, specific impacts and interactions within the context of India remain limited.

The present study was undertaken for Nagpur, located in the drought-prone Vidarbha region in the Indian State of Maharashtra. The city with a population of 24.48 lakh (Census of India 2011), is the district headquarter of Nagpur district. The city lies in the Tropical Savanna climate (Koppen's Classification) with dry conditions prevailing for most of the year. Nagpur experiences extremely hot summers between March to May with temperatures rising up to 48 °C during April/ May which is locally referred to as "Nava Tappa" or "Nau Tapa" meaning extreme heat days (heat waves), and the average annual rainfall is 1161.54 mm. The city has large areas of greens concentrated in the western part of the city, mainly in the form of institutional greens. The major lakes include Ambazari, Futala and Gorewada while the rivers Nag, Pilli, and Pora traverse the city landscape in a west-to-east direction. Development Plan of Nagpur (2011) mandates the protection of UGS (Urban Green Spaces) and waterbodies in the city, but many of its recommendations have not been implemented (Lahoti et al. 2019).

The authors argue that since the effect of BGI on heat mitigation varies based on different factors such as shape, size, canopy cover of BGI and surrounding land cover, BGI has to be planned for effective heat mitigation taking into account all those factors for specific climatic regime. The inclusion of planned BGI in master plans should be based on the quantitative assessment of the effect of BGI on temperature in the surrounding areas. Moreover, developing spatial strategies to build the necessary capacity to handle present and future climate change scenarios is necessary for heat resilience at the city level. The study aims to support the urban planning potential through the inclusion of planned BGI in master plans based on the quantitative assessment to improve microclimate in cities.

2.2 Literature Review

Urbanisation, as a process, induces population migration, socio-economic changes and an increase in area under built up leading to the alteration of natural systems (Rahman et al. 2022). These changes in land cover patterns lead to modifications in atmospheric composition in the near-surface layer and thermal stress in urban areas (IPCC 2007; Oke et al. 2017; Yu et al. 2020). Significantly warmer urban areas compared to surrounding rural areas known as the Urban Heat Island (UHI) effect, arise from these changes and may also be affected by changes in water runoff, pollution and aerosols (IPCC 2007; Oke et al. 1972). This is further amplified by global warming and climate change (Oleson 2012). Heat waves are expected to be on the rise, likely to increase the frequency of warm days and nights with an increase in the duration, intensity, and spatial extent of heat waves (IPCC 2022). These extreme heat wave events result in a high number of fatalities and heat-related illnesses such as general discomfort, respiratory difficulties, heat cramps, heat exhaustion, and non-fatal heat stroke (WHO 2014; NDMA 2020).

Some of the crucial impact factors that can induce or reduce the UHI effect involve the urban underlying surfaces (Liang et al. 2021), thermal emissions/waste heat from anthropogenic activities (Shahmohamadi 2011), urban green spaces, urban water bodies (Noëmie Probst 2022), land use, urban morphology, and surface geometry (Kotharkar et al. 2018), etc. However, in addition to these, the UHI effect might be influenced by more complex factors (Liu and Wang 2023).

For the mitigation of UHI, BGI is gaining interest globally as one of the spatial planning strategies under the umbrella of nature-based solutions (NbSs) (Sanusi and Bidin 2020); as it provides environmental and community benefits (IUCN 2016). European Commission and Wetlands International defines BGI as "*strategically planned interconnected networks of natural and semi-natural areas, including water bodies and green and open spaces that deliver a wide range of ecosystem services*" (EU 2019).

In the urban landscape, urban blue-green elements such as water bodies, urban parks, forests, grassland, playgrounds, and other UGS have cooling potential through evapotranspiration and shading, therefore, termed urban heat sinks (Santamouris et al.

2018; Yu et al. 2020; Noëmie Probst 2022). Passive cooling techniques such as green/cool roofs (Santamouris et al. 2018), increasing vegetation and permeable surfaces (NBC 2005), vertical greening, (Susca 2022) have also been mentioned in global literature.

Various methods including theoretical/empirical studies, observational techniques such as remote sensing and field measurements, experimental approaches, and numeric/computational methods have been commonly used to estimate the Urban Heat Island (UHI) effect in real time (Mirzaei et al. 2010; Sahu et al. 2014). Further, the studies based on the correlation between UHI, land cover and Land Surface Temperature (LST) explore the potential of urban blue-green to dissipate heat from their surroundings, resulting in urban microclimates (Noëmie Probst 2022; Yu et al. 2020).

In the context of India, remote sensing has become a widely employed approach to study UHI with comparatively fewer studies based on field measurement (Veena et al. 2019). This involves LST mapping using satellite datasets. Mondal et al. (2021) used a correlation between LST and different LULC indices to identify the heat-vulnerable zones in Imphal City; Pritipadmaja et al. (2023) studied spatial variations in the LST to assess the cooling effect of blue-green spaces in Bhubaneswar and Surawar et al. (2017) studied the effect of LULC change on daytime LST variation with focus of urban built-up density in Nagpur.

Size, shape, connectivity, and complexity (composition and configuration) of the blue-green space, and the greenness of the vegetation, etc. determine the cooling capacity (Santamouris et al. 2018; Yu et al. 2020; Antoszewski et al. 2020). UHI effects are often localized and depend on local climate factors (IPCC 2007), the cooling potential of BGI also differs with urban morphology and climatic conditions such as built density, wind intensity, direction, etc. (Santamouris et al. 2018; Kotharkar et al. 2018). Different methods have been used to analyse the effect of BGI factors responsible for heat mitigation. A study done in the tropical rainforest climate (Af) of Putrajaya, Malaysia by Sanusi and Jalil (2021); investigated the role of BGI on microclimate modifications in an urban park using a systematic random sampling method at four sites including waterbody, trees, grass and open areas. For each site, 16 sampling points were selected at a distance of 10 m apart, with a total of 64 sampling points. Another study conducted in subtropical Mediterranean climate, (Csa) at Tel-Aviv urban complex, Israel by Shashua-Bar et al. (2000); investigated the cooling effect at 11 small urban green sites with trees including 2 gardens, 4 avenues, 1 green square, 2 courtyards and 2 streets. For each site, several points were chosen at 20 m intervals and statistical analysis was carried out using 714 experimental observations gathered each hour. Another study by Taleghani (2018), considered six urban parks to demonstrate the role of effective UGS.

Existing studies in India have often overlooked the potential of urban blue-green spaces to mitigate the urban heat island effect in spatial planning efforts. A recent study in the tropical savanna climate of Bengaluru by Shah et al. (2021), analysed cooling effects at the level of individual UGS using remote sensing and found that on average, UGS provided local cooling effects within 347 m beyond its boundary, with an average temperature difference of 2.23 °C. According to the study, the level of greenness, size, and shape of UGS were found to influence their ability to provide cooling effects. Furthermore, academic researchers (Kotharkar and Badge 2018; Surawar and Kotharkar 2017, etc.) in Nagpur City have extensively investigated the UHI phenomenon using remote sensing and mobile survey techniques to analyse urban morphology and land use planning. However, there has been limited focus on investigating the specific influence of BGI on the microclimate in Nagpur's climatic region through in situ temperature measurements, underscoring the need for a comprehensive quantitative analysis.

In India, to address the challenges posed by rising temperatures, city-specific and regional Heat Action Plans (HAPs) have been framed. The criticism of these HAPs has been the absence of a local context, poor at identifying vulnerable populations, weak legal linkages with the existing disaster management and environmental governance among others (Pillai and Dalal 2023). HAP for Nagpur (2020) focuses on short-term health interventions. It has also suggested the objective strategies for spatial coverage such as tree plantation, change in urban planning approach, air pollution mitigation measures, deployment of shading structures for traffic police and workers, drinking water outlets water however, the plan fails to give a systematic approach for implementation of these strategies.

The study involves an attempt to answer the research question of 'how do the characteristics of urban BGI affect temperature in the surrounding areas in the context of a particular climatic zone (Aw), for the city of Nagpur'. Broadly, the independent variables were the characteristics of BGI, and the dependent variable was the temperature within and around the BGI.

2.3 Material and Methods

The present study was undertaken at two spatial scales. The city-level assessment involved spatio-temporal analysis of LST based on satellite imagery data for the years 2005, 2015 and 2021, and at the individual UGS/Blue space level, temperature readings collected through a primary survey conducted in April 2022 were examined.

For the city-level assessment based on satellite imagery, datasets of LANDSAT 5 TM (Thematic Mapper), and LANDSAT 8 OLI/TIRS (Optical Land Imager/ Thermal Infrared Sensor) were used to investigate LULC change and spatio-temporal variation of LST in Nagpur. The analysis made use of thermal infrared bands—specifically Band 6 for LT5 and Band10 for LT8—to calculate the LST following the steps described by Jose´ et al. (2004) and Avdan and Jovanovska (2016) respectively. Daytime satellite datasets (Path/Row—144 / 45) were acquired for April month of 2005, 2015 and 2021 from the USGS website (https://earthexplorer.usgs.gov/. Accessed on March 2022) shown in Table 2.1.

For detailed study of characteristics that determine temperature in the areas around BGI, temperature measurements were taken at different locations with a hand-held temperature measuring device (G-Tech Digital Anemometer). The surveys were conducted in April 2022 from 2nd april–5th april at solar noon time (1200–1600).

Table 2.1 Details of LANDSAT datasets used in the study

SN	Satellite	Dataset	Resolution (m)	Date of acquisition	Sun elevation	Cloud cover
1	LT_5	L1TP_144045_20050407_20200902_02_T1	30	2005-04-07	59.65	1.0
2	LT_8	L1TP_144045_20150419_20200909_02_T1	30	2015-04-19	64.99	0.0
3	LT_8	L1TP_144045_20210403_20210409_02_T1	30	2021-04-03	61.21	0.07

Temperature data was measured at 1.2 m above the ground level. Humidity and wind speed were not captured and studied.

For UGS, a systematic random sampling with a random start was used where the first point was randomly selected at the edge of parks, 10, 30 and 60 m from the boundary. Along with the measurement of temperature (°C) at the decided distance, characteristics (independent variables) derived from literature study such as percentage of UGS under tree coverage or canopy cover, percentage of ground vegetation and shape and size of UGS were recorded at each sampling location of urban greens. Throughout the city, 15 UGS were selected for the study. For each UGS, 3 sampling points were selected, thus a total of 45 sampling points were taken into consideration for the study of the thermal effect of UGS.

Similarly, a survey was conducted for surface waterbodies; Ambazari Lake, Futala Lake, Sakkardara Lake, Sonegaon Lake and Naik Lake. A total of 25 sampling points at a distance of 10, 20, 30, 50 and 100 m from the lakes were considered for the study of the impact of urban blue spaces on temperature.

ArcMap 10.8 was used for spatial analysis while IBM Statistical Package of Social Science (SPSS) version 26 software and MS Excel 2007 were used for non-spatial analysis. Pearson's correlation coefficient analysis has been used to assess the relationship between BGI characteristics and microclimate conditions.

2.4 Results and Discussions

2.4.1 Study Area

Nagpur has a geographical area of 225 Sq.Km and is located between 78°30″ to 79°30″E and 20°30″ to 21°45″N latitude, at an altitude of 310.5 m above mean sea level. Nagpur is the third largest urban agglomeration in the state of Maharashtra in terms of population with a population of 24.48 Lakh, with a decadal growth rate of 19.3% in 2001–2011 (Census of India 2011).

Typical to the central part of Deccan plateau (arid zone), the climate of Nagpur city is classified into four seasons namely the summer, monsoon, post-monsoon and winter (IMD 2011). It was suggested to set a threshold temperature of 43 °C (109.4°F) for a heat alert day and a threshold temperature of 45 °C (113°F) or higher for an extreme heat alert day during summer months (NDMA 2019). Nagpur has experienced a 1.5 °C increase in the average annual temperature and a drastic rise in the frequency of heatwaves over the past 50 years (NEERI 2020). Moreover, the Maharashtra State Action Plan for Climate Change (TERI 2014) has projected an increase in temperature based on past trends in Nagpur by 1.95–2.2 °C till 2050. The impacts of heat stress in the city have been observed as health impacts such as heat exhaustion and stroke, dehydration, diarrheal diseases, and worsening of environmental factors like water stress and deteriorating water quality (HAP 2020). Socio-economically disadvantaged communities with less protective measures and workers in outdoors or high-risk industries are the vulnerable groups with a higher risk of heat-related illnesses (Vasudha Foundation 2022).

In the above context of extreme heat conditions prevailing in certain months, it is critical and imperative to understand and assess BGI in the city. Overall, 94.53 Sq. Km. area, i.e., 42% of the total city area is under natural assets of which approximately 30% is under greens (park, grassland, forest, agriculture, avenue trees, etc.), and about 12% accounts for water bodies and open spaces (IGCMC WWF 2020). Nagpur has 136 urban parks covering around 142 hectares of land and a total of 12 lakes with three rivers within the city limits (City Development Plan for Nagpur-2041 2015). The major lakes include Ambazari, Futala, and Gorewada, while the rivers Nag, Pilli, and Pora traverse the city landscape from west to east. However, the Development Plan of Nagpur (2011) indicates a significant gap of about 10% in the provision of green coverage compared to URDPFI Guidelines, 2014. The City Development Plan for Nagpur-2041 (2015) outlines the capital investment plan for the rejuvenation of lakes and rivers. The city also has a Regional Heat Action Plan, 2019 prepared by Nagpur Municipal Corporation and has identified town planning measures in the long-term strategies (Vasudha Foundation 2022).

Table 2.2 LULC change in Nagpur: 2005–2021

LULC Categories	2005 Area		2015 Area		2021 Area		2005 to 2015	2015 to 2021	2005 to 2021	
	(Sq. Km.)	(%)	(Sq. Km.)	(%)	(Sq. Km.)	(%)	% Change	% Change	% Change	
Built up	93.07	42.22	113.07	51.29	123.77	56.14	9.07	4.85	13.92	13.92
Agriculture	64.94	29.46	53.12	24.09	39.09	17.73	−5.36	−6.36	–11.72	–20.13
Vegetation	50.87	23.07	37.73	17.12	32.33	14.66	−5.95	−2.46	–8.41	
Water bodies	2.82	1.28	3.37	1.53	3.1	1.41	0.25	−0.13	0.13	0.13
Bare land	8.76	3.97	13.17	5.97	22.17	10.06	2	4.08	6.08	6.08
Total	220.46	100.00	220.46	100.00	220.46	100.00	0.01	−0.02	0	0

Negative sign indicates reduction

2.4.2 Analysis

2.4.2.1 Spatio-Temporal Trend of LULC Change—2005–2021

In 2021, more than half (56.14%) of the geographical area of Nagpur was under built up, followed by 17.73% under agricultural use, 14.66% under vegetation, 10.06% of barren land, and 1.41% underwater bodies. More than half the city's UGS, in the form of recreational and institutional greens, are in the western and north-western parts.

As per LULC analysis for Nagpur city (Table 2.2, Fig. 2.1), urban green (UG) cover has reduced by 18.57 sq.km. i.e., 8.4% as a result of an increase in area under urban built up in the period 2005–2021. If the decrease in the area under agriculture in the peripheral areas of Nagpur City is also considered, the overall decline is nearly 20%. There is a significant increase (13.92%) in built up towards the north, north-west and north-east areas of the city. Bare land has also increased by 6.08% towards the western part of Nagpur, replacing agricultural land in the period 2005–2021 (Fig. 2.2).

2.4.2.2 LULC and Associated Temperature Profile

The spatio-temporal LST maps for Nagpur city derived for the pre-monsoon season of 2005, 2015 and 2021 are shown in Fig. 2.3. LST has increased from a minimum of 32.4 °C in 2005 to 34.5 °C in 2021 and a maximum of 47.5 °C in 2005 to 54.5 °C in 2021. Areas with high population and built-up density in the eastern, northern and south-eastern parts of the city experience higher temperatures as compared to

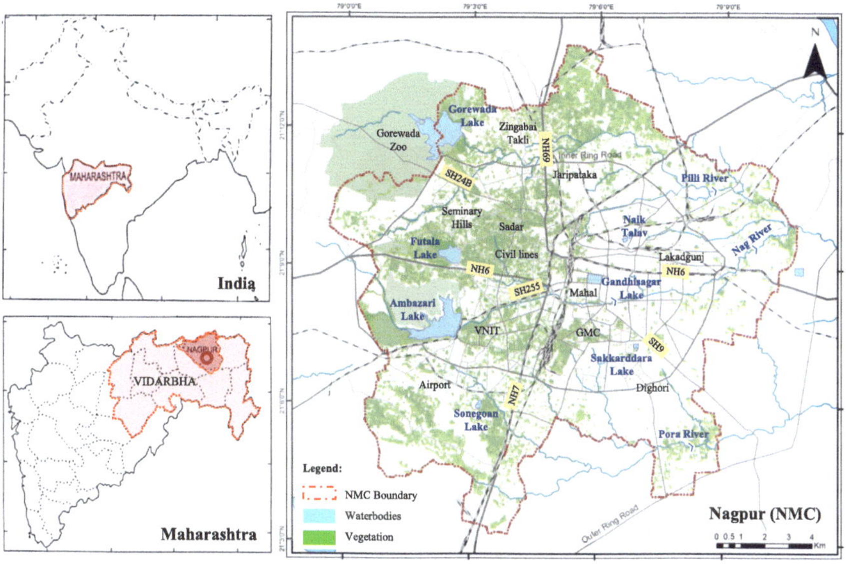

Fig. 2.1 Location of Maharashtra, Vidarbha region and Nagpur city (Reproduced by authors, 2022 using USGS landsat 8 dataset)

the rest of the city. These old areas of Nagpur such as Mahal, Itwari and Lakadgunj witness higher temperatures during night-time as well (Kotharkar et al. 2018).

Surface waterbodies show lowest temperature range of 32.4–40.0 °C followed by areas with vegetation in West and North-West areas, with LST range of 34.6–43.5 °C. These green areas range from medium dense forests to cultivated lands, parks and gardens such as Seminary Hills, Maharaj Bagh, Telankhedi Garden and Ambazari Garden stretch, marked with dense lush vegetation. There are some land pockets under institutes such as agricultural universities, engineering colleges (VNIT) and medical colleges (GMC) present in the city which also contribute to green cover. Areas under natural vegetation and water bodies witness temperature lower than the average LST range and act as "cool islands" (Yu et al. 2020) (Fig. 2.4).

Two cases, one of Ambazari Lake and the other of institutional greens in Nagpur, were examined to study the cooling effects of the selected blue-green infrastructure. The influence of BGI extends beyond the boundaries of the land cover, as indicated by the generated maps. Specifically, the study looked at the urban blue of Ambazari Lake (155Ha) and the dense institutional green located in the civil lines area to observe the effect on the temperature of the nearby areas. For measuring the distance, the land surface temperature maps in of Figs. 2.5 and 2.6 were juxtaposed with a 30 × 30 m grid.

The influence of Ambazari Lake clearly shows a lower range temperature up to 3 grids (90 m) beyond the lake boundary, which further varies depending on the surrounding LULC. The vegetation adjacent to the lake showed lower temperatures up to 5 grids (150 m). On the other hand, the LST of bare land was observed as

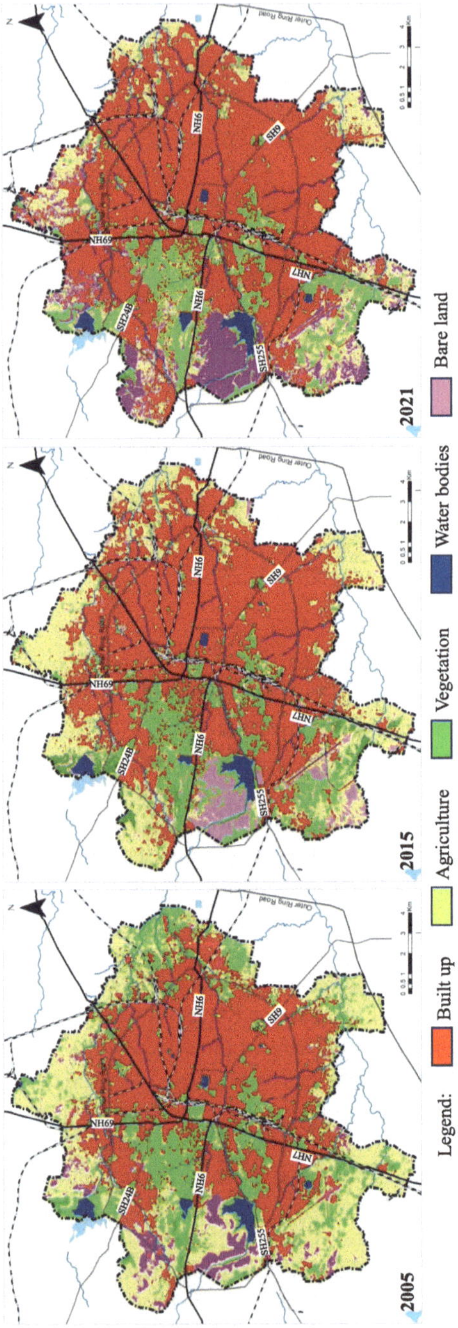

Fig. 2.2 LULC map of Nagpur for 2005, 2015 and 2021 (Maps derived by the authors based on landsat 5 and 8, USGS datasets)

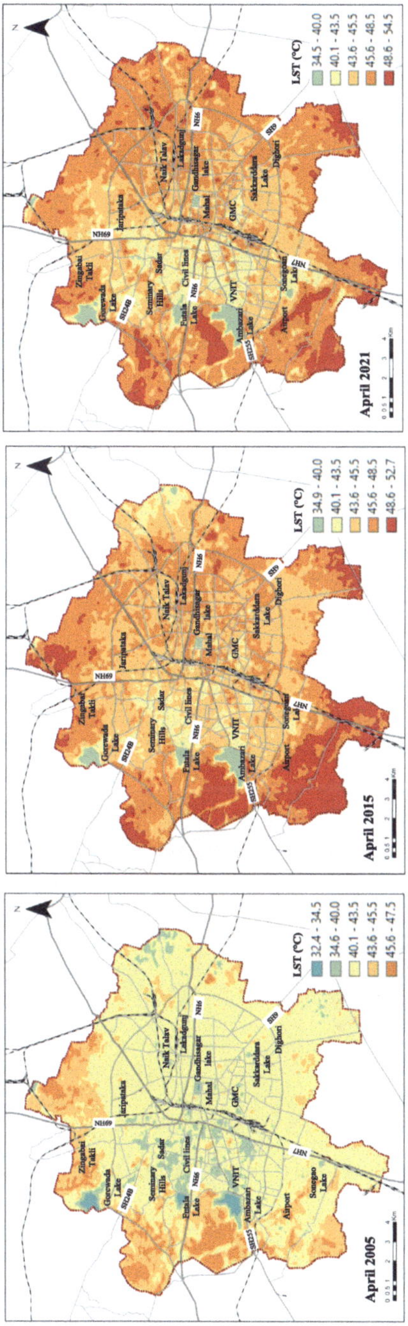

Fig. 2.3 Land surface temperature map for Nagpur city for years 2005, 2015 and 2021 (Maps derived by authors based on Landsat 5 and 8, USGS dataset)

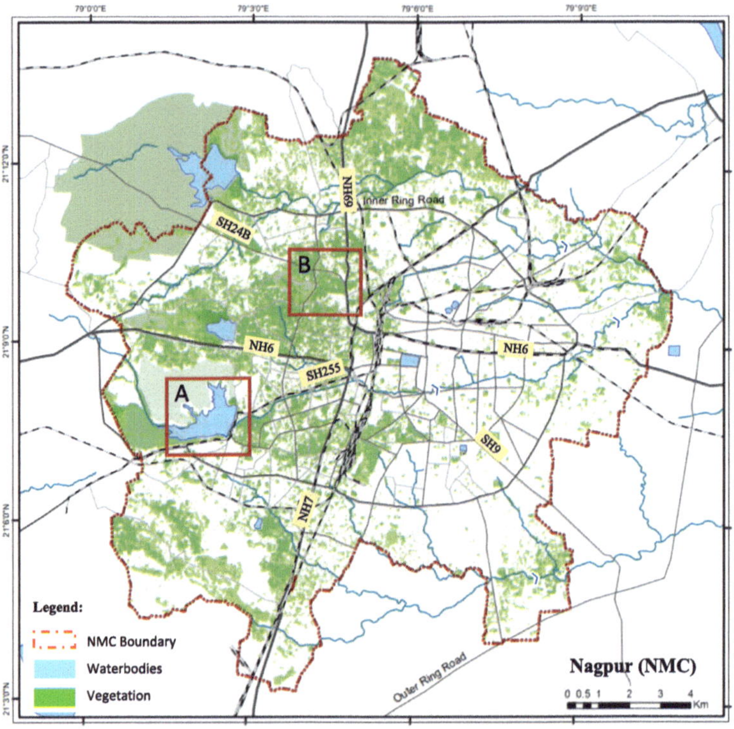

Fig. 2.4 Nagpur map showing locations—Patch **a** Ambazari lake and patch, **b** area under institutional green at civil lines Reproduced by authors, 2022 using USGS landsat-8 dataset

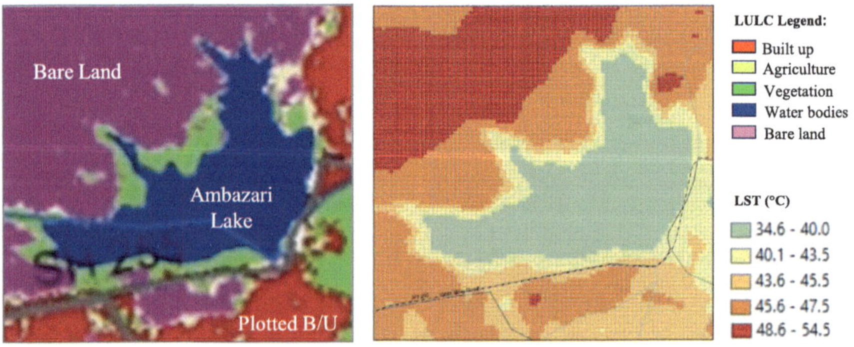

Fig. 2.5 Detail at Patch A: Ambazari Lake (Urban blue), LULC (on left) and LST (on right) for April 2021

Fig. 2.6 Detail at Patch B: civil lines (Institutional green), LULC (on left) and LST (on right) for April 2021

the highest (above 45.5 °C) and increased as the distance from the lake increased. The significant influence of dense green on LST was observed till 30 m beyond the boundary.

The temperature at Civil Lines under institutional green cover was observed to be in the lower LST range (between 40.1 to 43.5 °C), showing lower temperatures up to 1–2 grids beyond the boundary, which varied depending on the adjacent land cover. Areas with partial vegetation continued to exhibit lower temperatures, while land covered by grass showed an increase in temperature above 48.5 °C. For built up areas, the surface temperature was observed to vary with building characteristics and density, such as compact mid-rise, open/plotted mid-rise dense development, and sparsely built-up areas with greenery. The built up areas (north-east corner) showed surface temperatures in the same range as the institutional greens, as the buildings were low-density and low-rise, surrounded by scattered trees. On the contrary, the built-up areas (south-eastern side) exhibit higher temperature ranges exceeding 43.5 °C, due to the compact spacing of structures, the use of highly reflective surfaces, and the lack of vegetation.

2.4.2.3 Relationship Between Characteristics of Urban Green Spaces (UGS) and Temperature

UGS has a notable effect on reducing the temperature and creating a cooler surrounding environment within urban areas, as already noted in the previous section. However, even under the same climatic conditions, the impact of UGS varies corresponding to different characteristics of UGS. Pearson's Correlation Analysis was used to test the association between the nature of UGS such as size (area), shape, percentage of tree canopy cover, percentage of grass cover and temperature around the UGS. The present study in Nagpur city focused on 15 UGS with different configurations and levels of vegetation to measure their influence on ambient air temperature at 10, 30 and 60 m from the boundary of the UGS.

Increase in temperature with increasing distance from the boundary of the UGS was seen for all UGS. Since the influence of UGS changes with respect to distance from the UGS, a separate analysis of the association of determining factors such as size (area), shape, percentage of tree canopy cover, and percentage of grass cover and temperature was undertaken with respect to each of the distances at which temperature was measured. Pearson's correlation analysis was done, and the findings have been presented in Table 2.3.

Correlation between all factors besides area under grass cover and temperature at the select distances is statistically significant. The results suggest that there is a negative correlation between the size of the UGS and the temperature recorded at the edge, 10, 30 and 60 m from the boundary of the UGS. Thus, most UGS with larger areas record cooler temperatures as compared to their smaller counterparts. In addition, stronger Pearson's correlation coefficient for size and temperature at 60 m compared to temperature at 30 and 10 m imply that the effect of larger UGS on temperature is also for longer distances.

Similarly, as the value of the shape index (indicates complexity of the patch shape, with a ratio of perimeter to area, Shah et al. 2021) increases, the temperature decreases as seen in Table 2.3, Pearson's correlation coefficient is negative but statistically significant for shape index and temperature. This implies that UGS with irregular shapes have a higher cooling effect as compared to regular shapes such as squares and rectangles. Pearson's correlation coefficient is strongest at a 30 m distance from the edge beyond which it weakens.

Analysis was done for the percentage area under tree canopy cover within a UGS and its influence on cooling distance and cooling magnitude. Pearson's correlation coefficient was found to be negative (statistically significant); −0.726 for 10 m, −0.732 for 30 m and −0.717 for 60 m. Area under grass cover was not found to have any statistically significant influence on the temperature.

Table 2.3 Pearson's correlation coefficient between characteristics of UGS and temperature at varying distances

Distance from boundary of UGS (m)	Size (ha)	Shape Index	Area under tree canopy cover (%)	Area under grass cover (%)
0	−0.636*	−0.714**	−0.726**	−0.70
10	−0.636*	−0.714**	−0.726**	−0.70
30	−0.701*	−0.739**	−0.732**	−0.023
60	−0.779**	−0.659**	−0.717**	−0.30

* Correlation is significant at 0.05 level (2 tailed)
** Correlation is significant at 0.01 level (2 tailed)

2.4.2.4 Relationship Between Characteristics of Urban Blue Spaces and Temperature (T)

Urban waterbodies have the ability to absorb heat and serve as thermal buffers, effectively reducing the temperature of their surroundings. Past research has explored how the cooling effect of waterbodies on surrounding air temperature varies based on different factors such as area, river water quality (e.g., suspended particles), and albedo of the water surface (Noëmie Probst 2022; Yang et al. 2020). In the present case, the impact of five urban waterbodies of different sizes on temperature has been examined (Table 2.4).

For the small waterbodies like Sakkardara Lake and Naik Lake, the difference in temperature at the edge and 100 m from the edge is considerable, an increase of 3.4% and 1.9% respectively compared to the medium-sized waterbody of Sonegaon (2.4%) and the larger waterbodies of Ambazari (1%) and Futala (0.7%). Thus, it is clear that the zone of influence of larger lakes with respect to temperature effect is considerably greater. The rate of increase in temperature would also be influenced by the surrounding land cover and built density, besides the size of the water bodies.

2.5 Conclusion

Nagpur located in a region prone to drought and heat waves, needs careful consideration of integrating nature-based solutions into city planning to enhance heat resilience. The present paper sought to assess the contribution of BGI to urban cooling and mitigating heat in Nagpur. It is clear from both the literature review and the findings of the study, that the cooling effects of BGI vary based on climatic factors, its characteristics and surrounding land use/land cover.

Not only was temperature found to be lower for larger UGS and waterbodies, but the effect of these UGS and waterbodies was also observed to extend to a longer distance. Larger water bodies (exceeding 40 hectares) were found to have an impact extending over 100 m from the edge of the waterbodies, while UGS have shown influence up to 30 m. Similarly, UGS with irregular shapes were seen to have a higher influence on temperature within the UGS and the adjoining areas, thus exhibiting lower temperatures compared to UGS with regular shapes. Furthermore, UGS with a higher percentage area under tree canopy cover were also observed to have lower temperatures in and around the green space as compared to UGS with less area under tree canopy cover.

The findings have important implications for the type of UGS that may be proposed in a master plan. At present, the focus is largely on meeting the recreational needs of the population but to improve a city's adaptive capacity for heat mitigation and future heat resilience, it is imperative to plan BGI taking into account their cooling efficiency based on various characteristics. Thus, in the case of Nagpur, it is recommended to include larger green areas with irregular shapes with green fingers permeating into the built areas and a high percentage of area under tree canopy cover in the proposed green

Table 2.4 Temperature recorded around waterbodies in Nagpur (Primary Survey, April 2022)

Waterbodies	Area (Ha)	Temperature at edge	Temperature at 10 m	Temperature at 20 m	Temperature at 30 m	Temperature at 50 m	Temperature at 100 m
Ambazari	155	39.2	39.2	39.2	39.4	39.4	39.6
Futala	40	40.5	40.5	40.6	40.6	40.7	40.8
Sonegaon	16.42	40.7	40.7	40.8	41	41.3	41.7
Sakkardara	3.68	41	41	41	41.1	41.8	42.4
Naik	3	41.8	41.8	42	42.1	42.5	42.6

areas of the master plan. URDPFI (2014) suggests five categories in organized green for plain areas ranging from housing area park (0.50 ha) to district park (50 ha) and sub-city park (100 ha) catering to 5000 and 10,000 population, respectively. Within this recommendation, the characteristics of greens may be selected such that there are more urban forests, biodiversity parks in the category of district park and sub-city park categories. It is also clear that the effect of waterbodies is significantly more than UGS of a similar size, reiterating the need for the conservation of waterbodies.

The present paper also attempts to address the major limitation observed in current research with its narrow focus on satellite-based analysis for surface temperature evaluation and limited exploration through field measurement studies (Yu et al. 2020). It recommends using LST-based analysis at a city scale combined with detailed zone-scale field examinations that can be replicated in other locations within cities to further explore BGI's potential cooling effects, thus suggesting a more comprehensive approach.

It is anticipated that this study will find application in spatial planning by offering valuable information about how BGI can contribute to urban cooling, thus leading to the optimization of BGI by urban planners and policymakers during future city planning endeavours.

References

Almaaitah T, Appleby M, Rosenblat H, Drake J, Joksimovic D (2021) The potential of blue-green infrastructure as a climate change adaptation strategy: a systematic literature review. Blue-Green Syst 3(1):223. https://doi.org/10.2166/bgs.2021.016

Antoszewski P, Swierk D, Krzyzaniak M (2020) Statistical review of quality parameters of blue-green infrastructure elements important in mitigating the effect of the urban heat island in the temperate climate (C) zone. Int J Environ Res Public Health 17:7093. https://doi.org/10.3390/ijerph17197093

Avdan U, Jovanovska J (2016) Algorithm for automated mapping of land surface temperature using LANDSAT 8 satellite data. J Sens 1480307. https://doi.org/10.1155/2016/1480307

Blocken B (2015) Computational fluid dynamics for urban physics: importance scales possibilities limitations and ten tips and tricks towards accurate and reliable simulations. Build Environ 1–27

Census of India (2011) Census of India website: office of the registrar general and census commissioner, India. http://censusindia.gov.in/

City Development Plan for Nagpur-2041 (2015) Ministry of urban development, government of India capacity building for urban development. World Bank

Cohen-Shacham E, Janzen C, Maginnis S, Walters G (2016) Nature-based solutions to address global societal challenges. IUCN. https://portals.iucn.org/library/node/46191

CSIR-National Environmental Engineering Research Institute, Nagpur (NEERI) (2020) Environmental status Report 2019–20, Nagpur. NMC

Development Plan of Nagpur (2011) Nagpur Improvement Trust (NIT), Nagpur

European Commission and Wetlands International EU (2019) https://eur-lex.europa.eu/legal-content/EN/TXT/PDF/?uri=CELEX:52019DC0236&qid=1562053537296&from=EN

Gunawardena KR, Wells MJ, Kershaw T (2017) Utilising green and blue space to mitigate urban heat island intensity. Sci Total Environ 584–585:1040–1055, ISSN 0048-9697, https://doi.org/10.1016/j.scitotenv.2017.01.158

Heat Action Plan (HAP) (2020) Nagpur municipal corporation (NMC), Nagpur

India Meteorological Department (2011) IMD Data
Indira Gandhi Conservation Monitoring Centre (IGCMC) (2020) WWF-India. https://urban-leds.org/nagpur-advocates-for-actions-to-conserve-and-enhance-biodiversity/
IPCC (2007) Fourth assessment report of working group II. UN. https://www.ipcc.ch/site/assets/uploads/2018/03/ar4_wg2_full_report.pdf
IPCC (2014) Fifth assessment report of working group II. UN
IPCC (2022) Sixth assessment report of working group I. UN
Sobrinoa JA, Jime´nez-Mun˜oza JC, Paolinib L (2004) Land surface temperature retrieval from LANDSAT TM 5
Kotharkar R, Bagade A (2018) Evaluating urban heat island in the critical local climate zones of an Indian city. Landsc Urban Plan. 169:92–104. https://doi.org/10.1016/j.landurbplan.2017.08.009
Lahoti S, Kefi M, Lahoti A, Saito O (2019) Mapping methodology of public urban green spaces using GIS: an example of Nagpur City, India. Sustainability 11(7):2166. https://doi.org/10.3390/su11072166
Liang T, He J, Chen L, Yao Z, Zhang L, Che H, Gong S (2021) Simulation of the influence of a fine-scale urban underlying surface on the urban heat island effect in Beijing. Atmos Res 262:105786. https://doi.org/10.1016/j.atmosres.2021.105786
Liu Y, Wang Z (2023) Research progress and hotspot analysis of urban heat island effects based on cite space analysis. Land 12:1154. https://doi.org/10.3390/land12061154
Mirzaei PA, Haghighat F (2010) Approaches to study urban heat island—abilities and limitations. Build Environ 45:2192–2201
Taleghani M et al (2018) Outdoor thermal comfort within five different urban forms in the Netherlands. Build Environ 83:65–78. https://doi.org/10.1016/j.buildenv.2014.03.014
Mondal A, Guha S, Kundu S (2021) Dynamic status of land surface temperature and spectral indices in Imphal city, India from 1991 to 2021. Geomat Nat Haz Risk 12(1):3265–3286. https://doi.org/10.1080/19475705.2021.2008023
Nakata-osaki CM, Lucas de Souza LC, Rodrigues DS (2015) A GIS extension model to calculate Urban Heat Island intensity based on urban geometry. Comput Urban Plan Urban Manag, 1–16
National Building Code (NBC) (2005) Bureau of Indian standards, ICS 0.120; 91.040.01, ISBN 81-7061-026-5
NDMA (2019) National guidelines for preparation of action plan-prevention and management of heatwave. National Disaster Management Authority. https://ndma.gov.in/sites/default/files/PDF/Guidelines/heatwaveguidelines.pdf
NDMA (2020) Beating the heat. How India successfully reduced mortality due to heatwaves. https://ndma.gov.in/sites/
Oke TR (1982) The energetic basis of the urban heat island. Q J R Meteorol Soc 108:1–24. https://doi.org/10.1002/qj.49710845502
Oke TR, Mills G, Christen A, Voogt JA (2017) Urban climates
Oke et al (1972) Comparison of Urban Rural Cooling Rates at Night. In: Conference on urban environment and second conference on biometeorology, Philadelphia
Oleson K (2012) Contrasts between urban and rural climate in CCSM4 CMIP5 climate change scenarios. J Clim 25(5):1390–1412
Ozone Cell Ministry of Environment, Forest and Climate Change Government of India (2019) India cooling action plan; ministry of environment, forest and climate change: New Delhi, India
Perkins SE, Alexander LV, Nairn JR (2012) Increasing frequency, intensity and duration of observed global heatwaves and warm spells. Geophys Res Lett 2012(39):10
Pillai A, Dalal T (2023) How is India adapting to heatwaves? An assessment of heat action plans with insights for transformative climate action. CPR report. https://cprindia.org/wp-content/uploads/2023/03/Heat-Report_27March-23_Updated-Table.pdf
Pritipadmaja GRD, Sharma AK (2023) Assessing the cooling effect of blue-green spaces: implications for urban heat island mitigation. Water 15(16):2983. https://doi.org/10.3390/w15162983

Noëmie Probst et al. (2022) Blue green systems for urban heat mitigation: mechanisms, effectiveness and research directions. Blue-Green Syst 4(2):348. https://doi.org/10.2166/bgs.2022.028

Rahman MN, Rony MRH, Jannat FA, Chandra Pal S, Islam MS, Alam E, Islam ARMT (2022) Impact of urbanization on urban heat island intensity in major districts of Bangladesh using remote sensing and geo-spatial tools. Climate 10:3. https://doi.org/10.3390/cli10010003

Sahu S, Pawar A, Mukherjee M (2014) Urban Heat Island investigation: techniques and methods so far. Arch—Time Space People 16–22

Sánchez FG, Govindarajulu D (2023) Integrating blue-green infrastructure in urban planning for climate adaptation: lessons from Chennai and Kochi, India. Land Use Policy 124:106455, ISSN 0264-8377. https://doi.org/10.1016/j.landusepol.2022.106455.

Santamouris M, Ban-Weiss G, Osmond P, Paolini R, Synnefa A, Cartalis C, Muscio A, Zinzi M, Morakinyo TE, Ng EJJoCE Management (2018) Progress in urban greenery mitigation science–assessment methodologies advanced technologies and impact on cities. J Civ Eng Manag 24:638–671

Sanusi R, Jalil M (2021) Blue-green infrastructure determines the microclimate mitigation potential targeted for urban cooling. IOP Conf Ser Earth Environ Sci 918 012010. https://doi.org/10.1088/1755-1315/918/1/012010

Sanusi R, Bidin S (2020) Re-naturing cities: impact of microclimate. Hum Therm Comf Recreat Particip. https://doi.org/10.1007/978-3-030-37425-9_28

Shah A, Garg A, Mishra V (2021) Quantifying the local cooling effects of urban green spaces: evidence from Bengaluru, India 209:104043. https://doi.org/10.1016/j.landurbplan.2021.104043

Shahmohamadi P, Che-Ani AI, Maulud KNA, Tawil NM, Abdullah NAG (2011) The impact of anthropogenic heat on formation of urban heat island and energy consumption balance. Urban Stud Res 2011, Article ID 497524. https://doi.org/10.1155/2011/497524

Sharma D, Bharat A (2009) Urban Heat Island effect—causes impacts methods of measurement and mitigation options. Institute of Town Planners, India 6(2):69–77

Shashua-Bar L, Hoffman ME (2000) Vegetation as a climatic component in the design of an urban street: an empirical model for predicting the cooling effect of urban green areas with trees. Energy Build 31(3):221–235. https://doi.org/10.1016/S0378-7788(99)00018-3

Stewart & Oke (2012) Local climate zones for urban temperature studies. Bull Am Meteor Soc 93(12):1879–1900

Surawar M, Kotharkar R (2017) Assessment of urban heat island through remote sensing in Nagpur urban area using landsat 7 ETM+ satellite images. World Acad Sci Eng Technol Int J Urban Civ Eng 11(7)

Susca T, Zanghirella F, Colasuonno L, Fatto VD (2022) Effect of green wall installation on urban heat island and building energy use: a climate-informed systematic literature review. Renew Sustain Energy Rev 159:112100. https://doi.org/10.1016/j.rser.2022.112100

TERI (2014) Assessing climate change vulnerability and adaptation strategies for Maharashtra: Maharashtra state adaptation action plan on climate change (MSAAPCC). The Energy and Resources Institute, New Delhi, p 306

URDPFI Guidelines (2014) Urban and regional development plans formulation and implementation guidelines. Ministry of Housing and Urban Affairs. Government of India. https://mohua.gov.in/upload/uploadfiles/files/URDPFI%20Guidelines%20Vol%20I(2).pdf

Vasudha Foundation (2022). Climate change and environment action plan of Nagpur District. https://www.vasudha-foundation.org/climate-change-and-environment-action-plan-of-nagpur-district/

Veena K, Parammasivam KM, Venkatesh TN (2019) Urban heat island studies: current status in India and a comparison with the international studies. J Earth Syst Sci 129:85. https://doi.org/10.1007/s12040-020-1351-y

WHO (2014) Quantitative risk assessment of the effects of climate change on selected causes of death, 2030s and 2050s. https://apps.who.int/iris/handle/10665/134014.

Yu Z, Yang G, Zuo S, Jørgensen G, Koga M, Vejre H (2020) Critical review on the cooling effect of urban blue-green space: a threshold-size perspective. Urban Urban Green 49:126630

Chapter 3
Urban Heat Island and Its Implication for Cities

Rubika Singh and Neetu Kapoor

Abstract Urbanization substantially influences the local climate of a city, giving rise to the urban heat island (UHI) phenomenon and affecting the quality of human life and urban environment. UHI is a phenomenon caused by urbanization, population pressures, industrialization, insufficient vegetation, and transport systems which result in urban regions experiencing warmer temperatures than neighbouring outlying areas. As a result, the UHI effects have significantly risen, causing elevated temperatures, and altering the microclimate of the urban region. Numerous factors contribute to the development of UHI, including replacing natural land cover with impervious areas, buildings, and other infrastructures. The UHI effect is also exacerbated by heat from factories, cars, and air conditioners, which warm the surrounding air. As a result, urban surfaces experience slower cooling through nocturnal radiation, resulting in a delayed decrease in temperature during the night, whereas surrounding areas cool faster. The consequence of the UHI is thermal discomfort experienced by urban dwellers, which can lead to serious medical disorders, in some cases, it also increases the mortality rate caused by urban heat stress. The chapter consists of detailed insights into the UHI, various forms of UHI, techniques used to identify and measure UHI using remote sensing techniques and ground-based measurements, their impacts and factors contributing to the formation of UHI, as well as exploring the nature-based solution to minimize the UHI effect and improve the urban microclimate. Furthermore, it also employs a detailed case study carried out in Jhansi, India, to develop a comprehensive understanding of the UHI phenomenon. By employing a multifaceted approach, the study examines the influence of land use land cover (LULC) on UHI, and fluctuations in urban surface temperatures are determined through meteorological data remote sensing, and geographic information system (GIS) analysis. The study provides valuable insights for urban planners and policymakers, aiding in the development of mitigation and adaptation strategies for urbanization and climate change, promoting resilience and sustainability.

Keywords Urban heat island (UHI) · Urbanization · Climate change · Remote sensing techniques · Land use and land cover

R. Singh · N. Kapoor (✉)
Department of Architecture, National Institute of Technology, Hamirpur, Himachal Pradesh, India
e-mail: neetu@nith.ac.in

M. Firoz C et al. (eds.), *Nature-Based Solutions for Urban and Peri-Urban Areas*, Disaster Risk Reduction, https://doi.org/10.1007/978-981-96-5933-3_3

3.1 Introduction

Urbanization has been increasing significantly over the last century. The term "urbanization" refers to the movement of people from rural to urban areas, which increases the population of urban areas (Mcgranahan and Satterthwaite 2014; Silva and Mendes 2012; Sruthi Krishnan and Mohammed Firoz 2020). Cities are developing quickly due to increased urbanization, which changes the features and structure of the surrounding landscape (Ali et al. 2017). Globally, more than half of the population lives in urban areas than in rural ones, with 55% of the population living in urban areas as of 2018. In 1950, the percentage of the population living in cities worldwide was 30%, and that number is projected to increase to 68% by 2050 (Nations et al. 2018). Growing urbanization rates have resulted in substantial changes in land use, environment, economy, and population. Urbanization is often linked with a decline in environmental quality, including water, air, and noise quality. It is generally believed that the rise in global temperatures caused by climate change will be more severe in urban areas due to the Urban Heat Island effect (UHI), which is produced by the modification of natural surfaces into built-up, roads and impermeable surfaces (Cullis et al. 2019). UHI was initially coined by Luke Howard in 1818 when he noticed climate "artificial excess of heat" in London compared to the countryside (Gartland 2012). The term "UHI" refers to urban areas that experience increased temperatures when compared to their surrounding rural areas. Heat islands can be formed in towns and smaller cities, though this impact usually decreases with decreasing city size (EPA Office of Atmospheric Programs 2011). Urban cities with high UHI intensity experience reduced cooling impact at night and early morning due to built-up areas absorbing solar radiation and radiating the absorbed thermal energy slowly at night (Sharma et al. 2021). This causes an increase in energy consumption because of the use of air conditioning and refrigeration equipment. For every degree Celsius as temperatures rise, the maximum urban demand for power and energy-water consumption rises by 2–4% (Mathew et al. 2016). Maximum air and surface temperatures within the urban areas are also a result of artificial heat emitted by combustive processes from automobiles and industrial activity as well as heat escaping from commercial and home air conditioning. The main adverse effects of UHI are linked to rising temperatures in urban areas, particularly increased risks of heat waves and their effects (such as higher rates of urban resident mortality and morbidity), discomfort for people, and higher summertime energy consumption. UHI research can be classified as air UHI (AUHI) or surface UHI (SUHI) based on their formation mechanisms as displayed in Table 3.1. UHI has been measured at various scales for numerous cities worldwide as shown in Table 3.2. Researchers indicate that land use land cover (LULC) and land surface temperature (LST) are important factors to consider when studying UHI (Barat et al. 2018). LST of the earth's surface is a crucial measure that takes upward thermal radiation into account. It is one of the most important input parameters for many applications, including crop monitoring, greenhouse gas measurement, urban climatology, research on global warming, and the definition of the UHI phenomenon (Anderson et al. 2008; Li et al. 2013; Li

Table 3.1 Types of UHI

Types	Description	Typical identification method	Possible impact
SUHI	The SUHI refers to the higher LST in urban regions compared to rural areas. SUHI exists during the day and night but is most noticeable during the day. Seasonal variation in sun intensity, weather, and vegetation. SUHI is typically strongest in summer	Indirect measurerrent: remote sensing	• Urban Planning • Outdoor thermal comfort • Storm water runoff
AUII	When the urban atmosphere warms up relative to the surrounding rural atmosphere, it is referred to as AUHI (Harroodi, 2018). It generally appears in the evening, especially after sunset	Direct measurerrent: Fixed weather stations mobile traverses	• Thermal comfort • Building energy consumption • Ail quality • Urban ecology • Stormwater runoff
(a) CLUHI	CLUHI exists in the atmosphere surrounding human settlement, extending from the ground to the tops of the trees and roofs	Fixed weather station Traverse measurerrents Urban canopy models	
(b) BLUHI	The BLUHI exists above the canopy layer, which typically extends to a maximum elevation of 1.5 km above the surface of the earth. It reaches higher atmospheric altitudes and the tops of trees and roofs	Fixed weather stations Traverse measurerrents using aircraft meso-scale models	• Air mixing circulation • Ail quality • Atmospheric cloud formation • Atmospheric winds

and Becker 1993; Mathew et al. 2018; Proceedings-Lake; Ramachandra et al. 2016; Running et al. 1994; Schmugge and Becker 1991). As shown in Fig. 3.1, the chapter provides an in-depth understanding of UHI, its types, measuring methods, effects and factors contributing to its formation, along with a case study in Jhansi, India, using remote sensing data and GIS Software.

3.2 Factors Contributing to the Formation of UHI

As illustrated in Fig. 3.2, a wide range of factors contribute to the development of the UHI. The following section discusses the particular factors of the UHI as shown in Table 3.3.

Table 3.2 Previous studies on the assessment of UHI assessment

Author	Study area	Findings
(Halder et al. 2021)	Kolkata	There was an annual increase in temperature of 0.157 °C
(Sharma et al. 2021)	Noida	A primary factor contributing to the significant rise in urban hotspots in the southern and central regions of the city in 2019 is the transformation of natural and agricultural lands into developed zones
(Singh et al. 2017)	Lucknow	Land use types directly impact surface heat levels, with vegetation playing a significant role in reducing the impacts of UHI
(Singh et al. 2024)	Bhopal	Changes in LULC have resulted in a 2 °C increase in LST in Bhopal and the surrounding area
(Pramanik and Punia 2020)	Delhi	SUHI is a result of rocky exposed bare soil, fallow land, and built-up land
(Khare et al. 2020)	Delhi, Ahmedabad, Bangalore	Relationship between LST, NDVI, built-up areas, and green cover in significant Indian cities to ascertain how much urban greenery contributes to mitigating UHI
(Shirani-bidabadi et al. 2019)	Ishfan, Iran	Isfahan's UHI ratio increased from 0.16 in 1999 to 0.3 in 2016
(Abutaleb et al. 2015)	Greater Cairo, Egypt	LULC now significantly impacts the 0.5 °C to 3.5 °C temperature variations found in the heat islands
(Dan et al. 2009)	Beijing	The LST is highest on bare land and lowest in water, depending on the kind of land usage
(Srivanit and Hokao 2018)	Bangkok	Remote sensing technologies were found to be an efficient method for monitoring and analyzing urban growth trends and their impact on urban climates
(Fadhil et al. 2023)	Not Defined	Increasing the amount of urban greenery is a good way to lessen the effects of UHI
(Elsayed 2012)	Kaula Lumpur	Land management and plant cover are the key tactics for reducing the city's UHI intensity

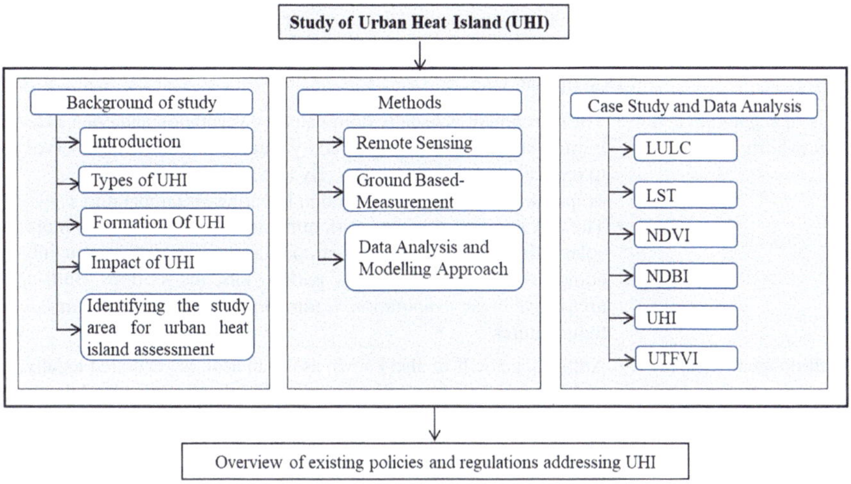

Fig. 3.1 Conceptual framework

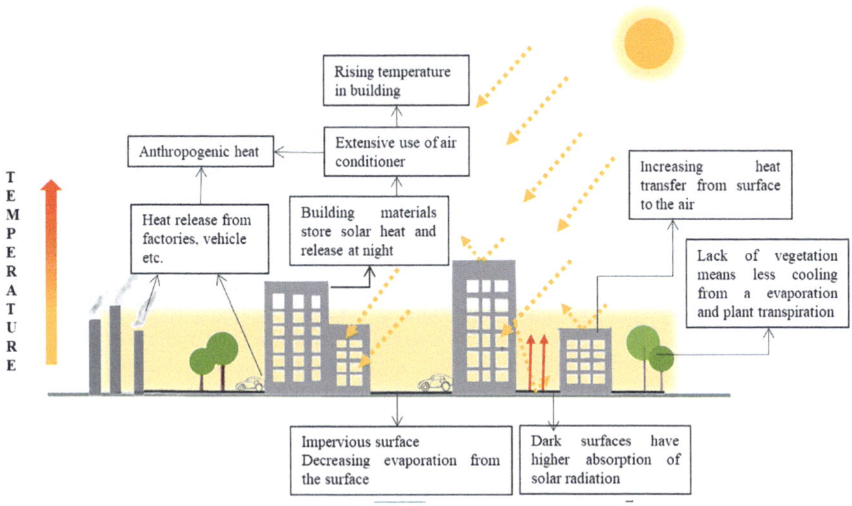

Fig. 3.2 Formation of UHI

Table 3.3 Factors contributing to the formation of UHI ; ; ; Sastry et al. 2017 (Source:Kaloustian 2015Imran et al. 2021Moyer 2016)

Factors	Description
Lack of vegetation and pemreability	• ThEnvironment is usually dominated by vegetation and open space in rural areas. Shade is produced by plants and trees, which serves to reduce surface temperatures. By a mechanism known as evapotranspiration, they also aid in lowering air temperatures • The dry, impermeable, and dark mineralized surfaces absorb more solar radiation that make up metropolitan environments are mainly composed of pavements, roads, parking lots, and rooftops. Built-up areas have lower evaporation, leading to higher surface and air temperatures
Anthropogenic heat	• Anthropogenic heat also known as urban heat, is generated locally by human-induced activities in cities. This heat is generated from a variety of sources, such as automobile emissions, summertime air conditioning use, and industrial facilities • As a result, these activities have a negative influence on the environment and related ecosystems • Based on urban activity and infrastructure, different amounts of heat are created; more energy-consuming buildings and transportation systems produce more heat
Urban geometry	• Urban geometry is the term used to define how a city's building layout, size, and spacing affect the total amount of radiation that the urban infrastructure absorbs, reflects, and emits • In urban areas, structures and surfaces are frequently at least partially covered by surrounding buildings and other objects. As a result, they grow into huge themral masses that are slow to dissipate their heat • Urban geometry has two elements urban canyons and Sky View Factor (SVF);
Properties of urban materials	• Properties of urban material influence the formation of UHI, particularly heat capacity, albedo, and thermal emissivity, which control how solar radiation is reflected, emitted, and absorbed • Common building materials for roadways and roofing materials, like concrete, cement, and asphalt, have thermal characteristics that differ greatly from ambient rural settings. These types of materials tend to capture and hold sun radiation in urban areas, which is then gradually released from the urban surface at night
Weather	• The two primary meteorological elements that influence the development of UHI are cloud cover and wind. With clear skies and slow wind speeds, UHI will be more severe during anticyclonic conditions. The surroundings and urban structures can be heated to the maximum extent by clear skies, but low wind speeds enable the air to warm and stagnate because there is less turbulence
Geographic location	• The terrain and geographic location of an area influence not just the surroundings, but also the weather. Local wind patterns, as well as other regional or local meteorological conditions, can affect UHI

3.3 Impacts of UHI in the Cities

3.3.1 Environmental Impacts

3.3.1.1 Greenhouse Gases and Air Quality Deterioration

Summertime smog primarily formed by UHI, is a combination of microscopic particles and ground-level ozone, resulting from the mutual interaction between pollutants (nitrogen oxides) and organic compounds that are volatile in the presence of heat and sunlight (urban heat island mitigation strategies: 2021 update november 2021 synthesis of knowledge, n.d.). Fossil fuels used in power plants release greenhouse gases, contributing to global warming and negatively impacting microclimates in urban areas as compared to rural areas, particularly through the creation of ground-level ozone and altering local microclimates and macroclimates, such as wind patterns, humidity variations, storms and flooding (Heat Island Impacts | US EPA n.d.; O'Malley et al. 2015).

3.3.1.2 Water Quality

The heat from pavement and rooftop surfaces can cause stormwater runoff to warm up. This water is then released into rivers, ponds, lakes, and streams, where it increases temperature as it drains into storm sewers. Warm stormwater runoff can induce temperature fluctuations in aquatic habitats, which can be exceedingly harmful or even fatal for aquatic life (Heat Island Impacts | US EPA n.d.). Primarily due to thermal pollution, SUHI deteriorates water quality.

3.3.1.3 Increased Energy Consumption

Hot summer days in cities put greater stress on the electrical grid during peak demand hours, which are typically summer afternoons when households and offices use lighting, appliances, and air conditioning. For every 1°F (0.6 °C) that the summertime temperature rises, the maximum urban power demand rises by 1.5–2%. (EPA Office of Atmospheric Programs 2011).

3.3.1.4 Health Impacts

The population's health and other socioeconomic characteristics are often taken into account when calculating the consequences of the UHI and, more broadly, climate change. UHI is linked to higher air pollution levels, decreased nighttime cooling, elevated surface temperatures during the day, and heat stroke and mortality due

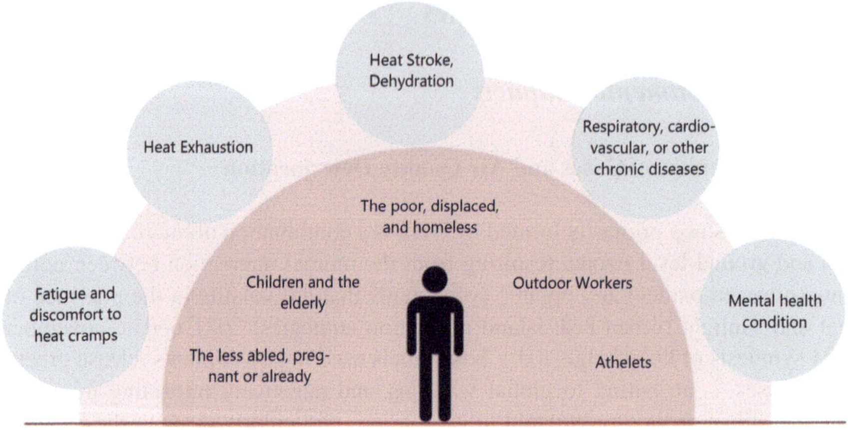

Fig. 3.3 Health-related impacts and vulnerable population groups

to heat risks. These factors can all have an impact on human health. Vulnerable populations are more at risk during these events as shown in Fig. 3.3.

3.3.1.5 Economic Impacts

Increased summer temperatures increase the need for air conditioning and, thus, the cost of power increases. On the other hand, higher ozone and carbon dioxide emissions are typically associated with an increase in hospital admissions, illnesses, and medical costs. Furthermore, people become less productive and less active during periods of intense heat. High temperatures, especially during heat waves, have a negative impact on outdoor labour (Kaloustian 2015). Presently, extreme heat costs New Delhi 4% of its entire economic output by lowering worker productivity; by 2050, this loss could rise to 5%. Labour productivity losses are expected to reach USD 6.1 billion (INR 475,013 crore) by 2050, accounting for 5% of potential output (Hot Cities, Chilled Economies: Economic Impacts of Heat on Cities n.d.).

3.4 Methods of Measuring UHI

3.4.1 Remote Sensing Method

The method of Remote Sensing can be used to study SUHI. This method enables us to map the SUHI pattern for an entire city or region. A location's infrared images can be obtained by satellites like LANDSAT, MODIS, ASTER, and TERRA or by taking photographs with thermal cameras mounted on the planes that are intended

Table 3.4 Types of ground-based measurement (*Source* Sahu et al. 2014; Sastry et al. 2017; World Bank 2020)

Ground-based Measurement methods	Description
Meteorological data	• To understand the effects of the urban climate, researchers typically compare the meteorological aspects that are measured and seen in urban and rural locations • National meteorological services operate observational systems and provide the necessary meteorological information (such as air temperature, wind, humidity, and precipitation). The main observational focus in this method is air temperature
Field measurement	• In this method, a region or area is chosen for UHI analysis. The region may surround a single structure or be as big as the entire city. This method involves measuring near-surface temperature variation in rural areas and adjacent urban areas using mobile or stationary instruments • This method has certain drawbacks, including the cost and time-consuming setup of measurement equipment in urban areas and the limited simultaneous measurement of certain parameters
Mobile survey	• In this kind of survey important pathways covering practically all land uses are identified. A vehicle can travel selected routes at a constant modest speed. To minimize weather variation, the survey must be finished in a duration of one to one and a half hours

to fly over that city (Sastry et al. 2017). The spatial structure of UHI can be better understood by remote sensing data, particularly in places with sparse monitoring networks (World Bank 2020). There are additional limitations to using remote sensing for UHI measurements, such as those related to cloud cover, geographical resolution, and temporal availability. They are unable to capture thermal images of surfaces that are vertical, like, outside walls of buildings (Sastry et al. 2017).

3.4.1.1 Ground-Based Measurement

Traditionally, UHI has been measured using ground-based approaches, which include either mobile transport of temperature sensors over a city and its countryside or a stationary network of sensors spread in and around a city. The type of ground-based measurement is discussed in Table 3.4.

3.4.1.2 Data Analysis and Modelling Approach

There are several ways to process the data that are gathered from the methods mentioned above. UHI intensity evaluation involves comparing meteorological factors, using mathematical formulas and computations, and integrating data into a GIS system for geographical distribution maps. There is a broad range of modelling

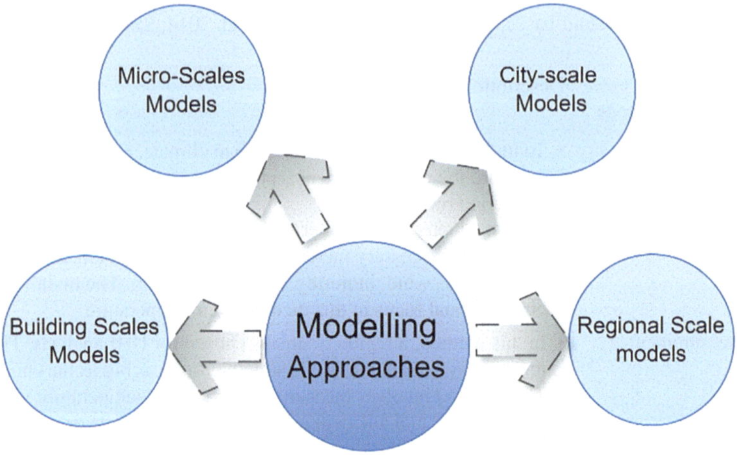

Fig. 3.4 Different types of modelling approaches

approaches in terms of their application to different issues and their spatial scales as shown in Fig. 3.4.

3.5 Material and Methods

3.5.1 Study Area

Jhansi city is situated in Uttar Pradesh, India (Fig. 3.5), located at 25.44 N, 78.56 E. Jhansi district is situated in the belt of drought-prone areas that runs through Uttar Pradesh. The study area of Jhansi City is 339 km^2. The demographic profile of Jhansi city is shown in Fig. 3.6. The city experiences a mild subtropical to semi-arid climate. With an average daily highest temperature of 24.10 °C and an average daily lowest temperature of 9.2 °C. Throughout the year, January is the coldest month. May is the hottest month, with an average daily highest temperature of 46 °C and an average daily lowest temperature of 28.80 °C. The district experiences 850 mm of annual rainfall on average, with the monsoon months of June, July, August, and September accounting for 91% of this total. The average wind speed is 4.8 km/h.

3.5.2 Methodology

The detailed methodology to prepare LULC, LST, Spectral indicator, Urban Thermal Field Variance Index (UTFVI) and UHI is shown in Fig. 3.7. For the study, Landsat 5 TM, Landsat 7 ETM+ , and Landsat 8 OLI/TIRS data imagery with minimum

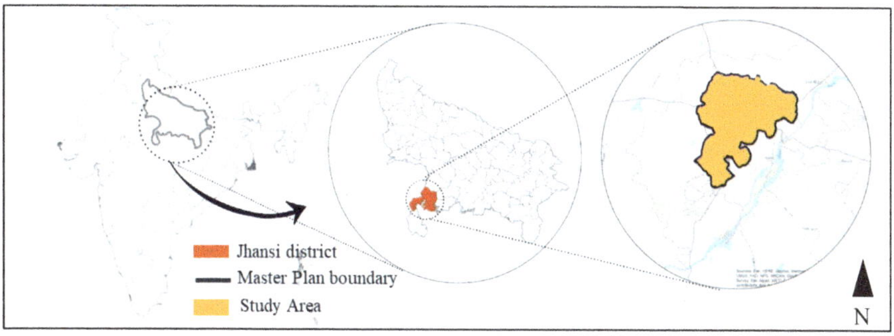

Fig. 3.5 Location map of the study area

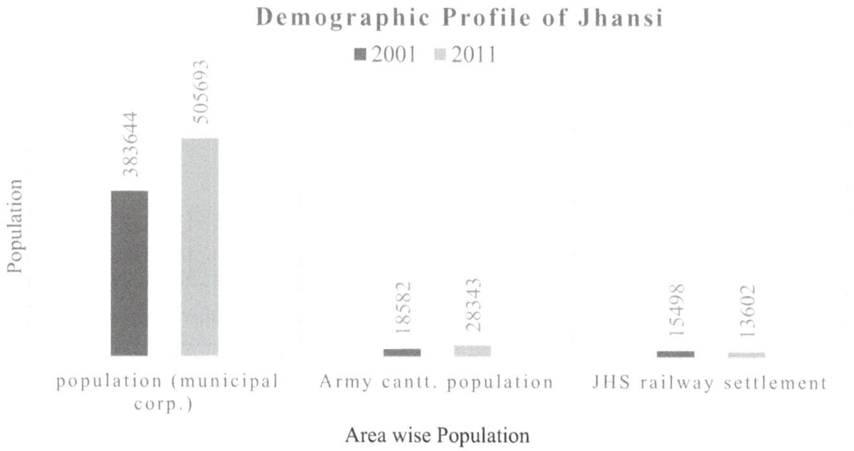

Fig. 3.6 Demographic profile of Jhansi

cloud cover for the years 2001, 2011, and 2021 was procured from the United States Geological Survey (USGS). Table 3.5 shows the satellite data specifications used in this study. Landsat satellite datasets were utilized to estimate the LULC and Landsat thermal band images were used for the retrieval of LST classes. UHI is computed using LST. In addition, spectral indices are generated to evaluate their correlation with LST, such as the Normalized Difference Built-up Index (NDBI) and the Normalized Difference Vegetation Index (NDVI). Furthermore, the thermal comfort and environmental quality of the urban area are also evaluated using the UTFVI. The boundary shape file was obtained from the Survey of India (SOI) and was clipped from the satellite data to define the study region.

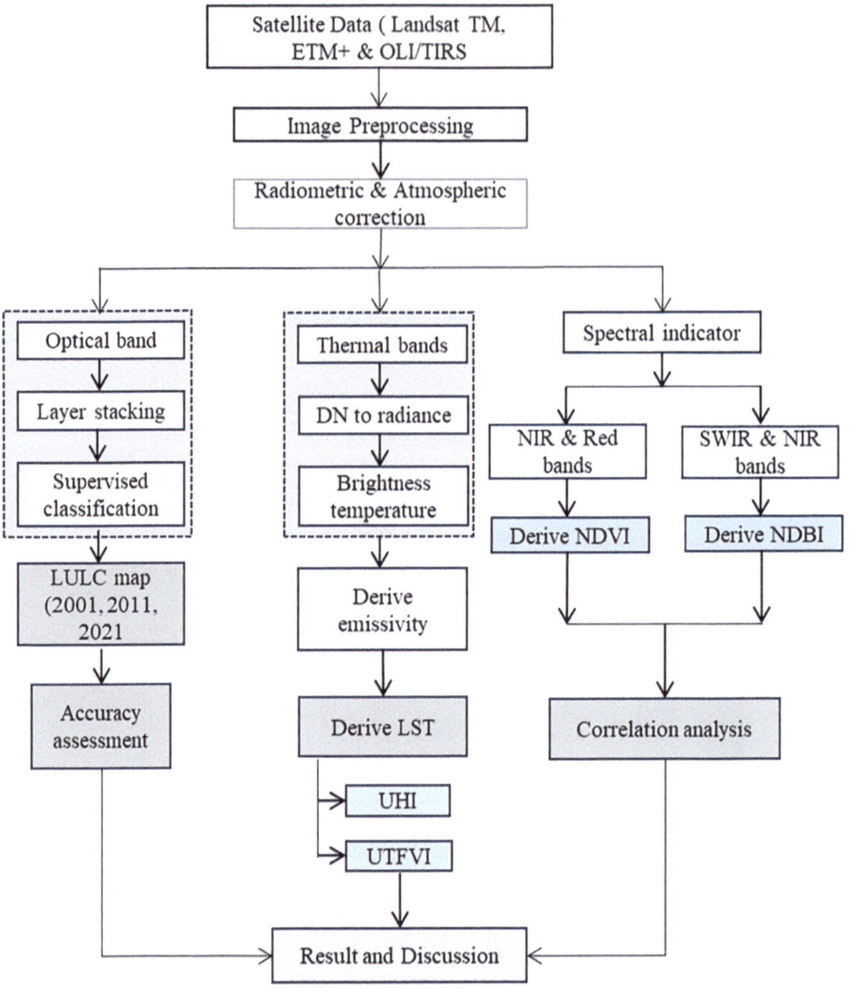

Fig. 3.7 Methodology

3.5.2.1 Image Pre-processed and Classification

The data was pre-processed using ENVI 5.3 software. Landsat 8 and Landsat ETM+ , Landsat TM images are composed of distinct band images that are layered and merged to create a multi-band image. These satellite datasets were used to classify LULC, LST, UHI, UTFVI and spectral indicators, then analyzed using ArcGIS 10.7.1 software.

To classify images, a supervised classification algorithm was used. The Maximum Likelihood Classifier (MLC) was used to select training sets for classification during this process. MLC is a statistical decision-making system that assigns pixels to classes

Table 3.5 Data specifications of different years

Satellite	Sensor	Date	Path and row	Data source
Landsat 5	TRR	10/08/2011	145.042	https://eai1hexpl oiei.usas.ao w
Landsat 7	**ETRR +**	18/09/2001 16/10/ 2011	145,042 145.042	
Landsat 8	OLfTIRS	3/10/2021	145.042	

based on their maximum likelihood. As indicated in Table 3.6, the study area was categorized into six categories: built-up land, cropland, vegetation, bare and fallow land, water bodies, and mining, and then a supervised classification technique and the MLC were employed.

3.5.2.2 Accuracy Assessment

After the classification procedure, the Kappa coefficient and overall accuracy were determined using an error matrix. The Kappa Index (K) can be computed using Eq. 3.1. For the accuracy assessment, it is suggested that at least 50 samples should be collected for each LULC class in the error matrix to reduce the possibility of a biased sample (Congalton 1991).

$$\mathrm{K} = (\text{Observed accuracy} - \text{Chance accuracy}/(1 - \text{Chance accuracy}) \quad (3.1)$$

3.5.2.3 Vegetation Index

Using remotely sensed data, the NDVI is used to determine vegetation density mapping. The NDVI scale ranges from −1 to + 1, where the green and vegetated area is represented by a value near + 1 on the NDVI scale, while an area without vegetation is represented by a value near −1 (Shukla and Jain 2021). It can be calculated as:

$$NDVI = \frac{(NIR - R)}{(NIR + R)} \quad (3.2)$$

where R is Visible Red Radiation and NIR is Near Infrared Radiation.

3.5.2.4 Built-Up Index

Using Eq. 3.3, the NDBI was calculated to detect urban areas. The values of the NDBI scale range from −1 and + 1. NDBI Values near + 1, indicate the presence of a built-up area, while NDBI values near −1 indicate water bodies.

$$\text{NDBI} = \frac{(\text{SWIR} - \text{NIR})}{(\text{SWIR} + \text{NIR})} \quad (3.3)$$

where SWIR is Short Wave Infrared.

3.5.2.5 Derivation of LST

The LST is a critical component of the Earth's energy budget and a key surface environment variable. Digital Number (DN) must be converted to radiations (L_λ) to estimate LST using ArcGIS 10.7.1.

Top of Atmospheric Spectral Radiance (TOA);

TOA calculates spectral radiance based on mean irradiance and solar zenith angle (Bagyaraj et al. 2023). Equation 3.4. is used to calculate TOA (L_λ)) radiance;

$$L_\lambda = \frac{(L_{max} - L_{min}).(Qcal - Qcal_{min})}{Qcal_{max} - Qcal_{min}} + L_{min} \quad (3.4)$$

The TOA radiance for Landsat 8 OLI was calculated using Eq. 3.5 (Bagyaraj et al. 2023; Mallick 2014).

$$L_\lambda = ML \text{ x } Qcal + AL \quad (3.5)$$

where L_λ = spectral radiance, L_{max} = spectral radiance to $Qcal_{max}$, L_{min} = spectral radiance to $Qcal_{min}$, Qcal = Level 1 pixel in DN, $Qcal_{max}$ = maximum quantized calibrated pixel value (1), $Qcal_{min}$ = minimum quantized calibrated pixel value (255), ML = band-specific multiplicative rescaling factor from the metadata (RADIANCE_MULT_BAND_n from the metadata), AL = band-specific additive rescaling factor from the metadata.

Conversion of Radiance to At-Sensor Temperature;

Equation 3.6 is used in the tool's algorithm to convert reflectance:

$$T_b = \frac{K_2}{\ln\left(\frac{K_1}{L_\lambda} + 1\right)} - 273.15 \quad (3.6)$$

The band-specific thermal conversion constants from the metadata are denoted by K_1 and K_2, and T_b is brightness temperature in °C.

Calculating LST;

Equation 3.7 evaluates LST based on At-sensor temperature (Avdan and Jovanovska 2016; Liu and Zhang 2011).

$$LST = \frac{T_b}{\left\{1 + \left(\frac{\lambda T_b}{\rho}\right) * \ln(LSE)\right\}} \quad (3.7)$$

where ρ is a constant that is determined by calculating h · c/σ (1.438 · 10 − 2 mK), where h is Planck's constant = 6.626 · 10 − 34 Js and c = light velocity = 2.998 · 108 m/s, LSE is Land Surface Emissivity and Tb is brightness temperature (°C) obtained using Eq. 3.6.

3.5.2.6 Estimating UHI Intensity

Equation 3.8 is used to estimate the UHI map, which shows the temperature variation in urban areas and the surrounding areas (Halder et al. 2021).

$$\mathrm{UHI} = \frac{T_s - T_{mean}}{SD} \tag{3.8}$$

where T_s is the LST in kelvin, T_{mean} is the mean LST in kelvin, and SD is the standard deviation of the computed LST map.

3.5.2.7 Estimating UTFVI

To describe urban health and heat dispersion, the UTFVI, or the degree of thermal comfort in an urban area, is computed using Eq. 3.9.

$$\mathrm{UTFVI} = \frac{T_s - T_{mean}}{T_{mean}} \tag{3.9}$$

3.6 Result and Discussion

3.6.1 LULC Change

According to Landsat images from 2001, 2011, and 2021, the LULC maps of the city of Jhansi generated as shown in Fig. 3.8. Figure 3.9a displays the year-by-year supervised classification area, whereas.

Figure 3.9b displays the increase and decrease of LULC classes over time. Results show that over time, the proportion of built-up land has increased, accounting for a total area of 25.13 km^2 in 2001, and 87.47 km^2 in 2021 as shown in Table 3.7. Further, it was found that the amount of vegetation decreased by 9% between 2001 and 2021 as shown in Table 3.8; in 2001, it covered 83.09 km^2 2001 and in 2021, it covered 54.14 km^2. In 2021, it was noticed that most of the bare and fallow land had been converted into built-up and cropland. It was observed that there has been a decrease in cropland from 27% in 2001 to 26% in 2021. Also, the overall area of bare and fallow land decreased by 18% in 2021, it was 130.35 km^2 in 2001 and reduced

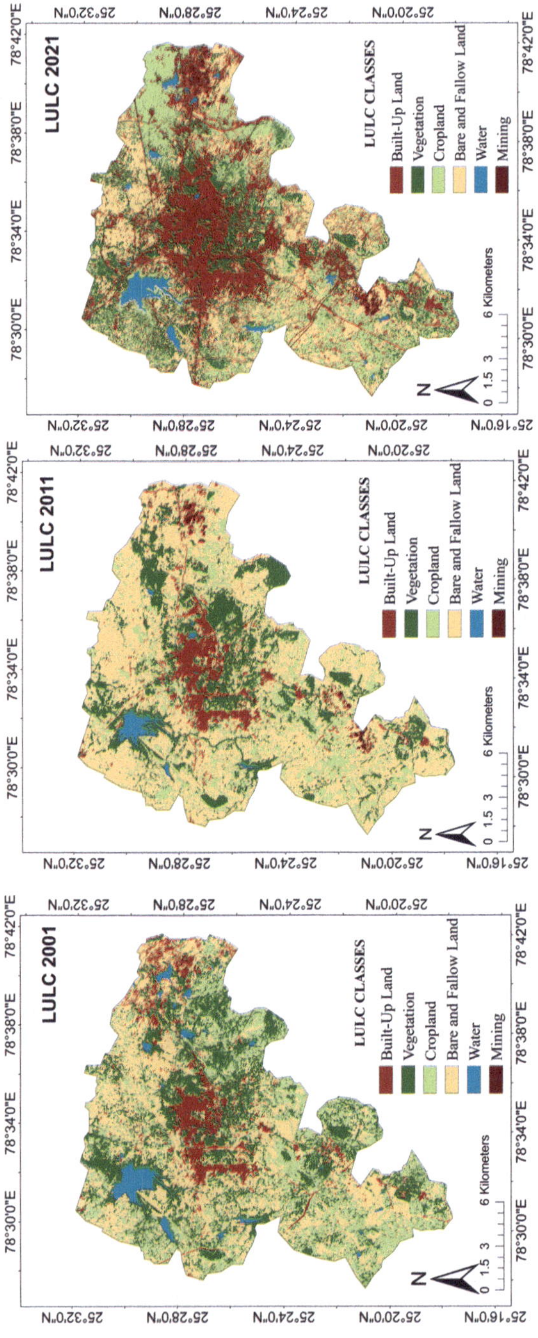

Fig. 3.8 LST maps of years 2001, 2011, and 2021

Fig. 3.9 **a** Area of different LULC classes, **b** area increased/decreased

Table 3.6 Description of the LULC classes

Sr. No	LULC classes	Description
1	Built-up land	Commercial area. Residential area. Transportations, Industrial area. Roads and Construction area
2	Vegetation	Natural growing forest, its including many types of trees. Glass area, open vegetated area
3	Cropland	Agriculture area
4	Bare land and fallow land	Exposed and Non developed area (soil, fallow land, rock)
5	Water bodies	Lakes, Ponds, Rivers and Open water bodies
6	Mining	Active mine (diaspore, soapstone, gypspum)

Table 3.7 Area of LULC classes

LULC classes	2001		2011		2021	
	Area (km^2)	%	Area (km^2)	%	Area (km^2)	%
Built-up land	25.13	7	24.02	7	87.47	26
Vegetation	83.09	24	63.68	19	54.14	16
Cropland	91.50	27	53.11	16	87.68	26
Bare and fallow land	130.36	38	192.31	57	98.21	29
Water	9.54	3	4.83	1	8.86	3
Mining	0.16	0	1.84	1	3.36	1
Total	339		339		339	

to 98.21 km^2 in 2021. The study's analysis shows that Jhansi has grown significantly during the last 20 years. The majority of the area under study is urbanized in its core regions, and as a result, there is less vegetation there. Many factors, including extensive urban development, unplanned urban population growth, and rural–urban migration in search of higher living standards, may be responsible for these changes.

Accuracy assessments for LULC classes have been performed in the current study. As a result, the total accuracy evaluation findings for 2001, 2011, and 2021 are, correspondingly 95%, 97%, and 95%, whereas the Kappa statistics assessment for the years 2001, 2011, and 2021 are 0.93, 0.95, and 0.94.

3.6.2 LST Variation Over Different LULC Classes

The study area observes severe temperature changes in different parts due to urban growth. Figure 3.10 shows the LST maps for the years 2001, 2011, and 2021. Table 3.9 illustrates the LST value, for the years 2001, 2011, and 2021. The maximum LST values were recorded for bare and fallow land in 2001 and 2011, respectively, at 34.8 °C and 38 °C. The existence of bare and fallow land in 2001 and 2011 resulted in higher LST values than in 2021. Here, "bare and fallow land" refers to uncultivated or openly exposed land to direct sunlight, which has a greater capacity to absorb solar radiation than agricultural and vegetative areas. Furthermore, the results demonstrated that the central region of Jhansi City has the majority of urban built-up areas, and has a high population density, Impervious surfaces such as pavement, and roads, and the extensive usage of tiles in urban areas can raise the surface temperature of that area (Reisi et al. 2019). The built-up land has been experiencing maximum LST values, 28 °C in 2001, 30.5 °C in 2011 and 31 °C in 2021 as shown in Table 3.10. Furthermore, the areas with the lowest LST were those with water bodies followed by vegetation and cropland. Vegetation has a lower temperature than built-up or bare soil because of transpiration, which reduces heat accumulation at the surface.

3.6.3 Correlation Between Spectral Indices and LST

The regional distribution of the NDVI and NDBI for the years 2001, 2011, and 2021 is shown in Figs. 3.11 and 3.12. The LST and NDVI have demonstrated a negative association; the corresponding R^2 values for the years 2001, 2011, and 2021 are, respectively, 0.01, 0.045, and 0.026. This is because the development damages the vegetated land and creates significant temperature variations in this area. Between the NDBI and LST, there is a positive relation found. The LST and NDBI correlation coefficients for the years 2001, 2011, and 2021 are, respectively, 0.19, 0.62, and 0.55.

3.6.4 UHI Intensity

The study found that UHI increased from the center of the city to all directions. In 2021, the city's central areas had significant variations in UHI due to increased development and vegetation degradation. Urban expansion has been linked to spatial and temporal variance in UHI. As shown in Fig. 3.13, it is observed that the city of Jhansi witnessed a higher intensity of UHI in 2021 compared to 2001 and 2011. It was observed that there was an increase in UHI intensity from 2001 to 2021 as shown in Table 3.11.

3.6.5 UTFVI

An analytical explanation for the UHI impact in the study region has been given by UTFVI. As indicated in Table 3.12, UTFVI is further classified into six phases utilizing six distinct ecological evaluation indices. The city's overall environmental

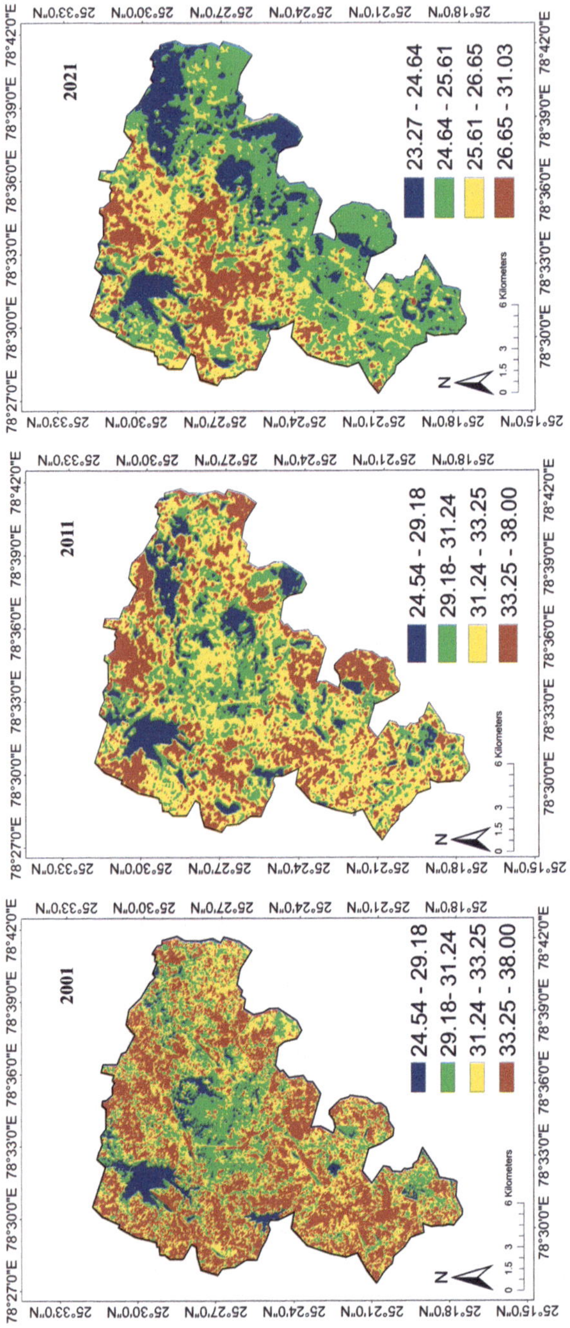

Fig. 3.10 LST of years 2001, 2011, and 2021

Table 3.8 LULC change dynamics in 2001, 2011, and 2021

	2001–2011		2011–2021		2001–2021	
	Change (km^2)	Change (%)	Change (km^2)	Change (%)	Change (km^2)	Change (%)
Built up land	−1.11	0	63.45	19	62.34	18
Vegetation	−19.41	−6	−9.54	−3	−28.95	−9
Cropland	−38.39	−11	34.57	10	−3.82	−1
Bare and fallow land	61.95	18	−94.10	−28	−32.15	−9
Water	−4.72	−1	4.03	1	−0.68	0
Mining	1.68	0	1.52	0	3.20	1

Table 3.9 LST over the years 2001, 2011, and 2021

Year	LST °C (MIN)	LST°C (MAX)	LST °C (MEAN)	SD
2001	19.57	34.8	27	1.99
2011	24.54	38	31.75	2.02
2021	23.24	31.03	25.56	0.99

quality and thermal comfort decreased between 2001 and 2021, which may indicate a rise in UHI phenomena. According to Fig. 3.14, the urban core area with the majority of built-up land had a weaker to higher UHI effect in 2021 compared to 2001 and 2011. Built-up regions absorb most of the radiation and raise the LST because of the different mineral composition of the concretized area (Pandey et al. 2022). The urban core area showed little to no UHI effect in 2001 and 2011, whereas UTFVI was higher on bare land and fallow land.

3.7 Overview of Existing Policies and Regulations Addressing UHI

According to the Addendum to Urban and Regional Development Plan Formulation (URDPFI) and Implementation Guidelines, 2014, some strategies can help to reduce the UHI effect (Ministry of Housing and Urban Affairs n.d.).

3.7.1 Cool Roofing

An easy and cost-effective strategy to reduce the quantity of heat that enters the buildings is to install a cool roof. Cool roofs enhance thermal comfort, conserve

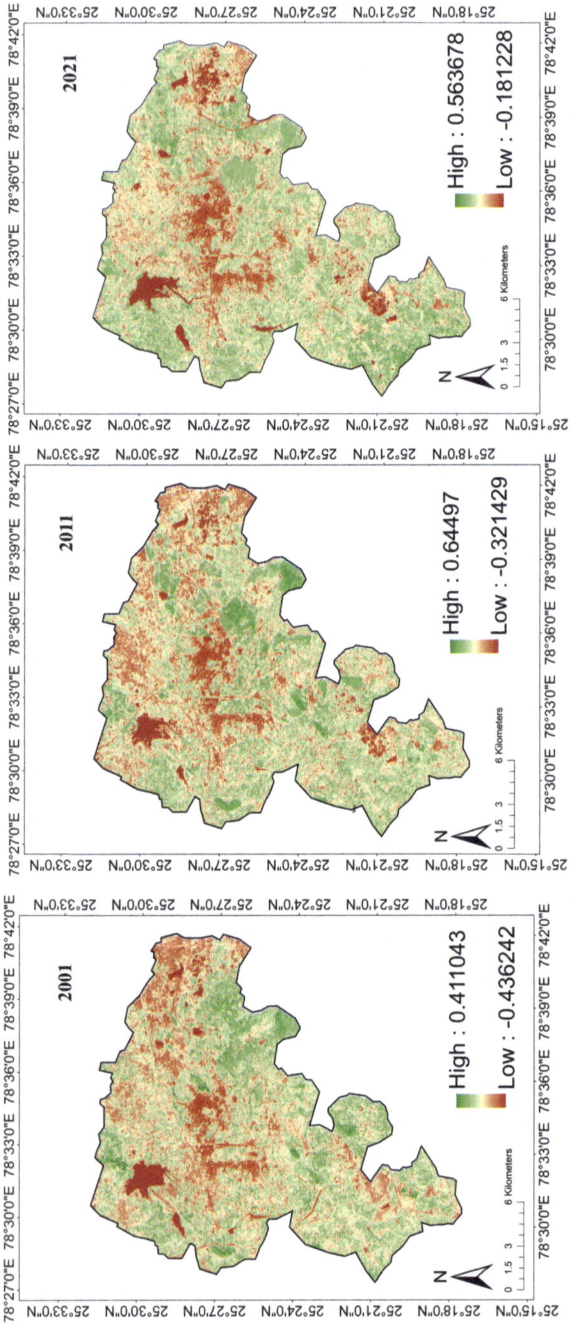

Fig. 3.11 NDVI maps of years 2001, 2011 and 2021

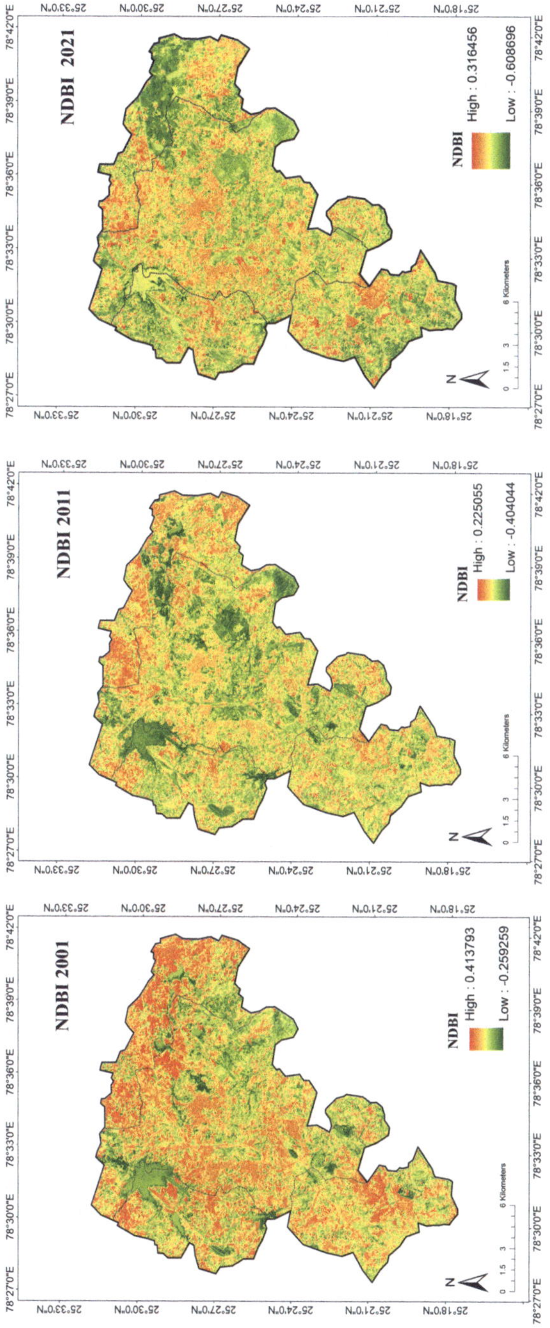

Fig. 3.12 NDBI maps of years 2001, 2011, and 2021

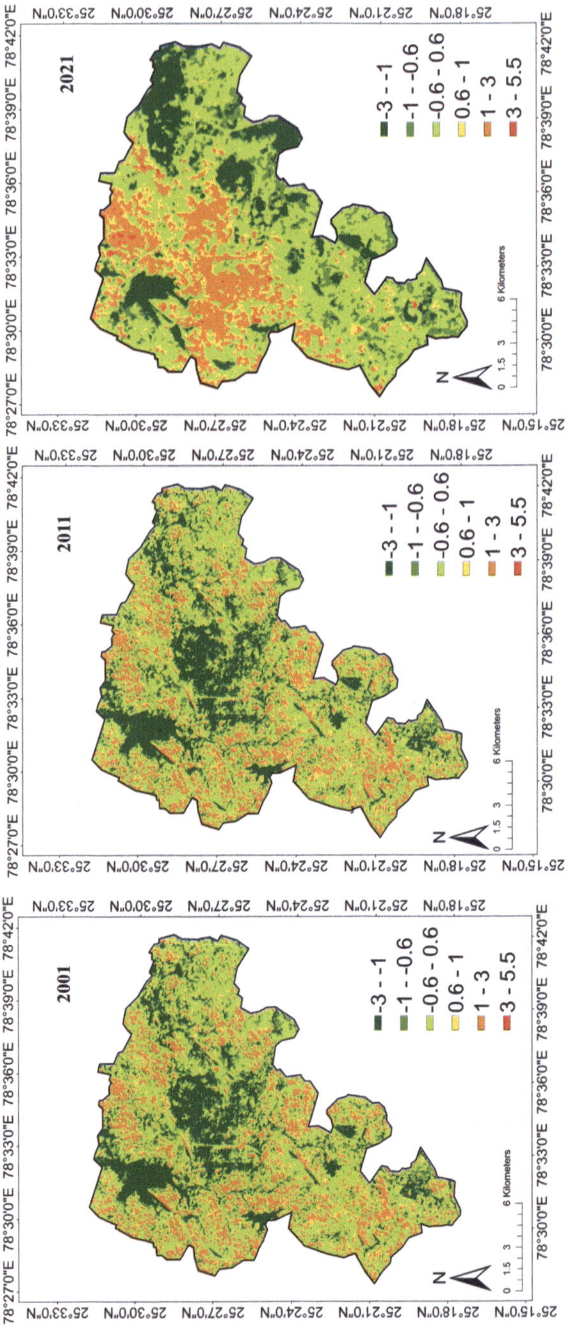

Fig. 3.13 UHI maps of year 2001, 2011 and 2021

Table 3.10 LST over different LULC classes for the years 2001, 2011, and 2021

LULC classes	2001		2011		2021	
	Min°C	Max°C	Min°C	Max°C	Min °C	Max°C
Built-up land	22	28	25	30.5	23.4	31
Vegetation	19.5	26.6	24.9	36.8	23.3	30.5
Cropland	21.1	27.6	25.8	37.6	23	29.7
Bare and Fallow land	21.1	34.8	25.4	38	23.3	31.9
Water	19.5	26.6	24.5	27.9	23.4	31
Mining	23.1	32.9	29.5	36.4	23.2	26.9

Table 3.11 UHI intensity

Years	UHI °C(max)	UHI °C (min)
2001	3.92	−3.72
2011	3.08	−3.56
2021	5.51	−2.30

Table 3.12 Scale of UTFVI (Halder et al. 2021)

LTFVI	T ~ HI phenomenon	Ecological evaluation index
< 0	None	Excellent
0.000–0.005	Weak	Good
0.005–0.010	Middle	Normal
0.010–0.015	Strong	Bad
0.015–0.020	Stronger	Worse
> 0.020	Strongest	Worst

energy, and reduce the need for cooling. Based on the surroundings, cool roofs can help keep indoor temperatures 2–5 °C (3.6–9°F) lower than normal roofs. Some of the Cool roofing materials; are membrane cool roofs, green roofs, mod roofs, coated cool roofs, and tiled cool roofs.

3.7.2 Water-Sensitive Urban Design

Urban evaporative cooling and misting systems, fountains, ponds, and man-made or natural wetlands are examples of water-inspired design (Ruefenacht and Acero 2017). In addition to these water-inspired designs, urban ambient temperatures can be lowered by 1–2 °C by using water-based landscape features including rivers, lakes, wetlands, and bioswales. To improve air quality, increase urban cooling, help

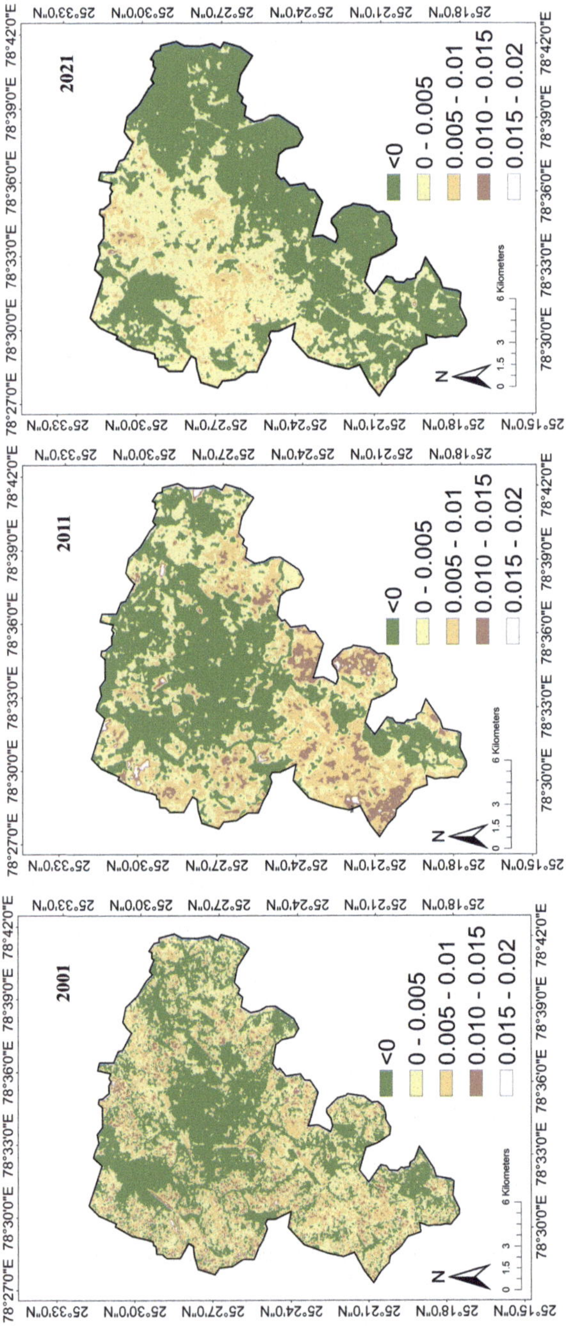

Fig. 3.14 UTFVI maps of years 2001, 2011 and 2021

manage flooding, and create visually appealing public areas, green infrastructure is typically integrated with water-based systems.

3.7.3 *Promoting Shading, Green Cover, and Vegetation*

Increasing vegetation is an important strategy to lessen the UHI effects, with planting trees and other vegetation carefully to shade roads and buildings and reducing runoff. Shading enhances outdoor thermal comfort by reducing radiant temperature, using vegetation to block radiations, and providing shade on streets, entrances, and public spaces. Additionally, a few ideas include creating additional mid-sized parks and open spaces for leisure and recreation in urban areas, creating green parking lots, and enhancing the façade of buildings with vertical greenery, a green roof, a vegetative layer at the top, and plant elements. Green roofs provide many advantages as; they lower the surrounding air temperature by 2.5 °C, which results in a 10% reduction in energy use (Refahi and Talkhabi 2015). The afternoon ambient temperature may decrease by up to 1–1.5°C with every 10% increase in the tree canopy, according to estimates (Ministry of Housing and Urban Affairs n.d.).

3.8 Conclusion

UHI, a local indicator of global climate change, negatively impacts energy consumption, carbon footprint, quality of life, human and other living things' health, and more. The chapter discusses UHI and its types, the factors contributing to the formation of UHI, the methods for measuring UHI, and the overview of the strategies used to reduce their effects and improve urban environmental conditions. Additionally, it employs a comprehensive study carried out in Jhansi, India. In the case study, analysis of LULC and its change, LST variation, UHI Intensity, and UTFVI index is investigated using remote sensing and GIS in Jhansi. The findings showed that urbanization affected the spatial distribution of LST, and the core region of Jhansi had an increase in temperature between 2001 and 2021 as a result of urban development, an urbanized structure, and a reduction in green spaces. The study also determined that there is a rise in UHI intensity in 2021 as compared to 2001 and 2011, due to factors such as developed land, bare or rocky surfaces, and a decrease in water bodies, vegetation, and Also, it was found that increased UTFVI was observed in urban core areas with an abundance of built-up terrain, suggesting a weak to significant UHI influence in 2021. The findings also showed that the LST and NDVI showed a negative correlation, with the UHI effect being less evident in vegetated areas. However, a positive relation between LST and NDBI was found, perhaps due to the warmer climate of built-up regions relative to agriculture and vegetation. Additionally, some of the following urban planning interventions and mitigating strategies are suggested in this study, which will help to reduce UHI, such as cool roofing, water-sensitive urban design, and promoting shading, green cover, and vegetation.

References

Abutaleb K, Ngie A, Darwish A, Ahmed M, Arafat S, Ahmed F, Abutaleb K, Ngie A, Darwish A, Ahmed M, Arafat S, Ahmed F (2015) Assessment of urban heat island using remotely sensed imagery over greater Cairo, Egypt. Adv Remote Sens 4(1):35–47. https://doi.org/10.4236/ARS.2015.41004

Ali SB, Patnaik S, Madguni O (2017) Microclimate land surface temperatures across urban land use/ land cover forms. Glob J Environ Sci Manag 3(3):231–242. https://doi.org/10.22034/gjesm.2017.03.03.001

Anderson MC, Norman JM, Kustas WP, Houborg R, Starks PJ, Agam N (2008) A thermal-based remote sensing technique for routine mapping of land-surface carbon, water, and energy fluxes from field to regional scales. Remote Sens Environ 112(12):4227–4241. https://doi.org/10.1016/J.RSE.2008.07.009

Avdan U, Jovanovska G (2016) Algorithm for automated mapping of land surface temperature using LANDSAT 8 satellite data. J Sens 2016(1). https://doi.org/10.1155/2016/1480307

Bagyaraj M, Senapathi V, Karthikeyan S, Chung SY, Khatibi R, Nadiri AA, Lajayer BA (2023) A study of urban heat island effects using remote sensing and GIS techniques in Kancheepuram, Tamil Nadu, India. Urban Clim 51. https://doi.org/10.1016/j.uclim.2023.101597

Barat A, Kumar S, Kumar P, Parth Sarthi P (2018) Characteristics of surface urban heat island (SUHI) over the gangetic plain of Bihar, India. Asia-Pac J Atmos Sci 54(2):205–214. https://doi.org/10.1007/s13143-018-0004-4

Beaun-Léa LM, Brunstein VM (n.d.) Urban heat island mitigation strategies: 2021 update November 2021 synthesis of knowledge. http://www.inspq.qc.ca/publications/2839

Congalton RG (1991) A review of assessing the accuracy of classifications of remotely sensed data. Remote Sens Environ 37(1):35–46. https://doi.org/10.1016/0034-4257(91)90048-B

Cullis JDS, Horn A, Rossouw N, Fisher-Jeffes L, Kunneke MM, Hoffman W (2019) Urbanisation, climate change and its impact on water quality and economic risks in a water scarce and rapidly urbanising catchment: case study of the berg river catchment. H_2 Open J 2(1):146–167. https://doi.org/10.2166/H2OJ.2019.027/862129/H2OJ2019027.PDF

Dan M, Huili G, Xiaojuan L, Wenji Z, Zhaoning G, Lin Z, Deyong H (2009) Study of urban heat island based on remote sensing in Beijing-capital zone. Jt Urban Remote Sens Event 2009:1–5. https://doi.org/10.1109/URS.2009.5137567

Elsayed ISM (2012) Mitigation of the urban heat island of the city of Kuala Lumpur, Malaysia. Middle East J Sci Res 11(11):1602–1613. https://doi.org/10.5829/IDOSI.MEJSR.2012.11.11.1590

EPA Office of Atmospheric Programs U (2011) Reducing urban heat islands: compendium of strategies—urban heat island basics. https://www.epa.gov/heat-islands/

Fadhil M, Hamoodi MN, Ziboon ART (2023) Mitigating urban heat island effects in urban environments: strategies and tools. IOP Conf Ser Earth Environ Sci 1129(1). https://doi.org/10.1088/1755-1315/1129/1/012025

Gartland L (2012) Heat Islands: understanding and mitigating heat in urban areas. Heat Isl Underst Mitigating Heat Urban Areas 9781849771559:1–200. https://doi.org/10.4324/9781849771559

Halder B, Bandyopadhyay J, Banik P (2021) Monitoring the effect of urban development on urban heat island based on remote sensing and geo-spatial approach in Kolkata and adjacent areas, India. Sustain Cities Soc 74. https://doi.org/10.1016/j.scs.2021.103186

Heat Island Impacts|US EPA (n.d.). https://www.epa.gov/heatislands/heat-island-impacts#emissions

Hot Cities, Chilled Economies: Economic Impacts of Heat on Cities (n.d.). https://onebillionresilient.org/hot-cities-chilled-economies/

Imran MH, Shammas MI, Rahman A, Jacobs SJ, Ng AWM, Muthukumaran S (2021) Causes, modeling and mitigation of urban heat island: a review. Earth Sci 10(6):244. https://doi.org/10.11648/j.earth.20211006.11

Kaloustian N (2015) On the urban heat island in Beirut [Doctoral Dissertation, Université Paris-Est]. https://theses.hal.science/tel-01305945

Khare S, Siber S, Singh SP, Amir SS (2020) A review and examination of urban heat island UHI in urban cities Delhi, Ahmedabad, Bangalore in India utilizing LANDSAT 8 and MODIS. Int Res J Eng Technol (IRJET) 7(6)

Li ZL, Becker F (1993) Feasibility of land surface temperature and emissivity determination from AVHRR data. Remote Sens Environ 43(1):67–85. https://doi.org/10.1016/0034-4257(93)90065-6

Li ZL, Tang BH, Wu H, Ren H, Yan G, Wan Z, Trigo IF, Sobrino JA (2013) Satellite-derived land surface temperature: current status and perspectives. Remote Sens Environ 131:14–37. https://doi.org/10.1016/J.RSE.2012.12.008

Liu L, Zhang Y (2011) Urban heat island analysis using the landsat TM data and ASTER data: a case study in Hong Kong. Remote Sens 3(7):1535–1552. https://doi.org/10.3390/rs3071535

Mallick J (2014) Land characterization analysis of surface temperature of semi-arid mountainous city Abha, Saudi Arabia using remote sensing and GIS. J Geogr Inf Syst 06(06):664–676. https://doi.org/10.4236/jgis.2014.66055

Mathew A, Khandelwal S, Kaul N (2016) Spatial and temporal variations of urban heat island effect and the effect of percentage impervious surface area and elevation on land surface temperature: study of Chandigarh city, India. Sustain Cities Soc 26:264–277. https://doi.org/10.1016/j.scs.2016.06.018

Mathew A, Khandelwal S, Kaul N (2018) Analysis of diurnal surface temperature variations for the assessment of surface urban heat island effect over Indian cities. Energy Build 159:271–295. https://doi.org/10.1016/j.enbuild.2017.10.062

Mcgranahan G, Satterthwaite D (2014) Urbanisation concepts and trends. https://www.iied.org

Ministry of Housing and Urban Affairs, G. of I (n.d.) Policies and guidelines : addendum to URDPFI guidelines, 2014—cooling action plan. https://mohua.gov.in/cms/policies.php

Moyer A (2016) Assessing the urban heat island of a small urban area in central Pennsylvania along the Susquehanna river [Doctoral Dissertation]. Shippensburg University of Pennsylvania

Nations U, of Economic D, Affairs S, Division P (2018) World urbanization prospects the 2018 revision

O'Malley C, Piroozfar P, Farr ERP, Pomponi F (2015) Urban heat island (UHI) mitigating strategies: a case-based comparative analysis. Sustain Cities Soc 19:222–235. https://doi.org/10.1016/j.scs.2015.05.009

Pandey A, Mondal A, Guha S, Upadhyay PK, Singh D (2022) Land use status and its impact on land surface temperature in Imphal city, India. Geol Ecol Landsc 1–5. https://doi.org/10.1080/24749508.2022.2131962

Pramanik S, Punia M (2020) Land use/land cover change and surface urban heat island intensity: source–sink landscape-based study in Delhi, India. Environ Dev Sustain 22(8):7331–7356. https://doi.org/10.1007/s10668-019-00515-0

Proceedings-Lake, Ramachandra TV, Chandran S, Alva M (2016) Conservation and sustainable management of ecologically sensitive regions in western Ghats. 560012. http://ces.iisc.ernet.in/energy://ces.iisc.ernet.in/energy

Refahi AH, Talkhabi H (2015) Investigating the effective factors on the reduction of energy consumption in residential buildings with green roofs. Renew Energy 80:595–603. https://doi.org/10.1016/j.renene.2015.02.030

Reisi M, Nadoushan MA, Aye L (2019) Remote sensing for urban heat and cool islands evaluation in semi-arid areas. Glob J Environ Sci Manag 5(3):319–330. https://doi.org/10.22034/gjesm.2019.03.05

Ruefenacht L, Acero JA (2017) Strategies for cooling Singapore a catalogue of 80+ measures to mitigate urban heat island and improve outdoor thermal comfort. https://doi.org/10.3929/ethz-b-000258216
Running SW, Justice CO, Salomonson V, Hall D, Barker J, Kaufmann YJ, Strahler AH, Huete AR, Muller JP, Vanderbilt V, Wan ZM, Teillet P, Carneggie D (1994) Terrestrial remote sensing science and algorithms planned for EOS/MODIS. Int J Remote Sens 15(17):3587–3620. https://doi.org/10.1080/01431169408954346
Sahu S, Pawar AS, Mukherjee M (2014) Urban heat island investigation: techniques and methods so far. Arch—Time Space People 14(3):16–22
Sastry M, Varma KH, Halve V (2017) Urban planning characteristics to mitigate climate change in context of urban heat island effect. www.teriin.org
Schmugge TJ, Becker F (1991) Remote sensing observations for the monitoring of land-surface fluxes and water budgets. Land Surf Evaporation 337–347. https://doi.org/10.1007/978-1-4612-3032-8_20
Sharma R, Pradhan L, Kumari M, Bhattacharya P (2021) Assessing urban heat islands and thermal comfort in Noida city using geospatial technology. Urban Clim 35. https://doi.org/10.1016/j.uclim.2020.100751
Shirani-bidabadi N, Nasrabadi T, Faryadi S, Larijani A, Shadman Roodposhti M (2019) Evaluating the spatial distribution and the intensity of urban heat island using remote sensing, case study of Isfahan city in Iran. Sustain Cities Soc 45:686–692. https://doi.org/10.1016/J.SCS.2018.12.005
Shukla A, Jain K (2021) Analyzing the impact of changing landscape pattern and dynamics on land surface temperature in Lucknow city, India. Urban For Urban Green 58. https://doi.org/10.1016/j.ufug.2020.126877
Silva LT, Mendes JFG (2012) City noise-air: an environmental quality index for cities. Sustain Cities Soc 4(1):1–11. https://doi.org/10.1016/J.SCS.2012.03.001
Singh P, Kikon N, Verma P (2017) Impact of land use change and urbanization on urban heat island in Lucknow city, central India. A remote sensing based estimate. Sustain Cities Soc 32:100–114. https://doi.org/10.1016/j.scs.2017.02.018
Singh P, Verma P, Chaudhuri AS, Singh VK, Rai PK (2024) Evaluating the relationship between urban heat island and temporal change in land use, NDVI and NDBI: a case study of Bhopal city, India. Int J Environ Sci Technol 21:3061–3072. https://doi.org/10.1007/s13762-023-05141-y
Srivanit M, Hokao K (2018) Thermal infrared remote sensing for urban climate and environmental studies: an application for the city of Bangkok, Thailand. J Arch/Plan Res Stud (JARS) 9(1):83–100. https://doi.org/10.56261/JARS.V9I1.168598
Sruthi Krishnan V, Mohammed Firoz C (2020) Regional urban environmental quality assessment and spatial analysis. J Urban Manag 9(2):191–204. https://doi.org/10.1016/j.jum.2020.03.001
World Bank (2020) Analysis of heat waves and urban heat island effects in central European cities and implications for urban planning the international bank for reconstruction and development. https://documents1.worldbank.org/curated/en/740251596528336330/pdf/Analysis-of-Heat-Waves-and-Urban-Heat-Island-Effects-in-Central-European-Cities-and-Implications-for-Urban-Planning.pdf

Chapter 4
Nature-Based Solutions and Planning Strategies for Optimizing Peri-Urban Open Spaces in Varanasi

Somi Sareen, Nazish Abid, Vidhu Bansal, Sunny Bansal, and Mazharul Haque

Abstract The phenomenon of rapid urbanization is a pervasive and continuing challenge in our modern world. It entails the rapid growth and extension of metropolitan centers, encroaching upon previously underdeveloped regions. This process frequently leads to the unfortunate consequence of diminishing the extent of valuable land cover. The phenomenon of urban sprawl poses a significant threat not only to the preservation of the natural environment but also to the delicate ecological equilibrium of peri-urban areas. Peri-urban zones, due to their close vicinity to urban centers, serve a vital function in minimizing the negative consequences of urban expansion and present a potential avenue for harmonizing urban growth with the preservation of green and natural environments. Peri-urban open spaces possess the capacity to function as significant buffers that can safeguard natural ecosystems, offer recreational prospects, and contribute to the holistic welfare of urban inhabitants. This chapter explores the comprehensive range of planning strategies established in the current body of literature on peri-urban settings. The study examines a wide range of strategies, including the implementation of green infrastructure and the establishment of zoning regulations that prioritize ecosystem services. The chapter aims to advance the understanding of spatial sustainability, promote the adoption of nature-based solutions in peri-urban areas and emphasize the significance of flexible strategies that can accommodate the distinct attributes of peri-urban regions at various topographical scales, local contexts, and implementation. The chapter will assess planning interventions to ascertain their pertinence within the complex landscape of peri-urban areas

S. Sareen (✉) · M. Haque
National Institute of Technology, Patna, India
e-mail: khansareen8@gmail.com

M. Haque
e-mail: mazharul@nitp.ac.in

N. Abid
University of Bahrain, Zallaq, Bahrain

V. Bansal · S. Bansal
Anant National University, Ahmedabad, India

M. Firoz C et al. (eds.), *Nature-Based Solutions for Urban and Peri-Urban Areas*, Disaster Risk Reduction, https://doi.org/10.1007/978-981-96-5933-3_4

and highlight the importance of using adaptable and situation-sensitive strategies to effectively tackle the challenges presented by the peri-urbanization of Varanasi. The chosen urban area serves as a case study to exemplify the application of specific planning principles, by studying the feasibility of implementing selected interventions from literature in the peri-urban regions of Varanasi and considering the unique attributes, challenges, and potentialities of the city. The study demonstrates the feasibility of formulating planning strategies and interventions for peri-urban open spaces in the context of Varanasi by evaluating the possibilities for comparable actions.

Keywords Peri-urban open spaces · Nature-based solution · Planning intervention · Varanasi · Sustainable urbanization

4.1 Introduction

The wave of rapid urbanization has both pros and cons for the regions it impacts. The impact is starkly visible in the underdeveloped or developing regions. Some of the positive impacts can be better economic opportunities, scope for infrastructure development, access to better services, and growth in innovation and creativity (Turok and Mcgranahan 2015). Economic opportunities due to urbanization may emerge from the concentration of economic activities and consequently generate job opportunities for people who are migrating from rural to urban areas (Kookana et al. 2020). The urban centre thus becomes an economic hub and leads to the generation of more economic activities (Turok and Mcgranahan 2015). The influx of people also requires infrastructure to support the growing population. Development of roads, electricity, water supply, and sanitation supply. Migrating to urban areas also often means access to better health care, education, and other essential services compared to rural areas (Zhang et al. 2020). Economic activities fuel other activities, and hence, urban areas tend to be hubs of innovation and creativity, nurturing knowledge exchange, technological advancements, and entrepreneurship (Melo and Jenkins 2021).

On the other hand, many negative aspects are also attached to rapid urbanization when there is a lack of counter forces to check it (Castells-Quintana et al. 2019). Some of the most common issues associated with this are slums formation, environmental degradation, social inequality, and pressure on resources (Surya et al. 2018). Rapid urbanization brings with it a challenge to provide shelter for the ever-swelling urban areas. This leads to the formation of illegal settlements or areas where the shelter conditions lack the aspect of liveability (Paul 2024). The phenomenon leads to multiple other consequential issues like overcrowding and excessive pressure on infrastructure services like water availability, sanitation, and healthcare. The quest for providing shelter to all leads to stress on natural resources like forests and water bodies and may lead to increased vulnerability to natural disasters (Debsarma et al. 2023). In many impacted areas, it may also lead to an increase in social inequalities.

4.1.1 Urbanization and Diminishing Landcover in Peri-Urban Areas

The pressure of development on the inner cores of the city leads to the haphazard expansion of urban areas, which often results in urban sprawl (Turok and Mcgranahan 2015). In other words, it can also be called the phenomenon of peri-urbanization. Here, as opposed to core urban areas, a transitional zone can be experienced, which is a mix of rural–urban features formed by the expansion of urban areas into rural landscapes. The urban sprawl emerges as haphazard peri-urbanization that leads to rapid changes in land use and land cover (Paül and McKenzie 2013). Multiple issues emerge from it, like the loss of natural habitat for developing arable lands, which in turn leads to the loss of native flora and fauna, disrupting the ecosystem cycles of the region (Debsarma et al. 2023). Sometimes, even the arable lands are converted for the development of housing, industries, and other infrastructure. This leads to the formation of more impervious surfaces, which eventually leads to the urban heat island effect (Stache et al. 2022).

The declining land cover contributes most towards the surface runoff (Shahfahad et al. 2022). Without vegetation or land cover, the surface runoff increases, leading to water contamination that results in the deposition of harmful chemicals into the aquatic ecosystem. It also makes the area more vulnerable to flooding (Castells-Quintana et al. 2019). The decline in land cover is also equally proportionate to declining ecosystem services like air and water purification, carbon sequestration, recreational opportunities, and many others. The communities who reside in the peri-urban zones and rely on ecosystem services for their day-to-day survival are affected by it to a maximum level. These lands are not only their primary resources for survival but also a part of their heritage and identity. Hence, it is important to control the declining land cover in peri-urban areas through integrated land use planning approaches (Mortoja and Yigitcanlar 2020). Green infrastructure, reforestation, and inclusion of peri-urban communities in the decision-making process are a few of the initial steps that can mitigate the negative effects of the current scenario (Wei et al. 2018).

4.1.2 Ecological Equilibrium

Ecological equilibrium is achieved through continuous monitoring and management of the ecological conditions (Palang et al. 2011). It also requires being resilient and adapting to changing conditions by prioritizing long-term benefits for the generations to come. There are multiple dimensions to achieve ecological equilibrium, which may include biodiversity conservation, ecosystem connectivity, development of green infrastructure, applying sustainable land use practices, efficient water resource management, peri-urban community awareness and engagement, and efficient implementation of policy and planning strategies (Semeraro et al. 2021). Efficient governance plays a huge role in it as it not only helps in enforcing the regulations but also can play a key role in fostering collaborations between the governing agencies and the stakeholders.

Sustainable practices can be incorporated into daily routines to support ecological stability. Urban areas are embracing robust systems and technologies, including DEWATS (Decentralised Wastewater Treatment Systems) for irrigation. This was demonstrated in Gorakhpur, Uttar Pradesh, where it also enabled the construction of new restrooms and the provision of clean water (Gupta et al. 2021a). Sustainable agriculture has benefited from increased soil quality and the use of composting for solid waste management.

Using Varanasi as a case study, this chapter explores the complexity of peri-urbanization with an emphasis on planning methods for peri-urban areas. Our objective is to identify ways that improve urban resilience and preserve ecological integrity by examining case studies, literature, and planning technology. The study question is: How can the design and administration of Varanasi's peri-urban open spaces be optimized to enable sustainable urbanization while maintaining ecological balance and taking socioeconomic considerations into account?

Methods have been suggested for incorporating ecosystem-based adaptation, biodiversity preservation, and green infrastructure into urban development. The study seeks to: (i) Analyse the contributions of peri-urban open spaces to sustainable urbanization; (ii) Review current policies and initiatives in Varanasi; and (iii) Identify and analyze successful NbS case studies from other regions for applicability in Varanasi by critically examining policy frameworks, regulatory mechanisms, and community engagement strategies.

4.2 Methodology

This chapter analyses Varanasi's peri-urban open space planning for sustainable urbanization using a thorough review-based methodology as shown in Fig. 4.1. The literature study analyses these spaces' theoretical and practical contributions to sustainable urban development. The case study explores the city's distinct topographical, cultural, and socioeconomic context and maps out its peri-urban zones. Policy documents, project reports from local authorities, and master plans such as the Varanasi Master Plan 2031 are reviewed to evaluate the current practices in Varanasi's peri-urban districts. The analysis of successful NbS case studies, and local environmental reports are all part of the section on the possibility of effective implementation of NbS. Considering the potential and challenges mentioned, the conclusion synthesizes the findings from the preceding sections to offer practical recommendations for the planning of peri-urban open areas.

Fig. 4.1 Methodological framework

4.3 Literature Review

4.3.1 Role of Peri-Urban Open Spaces in Sustainable Urbanization

As urbanization continues to accelerate worldwide, peri-urban areas have emerged as pivotal zones in the interface between urban and rural landscapes. These transitional areas, characterized by open spaces and greenery, play a vital role in minimizing the negative consequences of rapid urbanization by serving as buffers against unchecked urban sprawl. As cities expand, peri-urban areas absorb the pressure of urban development, thereby preventing the indiscriminate conversion of natural habitats into built environments. By retaining open spaces and green corridors, peri-urban zones help to mitigate environmental degradation, preserve biodiversity, and maintain ecological balance (Gupta et al. 2021a). For example, in cities like Nairobi, Kenya, peri-urban green spaces such as the Karura Forest act as important carbon sinks, mitigating the impacts of urban pollution and climate change (Shah and Irandu 2022). Similarly, in Brazil, peri-urban areas surrounding cities like São Paulo play a critical role in conserving biodiversity and protecting watersheds, thereby ensuring the provision of clean water for urban residents (Moreno-Peñaranda and Kallis 2010; Almeida et al. 2018). In Shanghai, China, peri-urban wetlands such as the Chongming Dongtan National Nature Reserve act as critical habitats for migratory birds and provide ecological services such as flood regulation and water purification (Bao and Gao 2021; Gou et al. 2021). Similarly, in Bangkok, Thailand, peri-urban agricultural lands help to mitigate the heat island effect, improve air quality, and provide food security for urban residents (Tsuchiya et al. 2015). Peri-urban open spaces play a vital role in sustainable urbanization by minimizing the negative consequences of rapid urbanization. It also acts as a buffer against urban sprawl, preserves natural ecosystems, and enhances the resilience of urban areas (Spyra et al. 2021).

Peri-urban open spaces can buffer natural ecosystems from urbanization impacts. These areas, including forests, wetlands, and agricultural lands, are vital for biodiversity and ecosystem services (Gottero et al. 2021). In India, the Aravalli Biodiversity Park in Delhi is a key biodiversity hotspot (Livesley et al. 2016). Malaysia's peri-urban mangrove forests protect coastal areas from erosion and storms, support fisheries, and sequester carbon (Mngumi 2020). In Berlin, Germany, the Grunewald forest provides wildlife habitats, carbon sequestration, and climate regulation (Spyra et al. 2021). Similarly, Tokyo's Tama Hills in Japan support native biodiversity and offer recreational spaces for urban residents (Tomitaka et al. 2022).

Furthermore, peri-urban open spaces play a crucial role in protecting water resources and mitigating natural hazards such as floods (Gupta et al. 2021b) and landslides Wetlands, floodplains, and riparian zones in peri-urban areas act as natural buffers against flooding, absorbing excess water and reducing the risk of downstream inundation (Goswami et al. 2020). In cities like Melbourne, Australia, peri-urban green belts such as the Yarra River catchment provide essential ecosystem services, including water filtration, flood control, and groundwater recharge (Van Den Bosch et al. 2018). By preserving peri-urban wetlands and riparian zones, cities can enhance water security, reduce the vulnerability of urban populations to natural disasters, and improve overall environmental quality.

Peri-urban open spaces contribute significantly to recreational opportunities and holistic well-being by providing urban residents with access to nature, outdoor activities, and cultural amenities. These spaces serve as valuable resources for physical and mental health, offering opportunities for exercise, relaxation, and social interaction (Žlender and Ward Thompson 2017). In Singapore, peri-urban parks such as the Dairy Farm Nature Park and the Singapore Quarry provide residents with opportunities for hiking, bird watching, and picnicking, enhancing quality of life and promoting community bonding. Similarly, in Japan, peri-urban green spaces such as the Tama Hills offer opportunities for outdoor recreation, nature education, and spiritual renewal, contributing to overall well-being and mental health (Mngumi 2020; Tomitaka et al. 2022).

In London, United Kingdom, peri-urban parks such as Hampstead Heath and Richmond Park offer extensive green spaces for recreational activities (Paudel and States, 2023). Similarly, in cities like Sydney, Australia, peri-urban reserves such as the Royal National Park and Ku-ring-gai Chase National Park provide opportunities for bushwalking, camping, and wildlife viewing, allowing urban residents to escape the hustle and bustle of city life (Van Den Bosch et al. 2018). By providing access to green spaces and recreational amenities, peri-urban areas help to alleviate stress, improve mood, and enhance the overall quality of life for urban residents, by providing venues for social interaction, cultural events, and community celebrations, strengthening social ties and promoting civic engagement (Gottero et al. 2021). For instance, in cities like Vancouver, Canada, peri-urban green spaces such as Stanley Park and Pacific Spirit Regional Park serve as popular destinations for outdoor recreation and community events, bringing people together and fostering a sense of belonging (Czekajlo et al. 2021).

4.3.2 Planning Strategies for Peri-Urban Areas

Efficient planning strategies are inevitable for the management of peri-urban areas. The planning approach is developed based on the local context so that maximum benefit can be achieved. Wherever possible, compact cities should be the planning approach as they help avoid urban sprawl. Since the Indian cities are already very compact and peri-urbanization is the development consequence hence, approaches like integrated land-use planning play a significant role in it as it seeks to balance the development and spatial planning with social, cultural, economic, and ecological aspects (Stoker et al. 2022; Peskett et al. 2023).

The peri-urban planning strategies should also include support for protecting agricultural land and farming so that food security can be maintained. Agricultural zoning, agroecological planning, and agricultural easements are a few of the ways to move forward in this regard (Paül and McKenzie 2013). All this should be done considering the ecological conditions of the particular region. Ecological resilience and holistic development can only be achieved if the local community is also taken into the decision-making process as they are more in sync with their environment. Participatory planning approaches work best in such kinds of areas as they help in community-based resource mapping, community-based knowledge bank development, and stakeholder forums, which are essential for comprehensive development (Waldman et al. 2017; Gomes et al. 2023). Here, it becomes important to adopt strategies that help conserve nature and make appropriate decisions for development. Focussing on the development of green infrastructure plays a significant role in it. Strategies like green belts, ecological corridors, parks, greenways, and wetlands can be incorporated to develop the peri-urban areas more sustainably (Waldman et al. 2017).

As opposed to grey infrastructure made of urban concrete jungle, green infrastructure constitutes a network of natural or semi-natural materials that help provide various ecosystem services. It uses natural systems or engineered systems that resemble the natural process to carry on with the designated functions (Semeraro et al. 2021). The concern here lies in how the implementation of green infrastructure can be initiated to benefit the peri-urbanization process. To initiate green development, it should be essential to assess and identify the priority areas, which may include ecologically fragile zones like riparian zones, wetlands, agricultural lands, and forested areas (Verdú-Vázquez et al. 2021). These areas are of utmost importance as they provide many kinds of ecosystem services and help with flood regulations, carbon sequestration, biodiversity conservation, and many other essential ecological functions. This can be done by devising zoning regulations that are specific to peri-urban areas and can augment their sustainable development. Eco-zoning principles contextually aligned with the peri-urban area's condition can be encapsulated in the development process (Honeck et al. 2020). Riparian setbacks, wildlife corridors, and biodiversity conservation areas can be some of the key areas where eco-zoning can be prioritized. Incentivizing for conservation can also be utilized to encourage sustainable development (Honeck et al. 2020).

Designation of dedicated green corridors can be another way of including green infrastructure in the peri-urban zones (Geneletti et al. 2017). They provide habitat connectivity and make people sustainably interact with their environment. NbS and multifunctional green spaces can also be included as part of green infrastructure. This may include rain gardens, green roofs, green streets, permeable pavements, and constructed wetlands. They will help not only enhance biodiversity and promote social well-being but also help the local communities (Smith 2018). At various levels, NbS offer a variety of advantages. In urban areas, they lessen the impact of heat islands, boost biodiversity, improve air quality, and manage stormwater using rain gardens, green roofs, and urban forests. Through restored wetlands and floodplains, NbS safeguards water supplies, lessens floods, and provides recreational areas near towns. Ecotourism also helps local economies by generating jobs in the area. Far from urban centres, NbS promote resilient supply chains, sustainable agriculture, and renewable energy sources while helping to control global warming through extensive initiatives like peatland restoration and reforestation. These solutions improve social, economic, and environmental resilience and are essential to urban sustainability (Thoidou 2021).

Flexible strategies must adapt to various scales and local contexts, as shown in Table 4.1, avoiding a one-size-fits-all approach. Policies should consider local socioeconomic conditions to respond appropriately (Sareen and Haque 2023). This involves tailoring green infrastructure designs and zoning regulations to local priorities (Tian 2015), and coordinating planning instruments and policies across scales, from regional to neighbourhood levels. Peri-urban planning impacts many levels of the supply chain for both human and natural environments, requiring sensitivity to multiple dimensions to create comprehensive, self-sustaining ecological hubs (Winarso et al. 2015). NbS and planning instruments should meet community aspirations in the built environment.

Table 4.1 Planning strategies discussion

Planning strategy	Examples/cases	Places	Scale	Mandates required	References
Green infrastructure	Implementation of green belts, ecological corridors, parks, greenways, wetlands, rain gardens, green roofs	Nairobi, Kenya (Karura Forest)—São Paulo, Brazil—Shanghai, China (Chongming Dongtan National Nature Reserve)—Bangkok, Thailand	Regional, city	- Zoning regulations specific to peri-urban areas—Incentives for conservation—Riparian setbacks and biodiversity conservation areas—Implementation of eco-zoning principles	(Moreno-Peñaranda and Kallis 2010; Tsuchiya et al. 2015; Bao and Gao 2021; Shah and Irandu 2022)
Eco-zoning	Riparian setbacks, wildlife corridors, biodiversity conservation areas		Regional, city, zonal level		
Green corridors	Creation of green corridors within and around urban areas, linking natural habitats		City		
Nature-based solutions	Rainwater harvesting, green streets, permeable pavements, constructed wetlands, green roofs	- Vancouver, Canada (Stanley Park, Pacific Spirit Regional Park)—London, United Kingdom (Hampstead Heath, Richmond Park)—Singapore (Dairy Farm Nature Park, Singapore Quarry	Zonal level	- Incentives for nature-based infrastructure in building codes and development regulations—Incorporation of nature-based solutions into urban planning guidelines	(Van Den Bosch et al. 2018; Mngumi 2020; Czekajlo et al. 2021)

(continued)

Table 4.1 (continued)

Planning strategy	Examples/cases	Places	Scale	Mandates required	References
Participatory planning	Community-based resource mapping, knowledge bank development, stakeholder forums, participatory design approaches	Nairobi, Kenya, São Paulo, Brazil, Bangkok, Thailand	City, zonal level	Establishment of stakeholder forums and community-based resource mapping processes—Inclusion of local communities in decision-making processes—Development of participatory planning frameworks and policies	(Tsuchiya et al. 2015; Shah and Irandu 2022)
Flexible strategies	Context-specific zoning regulations, incentive mechanisms for conservation, coordination of planning instruments	São Paulo, Brazil, Shanghai, China, Bangkok, Thailand	Zonal level	Tailored zoning regulations considering ecological sensitivity, land use patterns, and community preferences Conservation incentives: Implement tax breaks, grants, or subsidies to promote conservation practices	(Tsuchiya et al. 2015; Almeida et al. 2018; Gou et al. 2021)

4.4 Case of Varanasi

Varanasi, being one of the most ancient cities in continuous human habitation worldwide, presents an intriguing subject for examining the intricacies of peri-urban development. Varanasi, located in Uttar Pradesh, India, is a prominent centre of culture, religion, and education, situated along the banks of the Ganges River. The Varanasi Urban Agglomeration has seven urban sub-units, each of which plays a role in shaping its unique socio-economic environment. The urban population of the city, which is augmented each year by a significant number of pilgrims attracted to its spiritual places, highlights the urgent requirement for the implementation of sustainable urban management practices (Goswami and Kumari 2015). The Varanasi Development Authority has developed the Master Plan 2031 to tackle these difficulties, which outlines the Varanasi Development Region covering an area of 793 square kilometres (Patel et al. 2019). Notwithstanding these endeavours, the metropolitan area has escalating environmental challenges, and pervasive pollution, predominantly stemming from urban trash and industrial emissions.

Varanasi, the administrative center of Varanasi District, has seen significant urbanization within its Urban Agglomeration, comprising seven sub-units, including Varanasi, Ramanagar MB, and Banaras Hindu University (Ashraf and Munir 2019). Despite its cultural significance, the city faces major challenges, exacerbated by a 20% population increase since 2001 (Census 2011). This rapid growth has led to more vehicles, housing, and industrial activities, reducing green spaces to just 5.49% of the land area, well below the recommended 33% (Patel and Verma 2019). The Ganga Riverfront, a cultural symbol, suffers from severe pollution, with 80% attributed to urban waste (Patel and Shukla 2023). Varanasi's industrial zones, spanning over 600 acres, operate with little ecological consideration, worsening environmental degradation. Research indicates a 50.95% reduction in plants and trees and a 16.54% decrease in water bodies from 2000 to 2020 due to unchecked urbanization and land use changes (Singh et al. 2023) The Master Plan 2031 proposes interventions like regional parks and green belts, aiming to cover 12.87% of the total area, up from 5.49%. However, the plan's focus on quantitative goals overlooks the complex socio-ecological context of Varanasi's peri-urban regions.

Varanasi's peri-urban areas are seeing significant expansion, especially in the peripheral zones, due to increasing urbanization. Although these advancements are characterized by careful and systematic design and organization, they lack sufficient leisure and green areas. The growth patterns in peri-urban areas are mostly characterized by edge expansion typologies, which highlights the need for policy approaches that are sensitive to the various nature of urban growth (Patel et al. 2019). The presence of business and residential centres in Trans-Varuna regions highlights the necessity for all-encompassing town planning techniques that are specifically designed to suit local contexts as highlighted in Fig. 4.2 in the arrows across the Varuna River.

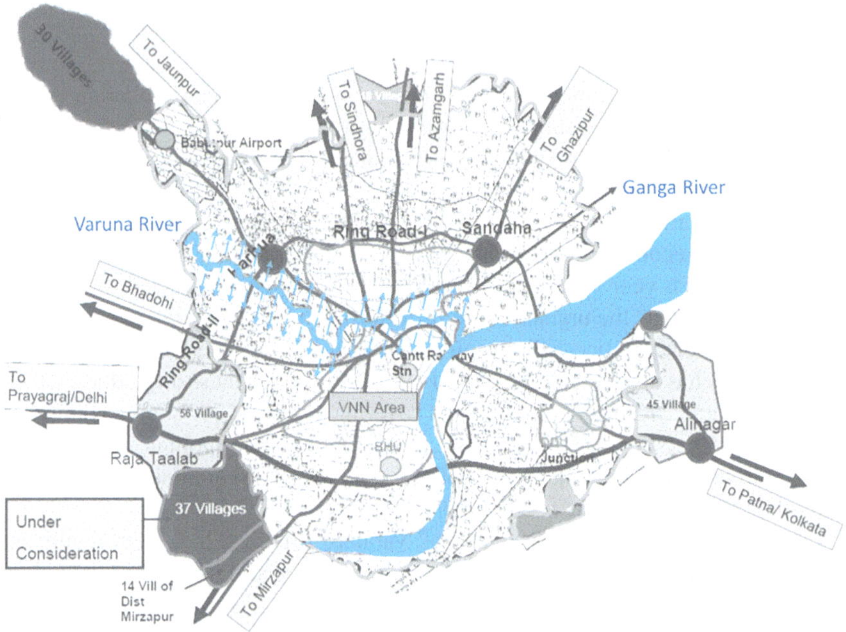

Fig. 4.2 Map showing Yamuna and Ganga rivers flowing through Varanasi

4.4.1 *Current Practices in Varanasi's Peri-Urban Areas*

The assessment of the peri-urban landscape in Varanasi further highlights a worrisome disparity between the suggested planning measures and the complex socio-ecological conditions present in the area. Notwithstanding the ambitious objectives delineated in the Master Plan 2031, the practical execution frequently falls short of effectively tackling pivotal challenges and guaranteeing fair and just results. For example, the process of artificializing *kunds* such as the *Mandakini Kund* and *Dudhdhama*. The rehabilitation initiatives of these previously crucial community water bodies have been spearheaded by third-party organizations, based on the analysis and recommendations of INTACH and Varanasi Smart City Limited. Nevertheless, the absence of active participation by Resident Welfare Associations (RWAs) or community groups has led to the erosion of the socio-cultural importance of these *kunds* (Fig. 4.3) (Mishra et al. 2022). Instead of functioning as central hubs for community involvement and water preservation, they have transformed into insignificant components within the environment, lacking significant ties to the local population.

Likewise, the conversion of wetlands into eco-parks at *Undi* (Fig. 4.4) exemplifies a fragmented strategy towards ecological restoration (Singh et al. 2023). The primary objective of these programmes is to improve the quality of green spaces in the form of wetland restoration, city forest development and eco-park construction. Being a standalone project, they frequently fail to consider the incorporation of these places

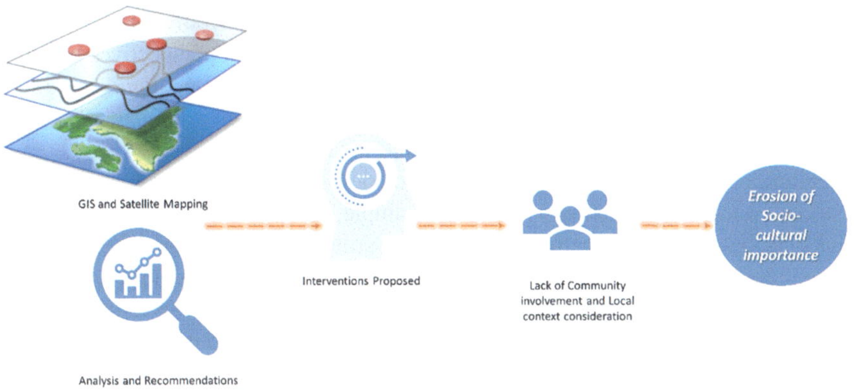

Fig. 4.3 Current planning intervention/implementation process

into broader ecological networks, such as the network connection with Varuna and Ganga Greens corridors at regional and sub-regional scales. Consequently, these ecological parks exhibit a deficiency in self-sustaining processes and demonstrate limited efficacy in their contributions to wider ecological conservation endeavours.

Furthermore, the sole reliance on Geographic Information System (GIS) analysis and satellite mapping in the planning procedure has resulted in erroneous decision-making, as evidenced by the implementation of initiatives such as the establishment of ring highways surrounding settlements like Sarai Mohana (Mishra et al. 2022). The plan places a higher emphasis on transit-oriented development (TOD) and infrastructure expansion, neglecting the socio-economic fabric and ecological integrity of

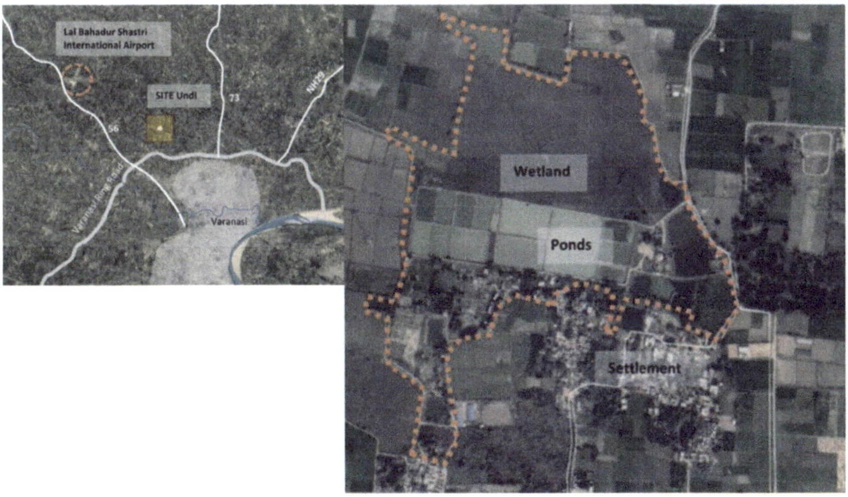

Fig. 4.4 Proposed city forest/wetland at Undi. *Source* Varanasi Development Authority

peri-urban communities, despite the agricultural and artisanal livelihoods that rely on these areas. The hierarchical approach not only excludes local perspectives but also worsens socioeconomic disparities and disturbs the fragile socio-ecological equilibrium. Because peri-urban areas are mere enablers of the built environment, and hence there is no prioritization of its open spaces and lands as assets.

Failure to consider temporal activities and socioeconomic backgrounds in interventions exacerbates existing issues. For example, informal communities on sandbanks lack sewerage infrastructure, creating pollution hotspots and health risks (Ashraf and Munir 2019). Planning often labels these areas as unoccupied, ignoring their importance to residents. This oversight misses the opportunities to use innovative solutions like bio-filtration systems to address pollution and enhance ecological resilience.

4.4.2 Potential for Successful Implementation of Nature-Based Solutions

To effectively NbS in Varanasi and guarantee the active engagement of the local population, it is imperative to adopt a comprehensive approach. It is essential to give precedence to the most urgent socioeconomic issues confronting the city, such as the contamination of water in the Ganges River and the intensification of the urban heat island phenomenon. Planning measures need to identify the scale of the challenges and customize NbS treatments by actively involving local stakeholders, such as citizens, community leaders, and policymakers. For example, restoration and preservation of riparian vegetation along the Ganges identified by zonation software and demarcating priority areas have the potential to alleviate water pollution while concurrently offering cooling properties to counteract urban heat.

NbS shall be integrated to tackle challenges encompassing enhancement of public health by expanding green spaces, mitigating the impacts of climate change by sequestering carbon through afforestation initiatives, and fostering biodiversity by establishing urban green corridors in the city and peripheral areas that are available. Effective collaboration, including urban planners, local government officials, environmental NGOs, and community groups in decision-making processes in Varanasi facilitates the attainment of cross-sectoral buy-in and commitment. Green bonds have the potential to be employed as a means of financing various projects, such as the rehabilitation of wetlands and urban forests. Additionally, the establishment of public–private partnerships can provide support for the advancement of green infrastructure in the new projects that are coming in the peri-urban areas group housing, educational institutes, etc. The inclusion and accessibility of financial channels should be ensured for all stakeholders, particularly marginalized groups that frequently experience the adverse effects of environmental degradation.

Figures 4.5 and 4.6 illustrate how Varanasi can approach urban planning with consideration for scale. A holistic approach is ensured by implementing NbS at multiple levels, such as rain gardens in neighbourhoods to lessen floods. Ecological resilience is increased by biodiversity conservation initiatives such as the Ganges Wildlife Corridor and green space preservation. Green infrastructure in cities and regions, such as ecological corridors and green belts, can improve socioeconomic conditions while halting environmental deterioration. Customized zoning regulations and conservation incentives can be used to protect riparian setbacks and biodiversity zones. Furthermore, utilizing artificial wetlands and rainwater collection can reduce water pollution and improve resilience. Adoption of NbS can be further encouraged by including its concepts in building incentives and urban planning guidelines.

In order to guarantee equitable benefits and safeguard marginalized populations, inclusive governance is crucial. Local communities can impact the design and execution of NbS through mechanisms such as participatory budgeting and community advisory committees. Planners also must weigh conflicting goals. For example, although river embankments can reduce floods in some locations, they may have a detrimental effect on ecosystems and communities downstream. Additionally, In Ramnagar, urban agriculture and community gardens can increase food security, develop green areas, and fortify social ties. There is plenty of space in this semi-urban location for gardening endeavours that will strengthen the local food supply and community ties. In this quickly urbanizing area, green infrastructure can reduce the heat island effect and increase building efficiency. The geography and hydrological attributes of Phulwaria make it the perfect location for blue-green infrastructure, like bioswales and artificial wetlands, which manage runoff, lessen flooding, and enhance water quality. Congestion can be decreased and healthy lives encouraged by establishing greenways and pedestrian routes, particularly along trans-river zones.

4.5 Conclusion

Varanasi is currently at an essential juncture in its peri-urban development, where measures at several levels have significant potential for enduring advancement. Peri-urban areas, which are frequently disregarded, offer distinct prospects as focal points for prioritization and the augmentation of ecological services. By acknowledging and protecting peri-urban regions as crucial assets. At the regional level, it is imperative to implement integrated efforts to preserve peri-urban open areas and harness their ecological importance. These areas, such as wetlands and riparian zones, play a vital role in mitigating environmental degradation and providing significant ecological benefits. By emphasizing the conservation of peri-urban areas, Varanasi has the potential to protect biodiversity, minimize pollution, and strengthen its ability to withstand the impacts of climate change. At the urban level, it is imperative to prioritize actions that effectively utilize the potential of peri-urban areas as valuable resources for sustainable development. Green infrastructure initiatives and nature-based solutions can potentially improve the quality of life in urban areas while preserving peri-urban ecosystems. Varanasi can boost biodiversity, improve air and

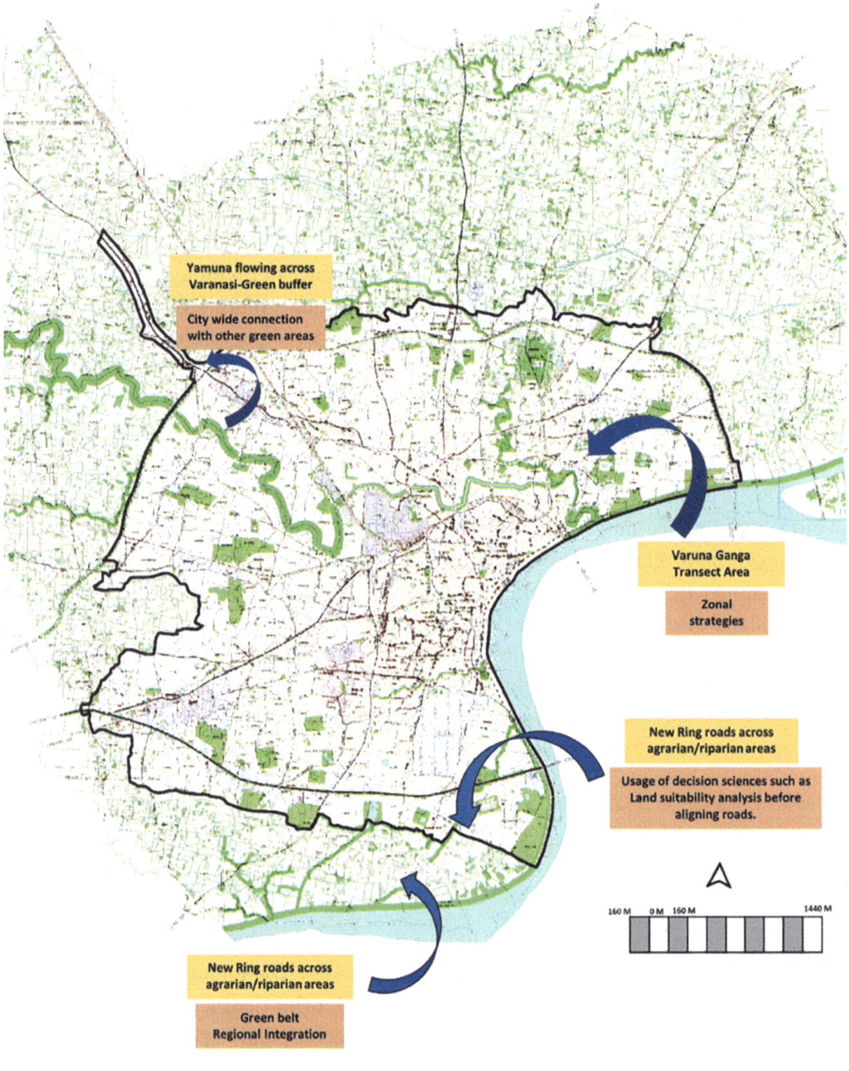

Fig. 4.5 Blue-green open area strategies in Varanasi

water quality, and improve the well-being of its citizens by including green corridors, parks, and wetlands in its urban development. At the local level, it is crucial to prioritize the needs of marginalized populations and alleviate the challenges posed by the impending population expansion. Open lands located on the outskirts of urban centres play a crucial role in providing communal spaces for recreational activities, livelihood pursuits, and cultural traditions, especially for marginalized communities.

Fig. 4.6 Scale informed approaches for NbS

Varanasi can guarantee that all inhabitants get the advantages of peri-urban interventions and could utilize green areas and ecological services by giving priority to social fairness and inclusive development.

References

Almeida CMVB, Mariano MV, Agostinho F et al (2018) Exploring the potential of urban park size for the provision of ecosystem services to urban centres: a case study in São Paulo, Brazil. Build Environ 144:450–458. https://doi.org/10.1016/j.buildenv.2018.08.036

Ashraf U, Munir A (2019) Urban Sprawl assessment and its effect on land use transformation in varanasi city using remote sensing and GIS. **8958**, 1056–1061. https://doi.org/10.35940/ijeat.B3647.129219

Bao J, Gao S (2021) Wetland utilization and adaptation practice of a coastal megacity: a case study of Chongming Island, Shanghai, China. Front Environ Sci 9:1–15. https://doi.org/10.3389/fenvs.2021.627963

Van Den Bosch CCK, Rodbell P, Salbitano F et al (2018) The changing governance of urban forests

Castells-Quintana D, Larrú JM, Lacalle-Calderón M (2019) Aid, microfinance and income inequality: a global view ayuda al desarrollo, microfinanzas y desigualdad en los ingresos: una visión global. Revista De Economía Mundial 52:143–174

Czekajlo A, Coops NC, Wulder MA et al (2021) Mapping dynamic peri-urban land use transitions across Canada using Landsat time series: spatial and temporal trends and associations with socio-demographic factors. Comput Environ Urban Syst 88:101653. https://doi.org/10.1016/j.compenvurbsys.2021.101653

Debsarma C, Sahu P, Kalubarme MH (2023) Influence assessment of urban expansion on groundwater level fluctuations in Gandhinagar, Gujarat, India. Environ. Monit. Assess. 195. https://doi.org/10.1007/s10661-023-11709-5

Geneletti D, La Rosa D, Spyra M, Cortinovis C (2017) A review of approaches and challenges for sustainable planning in urban peripheries. Landsc Urban Plan 165:231–243. https://doi.org/10.1016/j.landurbplan.2017.01.013

Gomes SL, Hermans LM, Butsch C et al (2023) A Delphi-based methodology for participatory adaptation pathways building with local stakeholders: methodological considerations and an illustrative application in peri-urban India. Environ Dev 46:100822. https://doi.org/10.1016/j.envdev.2023.100822

Goswami M, Nautiyal S, Manasi S (2020) Drivers and consequences of biophysical landscape change in a peri-urban–rural interface of Guwahati, Assam. Environ Dev Sustain 22:791–811. https://doi.org/10.1007/s10668-018-0220-1

Goswami KP, Kumari N (2015) Green spaces in the urban environment: a case study of Varanasi city **61**, 307–320

Gottero E, Cassatella C, Larcher F (2021) Planning peri-urban open spaces: methods and tools for interpretation and classification. Land 10. https://doi.org/10.3390/land10080802

Gou A, Li Y, Wang J (2021) Analysis of landscape microclimate and green space transformation strategy for old residential areas-case: Luoliu community. E3S Web. Conf. **271**, 2019–2022. https://doi.org/10.1051/e3sconf/202127102019

Gupta AK, Singh S, Agarwal M et al (2021a) Peri-urban ecosystems and urban resilience: knowledge compendium of case studies

Gupta AK, Singh S, Agrawal M (2021b) Peri-urban ecosystems and urban resilience

Honeck E, Moilanen A, Guinaudeau B, et al (2020) Implementing green infrastructure for the spatial planning of peri-urban areas in Geneva, Switzerland. Sustainability (Switzerland) 12. https://doi.org/10.3390/su12041387

Kookana RS, Drechsel P, Jamwal P, Vanderzalm J (2020) Urbanisation and emerging economies: issues and potential solutions for water and food security. Sci Total Environ 732:139057. https://doi.org/10.1016/j.scitotenv.2020.139057

Livesley SJ, Escobedo FJ, Morgenroth J (2016) The biodiversity of urban and peri-urban forests and the diverse ecosystem services they provide as socio-ecological systems. Forests 7:10–14. https://doi.org/10.3390/f7120291

Melo V de P, Jenkins P (2021) Peri-urban expansion in the Maputo city region: land access and middle-class advances. J. South Afr. Stud. **47**, 541–565. https://doi.org/10.1080/03057070.2021.1939499

Mishra M, Dugesar V, Raju KNP (2022) Flood risk and impact analysis of Varanasi city region, India. J. Sci. Res. **66**, 22–32. https://doi.org/10.37398/jsr.2022.660103

Mngumi LE (2020) Ecosystem services potential for climate change resilience in peri-urban areas in Sub-Saharan Africa. Landscape Ecol Eng 16:187–198. https://doi.org/10.1007/s11355-020-00411-0

Moreno-Peñaranda R, Kallis G (2010) A coevolutionary understanding of agroenvironmental change. a case-study of a rural community in Brazil. Ecol Econ 69:770–778. https://doi.org/10.1016/j.ecolecon.2009.09.010

Mortoja MG, Yigitcanlar T (2020) How does peri-urbanization trigger climate change vulnerabilities? An investigation of the Dhaka megacity in Bangladesh. Remote Sens 12:1–41. https://doi.org/10.3390/rs12233938

Palang H, Spek T, Stenseke M (2011) Digging in the past: new conceptual models in landscape history and their relevance in peri-urban landscapes. Landsc Urban Plan 100:344–346. https://doi.org/10.1016/j.landurbplan.2011.01.012

Patel SK, Verma P, Shankar Singh G (2019) Agricultural growth and land use land cover change in peri-urban India. Environ Monit Assess 191:10661. https://doi.org/10.1007/s10661-019-7736-1

Patel J, Shukla S (2023) Urban development in Uttar Pradesh: a historical study. Int. J. Multidiscip. Res. 5:1–11. https://doi.org/10.36948/ijfmr.2023.v05i05.7438

Patel SK, Verma P (2019) Agricultural growth and land use land cover change in peri-urban India

Paudel S, States SL (2023) Urban green spaces and sustainability: exploring the ecosystem services and disservices of grassy lawns versus floral meadows. Urban For Urban Green 84:127932. https://doi.org/10.1016/j.ufug.2023.127932

Paul A (2024) Assessment of metropolitan livability variations using objective-subjective approaches. World Dev. Sustain. 4:100135. https://doi.org/10.1016/j.wds.2024.100135

Paül V, McKenzie FH (2013) Peri-urban farmland conservation and development of alternative food networks: Insights from a case-study area in metropolitan Barcelona (Catalonia, Spain). Land Use Policy 30:94–105. https://doi.org/10.1016/j.landusepol.2012.02.009

Peskett L, Metzger MJ, Blackstock K (2023) Regional scale integrated land use planning to meet multiple objectives: good in theory but challenging in practice. Environ Sci Policy 147:292–304. https://doi.org/10.1016/j.envsci.2023.06.022

Sareen S, Haque M (2023) The dynamics of peri-urban spatial planning : an overview. J. Urban. Plan Dev. 149:03123002(1–15). https://doi.org/10.1061/JUPDDM.UPENG-4275

Semeraro T, Radicchio B, Medagli P et al (2021) Integration of ecosystem services in strategic environmental assessment of a peri-urban development plan. Sustainability (Switzerland) 13:1–26. https://doi.org/10.3390/su13010122

Shah PS, Irandu E (2022) Recreational green spaces as the future for sustainable cities: case of Karura forest in Nairobi, Kenya. J. Sustain. Environ. Peace 5:87–95. https://doi.org/10.53537/jsep.2022.06.001

Shahfahad TS, Rihan M et al (2022) Modelling urban heat island (UHI) and thermal field variation and their relationship with land use indices over Delhi and Mumbai metro cities. Environ Dev Sustain 24:3762–3790. https://doi.org/10.1007/s10668-021-01587-7

Singh NK, Nathawat MS, Delhi N (2023) Quantifying the characteristics and types of urban growth of Varanasi city using multi-temporal remotely sensed data and geospatial techniques 45

Smith JL (2018) The coproduction of peri-urban space. Geogr Rev 108:592–614. https://doi.org/10.1111/gere.12290

Spyra M, Kleemann J, Calò NC et al (2021) Protection of peri-urban open spaces at the level of regional policy-making: examples from six European regions. Land Use Policy 107. https://doi.org/10.1016/j.landusepol.2021.105480

Stache E (Eva), Schilperoort B (Bart), Ottelé M (Marc), Jonkers HM (2022) Comparative analysis in thermal behaviour of common urban building materials and vegetation and consequences for urban heat island effect. Build Environ 213. https://doi.org/10.1016/j.buildenv.2021.108489

Stoker P, Albrecht T, Follingstad G, Carlson E (2022) Integrating land use planning and water management in U.S. cities: a literature review. J Am Water Resour Assoc 58:321–335. https://doi.org/10.1111/1752-1688.13022

Surya B, Saleh H, Remmang H (2018) Economic gentrification and socio-cultural transformation metropolitan suburban of Mamminasata. J Eng Appl Sci 13:6072–6084. https://doi.org/10.3923/jeasci.2018.6072.6084

Thoidou E (2021) Spatial planning and climate adaptation: challenges of land protection in a peri-urban area of the mediterranean city of thessaloniki. Sustainability (Switzerland) 13. https://doi.org/10.3390/su13084456

Tian L (2015) Land use dynamics driven by rural industrialization and land finance in the peri-urban areas of China: "the examples of Jiangyin and Shunde." Land Use Policy 45:117–127. https://doi.org/10.1016/j.landusepol.2015.01.006

Tomitaka M, Iwachido Y, Sasaki T (2022) Spatial scales and urban greenspace types influence public conservation preferences. Land Use Policy 123:106403. https://doi.org/10.1016/j.landusepol.2022.106403

Tsuchiya K, Hara Y, Thaitakoo D (2015) Linking food and land systems for sustainable peri-urban agriculture in Bangkok metropolitan region. Landsc Urban Plan 143:192–204. https://doi.org/10.1016/j.landurbplan.2015.07.008

Turok I, Mcgranahan G (2015) Urbanization and economic growth : the arguments and evidence for Africa urbanization and economic growth : the arguments and evidence for Africa and Asia. https://doi.org/10.1177/0956247813490908

Verdú-Vázquez A, Fernández-Pablos E, Lozano-Diez RV, López-Zaldívar Ó (2021) Green space networks as natural infrastructures in peri-urban areas. Urban Ecosyst. 24:187–204. https://doi.org/10.1007/s11252-020-01019-w

Waldman L, Bisht R, Saharia R et al (2017) Peri-urbanism in globalizing India: a study of pollution, health and community awareness. Int J Environ Res Public Health 14:1–17. https://doi.org/10.3390/ijerph14090980

Wei J, Qian J, Tao Y et al (2018) Evaluating spatial priority of urban green infrastructure for urban sustainability in areas of rapid urbanization: a case study of Pukou in China. Sustainability (Switzerland) 10. https://doi.org/10.3390/su10020327

Winarso H, Hudalah D, Firman T (2015) Peri-urban transformation in the Jakarta metropolitan area. Habitat Int 49:221–229. https://doi.org/10.1016/j.habitatint.2015.05.024

Zhang S, Li Y, Zhang Y et al (2020) Does sanitation infrastructure in rural areas affect migrant workers' health? Empirical evidence from China. Environ Geochem Health 42:625–646. https://doi.org/10.1007/s10653-019-00396-2

Žlender V, Ward Thompson C (2017) Accessibility and use of peri-urban green space for inner-city dwellers: a comparative study. Landsc Urban Plan 165:193–205. https://doi.org/10.1016/j.landurbplan.2016.06.011

Chapter 5
Perspectives on Convergence of NbS and Agile City Principles in Tier II Industrial Cities of India

Nabanita Saha and Basudatta Sarkar

Abstract Nature-based solutions (NbS) have only recently entered the planning research and practice in India. Large cities like, Kochi, Chennai, Mumbai, Hyderabad, and Kolkata have taken action to revitalize their blue-green infrastructure. However, being on the cusp of the fourth industrial revolution and associated development, smaller tier II industrial cities also demand equal attention towards NbS. The rapidly changing urban landscape in these cities is leading to a shift of blue-green spaces to built spaces, making it vulnerable to climate change impacts. In the context of cities, agility involves adapting to rapidly changing circumstances, engaging citizens in decision-making, fostering innovation and resilience, prioritizing citizen participation and engagement and seeking input and feedback from citizens for an informed decision-making process. This research focuses on the convergence of components of agile governance and NbS. It aims to understand the stance of urban authority in Tier II industrial cities and analyses the factors that drive or impair the successful implementation of NbS. Following the analysis of surveys and interviews with citizens and city administrators, this research proposes a procedural framework as a solution towards embedding agile principles with respect to the implementation of NbS.

Keywords Nature-based solutions · Agile city principles · Tier II cities · Industrial cities · Indian context

---

N. Saha · B. Sarkar (✉)
Department of Planning and Architecture, National Institute of Technology Rourkela, Rourkela, India
e-mail: sarkarb@nitrkl.ac.in

N. Saha
e-mail: sahan@nitrkl.ac.in

M. Firoz C et al. (eds.), *Nature-Based Solutions for Urban and Peri-Urban Areas*, Disaster Risk Reduction, https://doi.org/10.1007/978-981-96-5933-3_5

5.1 Introduction

Since a long time, ideal cities have been described as those which are in balance with their natural environment (Taheri 1990). Urban ecosystem encompasses the complex interactions between society, economy, and nature, often considered analogous to the biological metabolism wherein resource from nature is converted to beneficial products and waste products by social and economic activities of humans (Decker et al. 2000; Zhang et al. 2006). The increase in the city size is increasing the demand for material input, infrastructure, and disposal of the waste thus generated. Socio-economic activities and the carrying capacity of nature are considered competing aspects of the urban ecosystem, implying that a reduction in material input will safeguard the natural environment (Zhang et al. 2006). However, converting the current way of urban life into a sustainable way with an increased degree of recyclability is a slow process. The rate of grey infrastructure development is much faster than the change in the urban way of life, especially in India. In this scenario, nature-based solutions (NbS) are considered sustainable solutions to counter issues of fast-paced urbanization and to supplement the grey infrastructure, thereby, supporting human well-being. Amid the concerns of climate change and declining human well-being, Russell (2011) emphasized on developing an 'urban culture of agility'. 'Agile urbanism' underscores the ability to react to fast-paced urban transformations and propagate temporary interventions in the city's infrastructure to solve immediate urban issues (Schwimmer and Schaufler, 2023). These interventions may subsequently be converted into permanent solutions depending on the communities' reception and acceptance. The World Economic Forum (2018) identifies eight key domains wherein agility plays key roles in managing the demands of the rapidly changing urban scenarios—building, land, security, energy, mobility, education, governance, and IT. At the core of urban agility is the government's appreciation and support to these changes and the communities' participation in agile administration. Therefore, installing NbS in an urban area would entail repurposing and retrofitting urban land, initially as temporary solutions and subsequently as permanent features depending on the success of such interventions and the adaptive capability of city administration, citizens and other stakeholders. In the current scenario, implementation of NbS in urban areas require flexibility in land use to install components like sustainable drainage, green–blue infrastructure etc., which demands an agile administration system to induce participation and incorporate feedback from the stakeholders.

The uncertain and rapid changes in the physical characteristics of urban centers require two types of adaptations. Type I adaptation (problem-oriented and preservation-oriented approach) refers to infrastructure and basic services addressing the absence or shortage of the same, and Type II adaptation (regeneration-oriented approach) addresses the future needs in a long-term basis resulting from changes in the urban environment such as socio-economic and climate changes (Burton 2004; Xiao et al. 2017). Researchers have suggested a holistic integration of adaptation responses on a spatial and temporal scale. However, the Tier II cities in India lack the necessary support to implement the adaptation responses as they are considered

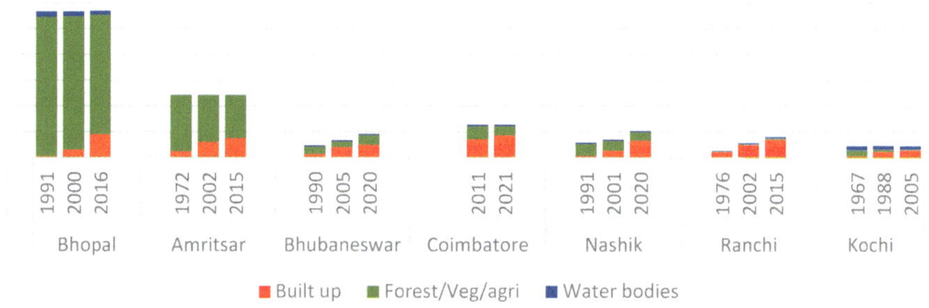

Fig. 5.1 Ratio of blue-green areas to built areas in Tier II cities of India. *Source* Authors' compilation[1]

as expensive in terms of both time and money along with various other challenges (Pathirana et al. 2018; Tessler et al. 2015; UNEP 2014).

5.1.1 Focus on Tier II Industrial Cities in India

The concept of NbS has only recently entered the planning research and practice in India. Large cities like, Kochi, Chennai (García Sánchez and Govindarajulu 2023; Tajuddin and Dabrowski 2021), Mumbai (Chitra 2022), Hyderabad (Kadaverugu et al. 2021), Kolkata (Basu et al. 2022; Roy-Basu et al. 2020), Delhi (Pawar and Dhote 2023), Bengaluru (Santhanam and Majumdar 2022; Shah et al. 2021) and others have taken action to revitalize and maintain their green spaces, parks, gardens, playgrounds, forests, lakes, and rivers. However, the smaller Tier II cities, especially the industrial cities, also demand equal attention towards NbS. Being in the cusp of fourth industrial revolution and associated development, they are eager to implement urban policies to improve livability and global visibility (Valsan 2017). These cities have also diversified their economic base to attract more investments and stay relevant in the globalized economic condition. Having the capacity to provide basic infrastructure at par with Tier I cities, they are attracting investment in smart infrastructure, real estate (Law 2023), transforming into startup hubs and technology centers (Balaji 2017; The Economic Times 2023), leveraging cultural heritage for economic and social benefit (Plaza 2016), and developing human capital and exploring collaborative advantage through regional networking (Hohmann and Roberts 2020). These cities are persuasively positing themselves as the next power centers, ready to capture a greater share of the country's growth. However, the steady increase in the built-up areas and decrease in the green and blue spaces in these cities, as represented in Fig. 5.1, increases their vulnerability to climate change imapcts.

[1] Source of data: Bhopal (Ghosh 2019), Amritsar (Minakshi et al. 2017), Bhubaneswar (Bose et al. 2023), Coimbatore (Nagarathinam 2022), Nashik (Bhailume 2021), Ranchi (Ahmad and Goparaju 2016), Kochi (Department of Town and Country Planning Government of Kerala 2010).

5.1.2 *Research Problem*

Indian government has been directing funds toward the development of small towns and cities through various schemes,[2] which pivot around urban service provisions, but do not explicitly mention the use and establishment of alternative practices like NbS to solve urban issues. Whereas lack of fund and resources, awareness of climate change, and skilled personnel in smaller cities have been commonly attributed to insufficient efforts toward sustainable strategies (Simon et al. 2021), the absent synergy between Type I and Type II adaption actions, lack of flexibility of adaptations, rejection of strategies by stakeholders, and lack of collaboration between different agencies have significantly contributed to failure in climate change adaptations (Pathirana et al. 2018). The Tier II cities have the unique opportunity to learn sustainable urban management from the Tier I cities which have already faced the negative impacts of rapid urbanization and are dealing with a distressed urban ecosystem through alternative practices. However, given the varied morphology, social norms, and culture, the issues related to NbS implementations in smaller cities must be contextualized. It thus becomes imperative to analyze the bottlenecks that impede the implementation of NbS specific to smaller cities. This research aims at understanding the stance of urban authority, specifically in Tier II industrial cities, considering Rourkela as a case in point. Accordingly, the following objectives have been formulated:

1. Investigation of the impediments and challenges faced by Tier II cities while initiating and implementing NbS in a rapidly changing urban scenario.
2. Exploration of possibilities to incorporate agility in governance to ensure seamless and efficient implementation of NbS in transforming urban areas.
3. Proposing a procedural framework as a solution towards embedding agile principles with respect to the implementation of NbS.

5.2 Convergence of NbS and Agile City Principles in Urban Planning

Particular to NbS, Wamsler et al. (2020) identified five strategies for successful implementation—"(i) targeted stakeholder collaboration; (ii) strategic citizen involvement; (iii) outsourcing; (iv) the alteration of internal working structures; and (v) concealed science policy integration". They also observed an "increasing need for relational approaches that, in turn, require individuals to develop the cognitive/emotional capacity to establish trust, communicate inclusively and promote social learning, while at the same time dealing with an increasingly complex and uncertain working environment". The first three strategies among these five have been considered the main focus of agility (Beck et al. 2001). Although originated from the field of software development, it is designed to achieve a higher level of adaptability by creating a dynamic organizational systems and methods of collaboration (Mergel

[2] See Infrastructurel National Portal of India for details.

et al. 2018). Agile governance must focus on individuals and interactions, development of a working model, stakeholder collaboration, and preparedness to update or change with time rather than following a plan. The bottom-up approach starts with empowering citizens to express their choices in the decision-making process and also connect with various other stakeholder groups. These activities lead to continuously updated public policies by designing and testing them to assess their ability to fulfill the needs of the citizens. Identification of needs emerging from the complexity, uncertainty, and urgency and consequent optimization of associated infrastructure and services result in improved citizens satisfaction and successful implementations (Mergel et al. 2018; Soe and Drechsler 2018; Xiao et al. 2017). Table 5.1 provides a brief account of agile governance being used in urban planning.

The broad objectives of the agile approach elaborate on the importance of stakeholders' involvement and collaboration and a digital base to streamline this collaboration. The developed interfaces respect stakeholders' needs emerging from the complexity, uncertainty, and urgency of the changing urban scenario resulting in a working model that is to be continuously updated and tested to determine a lasting outcome. The three broad components of agile governance can be listed as—physical, digital, and environmental. The physical component deals with the adaptive capacity of current infrastructure; the digital component deals with harnessing technology or use of technological advancement; the environmental component deals with the innovative mitigation measures to the environmental impacts. City administrators have deployed agility while addressing the issues of sustainable development and mitigation of negative environmental impact, especially through NbS, the brief account of which is presented in Table 5.2.

For efficient designing or planning of NbS, agile attributes are to be determined at the very nascent stage. The desired outcomes in the changing context of needs can be achieved through flexible adaptation responses, anticipation and management of probable or possible events, and participation of all stakeholders in the decision-making process. Thus, governance needs to be agile with a bottom-up approach where public policies are continuously updated to fulfill the needs of the citizens. The determinants of agile governance focus on multi-stakeholder engagement, flexible budgeting, multi-stage procurement process, consultative approach for new policies on the use of 4IR (Fourth Industrial Revolution) technologies and innovation, cross-entity digital services, and feedback system of citizen satisfaction. Given that Tier II cities have the unique opportunity to implement sustainable urban management, this research attempts to propose a procedural framework for the implementation of NbS, addressing their specific shortcomings with embedded agility.

5.3 Methodology

To formulate a framework for the implementation of NbS by incorporating agile principles, the industrial city of Rourkela, in the eastern Indian state of Odisha (Fig. 5.2), was taken up as a case example. Rourkela is actively improving its livability and

Table 5.1 Integration of agile principles in urban planning

Literature	Emphasis
(Schwimmer and Schaufler 2023)	Agile urbanism will be the key to maneuver through the temporarily changing functions and myriad infrastructural requirements of ever-changing urban environments. To incorporate the citizen's need for social and adaptive spaces, city administrators need to respond to short-term changes in urban flow and subsequent needs through end-to-end digital planning, introduction of short iteration cycles and flexible design components
(American Planning Association 2021)	Agile processes like design thinking and a "fail fast, fail small" approach have been the buoy for contemporary planners. Experimentation, prototyping, pilots, community engagement, and feedback loops generate innovative ideas for community-specific solutions through digital, automated, and technologically advanced approaches
(Salem 2016)	The key aspect of the 'smartness' of cities is the utilization of information and communication technologies (ICTs). As enablers of "smart" transformation, acceptance of digital technologies by society is imperative. Digitalization has been the foundation of Dubai's collaborative and networked style of governance incorporating agility which led to smart urban services
(Farias et al. 2020)	In the context of Brazil, the participatory approach to urban planning is often not accessible to the general public due to the digital divide. To accommodate community participation, visualization can be used as a tool for engaging the community for opinions on alternative scenarios
(Hämäläinen 2021)	It is argued that a digital twin can be used to create a virtual replica of a city, which can be used to explore different simulated planning scenarios and evaluate the impact of different decisions. Implementing pilot projects, embracing experimentation, learning, and adapting to new technologies and approaches through interactive platforms is vital for creating resilient and responsive cities
(Krämer et al. 2013)	Since the traditional urban planning methods are considered time intensive to effectively address the complex challenges of modern cities, the use of Domain-Specific Languages (DSLs) and ICT-enabled tools is proposed to create a more agile and iterative approach to urban planning. The need for increased collaboration and transparency among stakeholders, including citizens, developers, and city officials is emphasized

(continued)

Table 5.1 (continued)

Literature	Emphasis
(Abouaiana and Battisti 2023)	It formulates a transdisciplinary framework wherein stakeholder identification, community participation, support to decision-making, and conflict mitigation will eliminate uncertainties during organized stakeholder collaboration for a planning decision. It underscores the mass collaboration of the right stakeholders towards a common objective using agile tools like digital-based approaches and agile methodology
(Faber et al. 2018)	The emphasis is on collaboration, iterative and data-driven decision-making approaches in urban planning. Ultimately, the article argues that by leveraging agile governance principles and using digital tools like DSLs, cities can create more dynamic, responsive, and data-driven urban planning processes, leading to the development of more sustainable and inclusive smart cities

global visibility and is steadily moving towards 4IR. The rapidly changing urban landscape is leading to a shift from blue-green spaces to built spaces. Although Rourkela was built around the steel plant and its ancillary requirement, it is now showing a high degree of diversification in economic base and composition of population positing itself as an upcoming powerhouse of Odisha.

The methodological framework for this study is represented in Fig. 5.3. It employed a qualitative research method wherein the semi-structured interview provided a basis for conducting a detailed two-pronged self-administered focus group questionnaire survey—one for city administrators and the other for residents. Based on the result of the semi-structured interview, the questionnaire contained inquiry pertaining to eight domains—access to information, database and knowledge generation and acceptance of NbS, efforts toward uptake and advocacy, scope for policy integration, extent of stakeholder networking, opportunities for active community engagement, scope for monitoring and assessment, financing mechanism and future scope in terms of knowledge development and sharing. Questions were of two types—yes/no/maybe and descriptive. Interviews and surveys were conducted with authoritative personnel at Rourkela Municipal Corporation and Rourkela Smart City Limited belonging to different policy domains and directly engaged with developmental work. The resident focus group consisted of 44 respondents (51% male and 49% female; age group of 20–40). Tables 5.3 and 5.4 show examples of mapping the questions with the dimensions for the citizens and administrators respectively. In addition to the survey and interviews, policy documents were analyzed and insight through participatory observation was recorded.

Table 5.2 Implementation of NbS by incorporating agile principles in urban governance

Literature	Emphasis
(Salliou et al. 2023)	Considering that urban landscapes must deliver and maintain key ecosystem services, in spite of shocks, the landscape designers are expected to have constant input from scientists and engineers to maintain the benchmark of service provision. Simultaneously, stakeholders must be included to gain local knowledge and well as for co-production. Therefore, maintaining the science-design loop and energizing the designer stakeholder loop is essential for climate change adaptations
(Nur et al. 2021)	It recommended that a clear standard operating procedure need to be stated for the community to participate in any NbS-related project, and the government should disseminate adequate knowledge about the project for it to be acceptable to the community
(Kauark-Fontes et al. 2023)	Efficient implantation of NbS and its management depends on adaptive governance. It is found that NbS adaptive governance can be a global concept and yet incorporate specificities of the region, thereby highlighting the need for a "collaborative environment, composed of collective decision-making, co-management, and polycentricity"
(Hölscher et al. 2024)	Particularly for implementing NbS, it is required that collaborative governance practice must include a co-production approach suited to the local context. An inclusive environment is created through policies for co-production and collaboration and engendering networking between all stakeholders and experts
(Frantzeskaki et al. 2022)	For the long-term success of NbS implementation, the emphasis is on a collaborative approach which includes people in the co-creation process—from ideation to construction and management beyond the project period
(West and Jia 2023)	Addressing the challenges faced while forming multilevel collaboration, a Climate Sprint Model was developed to boost collaboration for climate change actions. Using the 'scrum technique', which has inherent feedback loops, it fosters rapid feedback, group discussions, and multi-disciplinary teams
(Vögt et al., 2023)	This article emphasizes the implementation of agile data workflows to integrate climate data into climate change adaptation planning. It identified the transdisciplinary and trans-sectoral co-development approach to enhance agility in urban management and emphasized on communication and feedback loops between various stakeholders
(Battistia et al., 2024)	NbS implementation is based on stakeholder collaboration and citizens' cooperation. Networking capabilities, integration of resources, co-responsibility, and knowledge transfer are among the key elements that ensure sustenance of the NbS

Fig. 5.2 Map showing the location of the city of Rourkela, Odisha in India. *Source* of base map: Modified from (Thevapalan 2021)

5.4 Results and Discussion

The literature on NbS has underscored the inefficiency of the 'one-size-fits-all' approach in its implementation and necessitated the understanding of specific capabilities and requirements of the locale where it is to be installed. It is, therefore, essential to contextualize and understand the bottlenecks that cities like Rourkela face in mainstreaming and implementing NbS while balancing the rapid changes in the city. The outcome of interviews, surveys, and observations of the urban processes by the authors provided insight into the degree of knowledge and acceptance of the concept of NbS by the administration and the general issues that hinder the operationalization of NbS. The outcome has been abridged below.

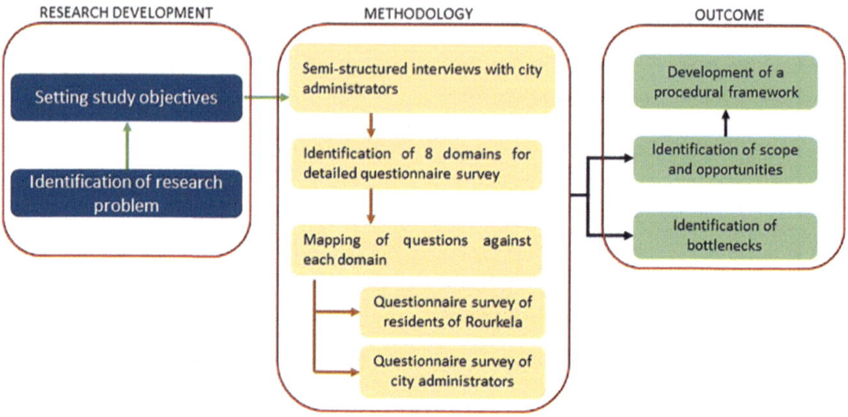

Fig. 5.3 Flowchart of methodological framework

Table 5.3 Questionnaire mapping for residents of Rourkela

Domain	Questions
Knowledge and acceptance	Have you heard the term Nature-Based Solutions (NbS)? Examples of NbS are parks, playgrounds, urban forest areas, green roofs, bioswales/swales, sustainable urban drainage, blue-green infrastructure, etc. Do you think such solutions are viable for Rourkela?
Uptake and advocacy	Do you think the existing green areas and water bodies are providing make the city look good?
Integration with policies and regulation	Do the city administrators have rules and regulations to protect green spaces and waterbodies of Rourkela?
Stakeholder networking and partnership and active community engagement	Is there any opportunity for you to express your opinion on any city-related decision?
Monitoring and assessment	Who do you think does the maintenance of the green spaces and waterbodies?
Financing mechanism	If the government provides incentives, would you be eager to implement NbS in your building or residential complex?
Knowledge development and sharing	What is your perception of the poorly maintained green spaces around the cities?

Divergence of green mandates of different development agencies: The overarching guidelines followed by industrial developments in India are the 'Guidelines for development of greenbelt' issued by the CPCB (Central Pollution Control Board of India) (CPCB 2000). The guideline recognizes the ecosystem services provided by a green belt with clear specifications regarding its elements. However, these green

Table 5.4 Questionnaire mapping for administration

Domain	Questions
Knowledge and acceptance	If NbS were to be implemented in Rourkela, how can awareness be created among the public?
Uptake and advocacy	Do you think the existing green parks, rivers, waterbodies can help in solving any type of urban issues present in Rourkela?
Integration with policies and regulation	Examples of NbS are parks, playgrounds, urban forest areas, green roofs, bioswales/swales, sustainable urban drainage, blue-green infrastructure, etc. Do you think such solutions are viable for Rourkela?
Stakeholder networking and partnership	Is it possible to form multi-stakeholder partnership in support of NbS?
Active community engagement	Would community engagement/participation play a role in the planning, design, and management of NbS in Rourkela?
Monitoring and assessment	Do any public institutions or city administration of Rourkela have a system in place to monitor and assess the NbS?
Financing mechanism	What could be the probable financial mechanism for the implementation of NbS?
Knowledge development and sharing	Do the public institutions or city administration of Rourkela aim to develop a knowledge base, database, and skill related to NbS?

belts are developed as isolated entities. Green belt development by the industries is, most of the time, a fulfillment of the CPCB requirement, rather than a city-wide nature-based integrated solution.

Mechanism for co-ordination between different city management agencies: Similar to industrial towns in India, Rourkela is divided into two parts in its maintenance and functions. The steel plant, its ancillary functions, and the residential area of its employees were managed by the steel pant authorities and the remaining area is maintained by the municipality. Rourkela Development Authority was created by the Government of Orissa in the year 1995 and more recently the Rourkela Smart City Limited was established to manage the smart city developments. The main activities of the Rourkela Development Authority pertain to different land use development and re-development programmes in a planned manner (Rourkela Development Authority: https://www.rda-rourkela.in/index.html). Although the city exhibits skilled coordination among administrators in implementing urban projects, there is little clarity on strategies for green development or river basin management, making the design and implementation of NbS problematic.

Urban lost spaces: The city has a large number of neglected, empty, unused areas that exist beside roads or in urban districts. Such spaces have been termed as 'urban lost spaces' in the literature of urban design (Franck and Stevens 2006; Loukaitou-Sideris 1996). These are the incidental spaces that might have lost their intended functions and are often undesirable. These lost spaces have a high potential to provide ecosystem services through urban greening practices and scaling them up to form a network of green stormwater infrastructure (Albro 2019). However, such spaces in Rourkela are not documented and accounted for, making it difficult to identify such land parcels and to bring them within the ambit of NbS.

Entrusting land: The smaller open spaces within individual plots can also contribute to NbS. However, residents more often are not forthcoming in allowing the usage of their land for public purposes. A more effective strategy would be to involve residents in every step of the decision-making process and quick information dissemination, thereby, building up their trust along with incentivizing their initiatives.

Singular objective for green space development: The green and blue spaces which are taken up for improvement in the city are developed as recreational amenities. Although they provide much-needed public spaces for the residents, the multi-dimensional benefits of such spaces are not perceived, and neither the scope for networking of these green areas are explored. This, essentially, stems from the lack of awareness about NbS and the lack of a mechanism to incentivize such initiatives.

Three major barriers to implementation of NbS were revealed—stakeholders' coordination, identification of multi-use land parcels, and lack of knowledge regarding the multi-functionality of blue-green spaces in the city. Further analyzing the survey responses from the residents, the following observations are made and discussed:

1. Air and water pollution, water shortage, ill-maintained parks, playgrounds, and water bodies, and lack of recreation amenities emerged as the main urban issues in Rourkela. Waste management, maintenance of sewer drainage, parking issues are additional issues mentioned by the residents. Residents find shortcomings in the provisions and maintenance of basic infrastructure. However, these deficiencies in the provision of urban infrastructure can be offset by using alternative practices like the NbS. The city has scope for implementing NbS to alleviate its infrastructural deficiencies and supplement the grey infrastructure.
2. Apart from recreation and aesthetics, residents have identified the purpose of green spaces and water bodies majorly as stormwater management, prevention of water logging, ecological management, controlling air pollution and healing properties. The residents are cognizant of the multi-benefits of NbS elements through information from various sources. However, structured understanding through formal workshops or programs by the city administrator has not been available. This hinders the mainstreaming of NbS.
3. Residents identified the municipality as responsible for the maintenance of green spaces and water bodies. Their perception of 'not used' green spaces is directed to the unidentified use or unplanned use of green spaces. With awareness and education, these 'not used' green spaces can be engaged for the provision of ecosystem services with shared responsibility for maintenance and management.
4. Residents get informed about new policy implementation or development projects mainly from social media and sometimes from the websites of municipal corporations. The residents are perceptive of the advantages of social media

and are attuned to various digital platforms. Opting for social media, or an exclusive digital platform shared by the city administrators and all stakeholders will facilitate the fast-paced information dissemination, community engagement, co-ordination, and collaboration between all stakeholders for NbS-related projects.

Surveys with administrators resulted in very few responses, responders include cluster coordinator, project assistant and senior system executive. Therefore, definitive results could not be obtained. However, it was evident that the city administrator must develop an intention toward implementing NbS since they have the opportunity to anticipate the urban problem that may arise in the near future and also take cue from other Indian cities on the implementation of NbS. Although insufficient responses were received, the following observations are made based on the response to the descriptive questions:

1. Main infrastructure-related issues in Rourkela are unplanned or non-systematic house development and drainage system. River banks and slum areas are the parcels of lands that are the first choice for infrastructure development.
2. City administrators are aware of the terms—green infrastructure, green corridor, restorative agriculture, urban forestry, sustainable urban drainage, and most importantly NbS. It is noted that they heard about NbS from social media and not from any structured source of information dissemination.
3. Awareness for NbS can be created by the active participation of different stakeholders like the administrative body, social media, NGOs, through community meetings, workshops, and awareness programmes.
4. There is an initiative for clean and green Rourkela however, it is not well known or popularised.
5. There is long-term plan for the preservation or creation and monitoring of NbS through third-party contractor, local bodies and NGOs, and there is also a database available related to green spaces and water bodies.

5.5 Proposal for a Framework

It is revealed that lack of information and knowledge about the scale, scope, and benefits of NbS is one of the major hindrances in mainstreaming it as an alternate practice to optimize grey infrastructure. Although residents and administration are aware of some benefits of NbS, lack of structure effort does not allow the initiation of NbS projects. It is also observed that, whereas co-ordination among various departments and citizen collaboration can be expected, awareness of the science-design loop is not present. In cities like Rourkela, it is recommended that implementation of NbS must be initiated as a top-down approach. After sufficient success, building of trust and community engagement in initial projects, the later projects can be initiated through a bottom-up approach wherein, ultimately the stakeholders can take control of maintaining and managing the NbS. Also, integration of Type I and Type

II adaptations is essential for the overall development. Although the observations are based on an analysis of Rourkela, these can be extended to other Tier II cities that are either at the nascent stage of NbS implementation or have not initiated any implementation. In such cities, an hybridized approach of top-down and bottom-up approach has to be opted for.

This research proposes a procedural framework (Fig. 5.4) as a solution toward embedding agile principles with respect to the implementation of NbS clearly emphasizing stakeholder networking and community at every stage of planning. It also suggests the utilization of digital technologies in data and feasibility assessment and a digital product to aid in stable and instantaneous community engagement. Based on Xiao et al. (2017) framework of the agile participatory process, the stages of planning for NbS implementation have been divided into four major sections. First, 'development of intention' wherein a knowledgebase is created and acceptability of NbS by stakeholders and the general public increases. Second, 'feasibility assessment' helps in an informed decision-making and master plan phasing emphasizing the advocacy and uptake of NbS and an assessment of the financing mechanism. The third stage—'projection, design and implementation' involves design iterations and incorporation of the strategies into development control tools through building permits and environmental permits. The fourth stage is 'operationalization and expansion' wherein post-implementation monitoring and quality assessment are to be done.

5.6 Conclusions

This research investigated the impediments and challenges faced by Tier II cities while initiating and implementing NbS and proposed a procedural framework to embed agile principles for the implementation of NbS. The framework is the intersection of standard planning processes and agile participatory processes. The framework utilizes the participatory process as a feedback mechanism with stable stakeholder networking and quick feedback from the community. Given that some form of digital platform is available to a large number of city residents it incorporates the use of digital technology in not just obtaining data and data management for analysis and evaluation, but also developing digital interface for quick dissemination of information and to receive instant response. This framework can be utilized in a hybridized top-down and bottom-up approach to NbS implementation. However, this does not imply foregoing the face-to-face interaction, which is still considered central to the framework. The framework also attempts to bridge the gap between technical and human-centric approaches through design-science and design-community loops.

Although Rourkela, an industrial city, has been studied as a case, this framework is expected to guide the implementation process of NbS through an agile participatory process in other Tier II cities of the country. The framework is expected to contribute towards long-term sustainability solutions in a rapidly changing urban scenario in smaller cities. The present research is based on the survey of a small group of people

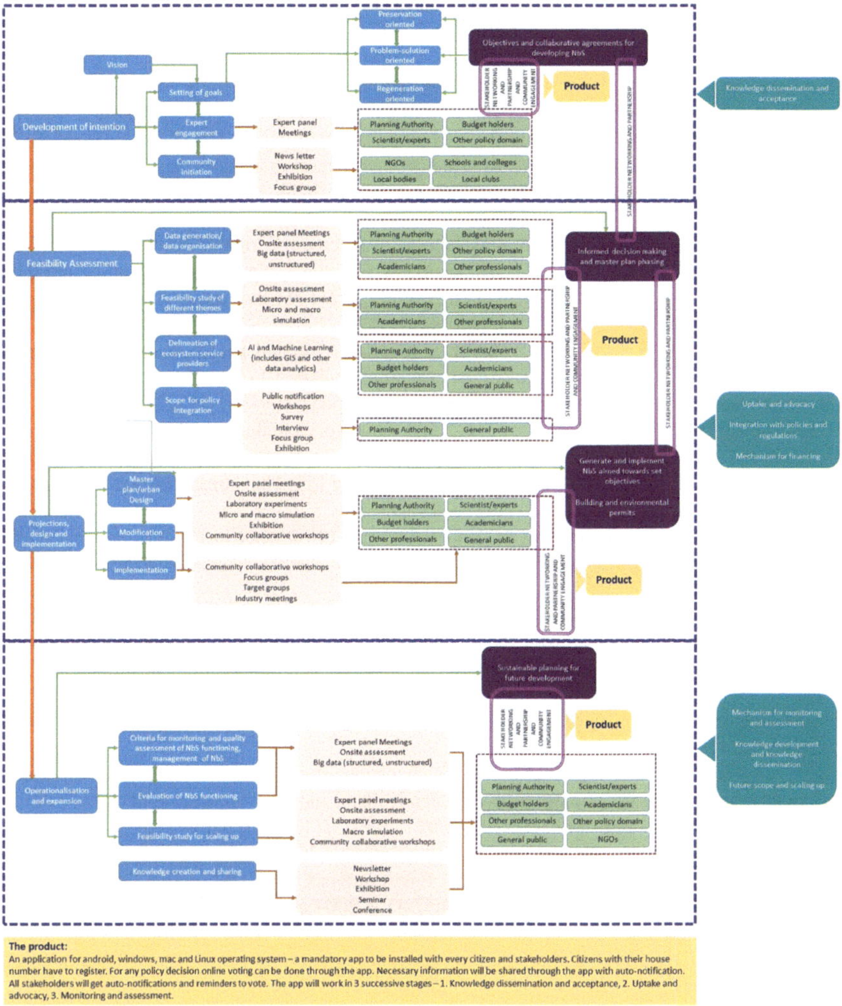

Fig. 5.4 Incorporating agile participatory process in methodological framework to implement NbS

residing in Rourkela. To further the research on NbS implementation in Tier II industrial cities, comparative studies of several other similar cities need to be conducted with robust surveys. Such comparative studies will lead to the formulation of national guidelines to implement NbS in Tier II cities in India and amendment in city-specific building by-laws.

References

Abouaiana A, Battisti A (2023) Agile-transdisciplinary conceptual framework for retrofitting mediterranean built environments. In: Sayigh A (ed) Mediterranean architecture and the green-digital transition. Springer, pp 127–146. https://doi.org/10.1007/978-3-031-33148-0_11

Ahmad F, Goparaju L (2016) Analysis of urban sprawl dynamics using geospatial technology in Ranchi City, Jharkhand, India. J Environ Geogr 9(1–2):7–13. https://doi.org/10.1515/jengeo-2016-0002

Albro SL (2019) Vacant to vibrant: creating successful green infrastructure networks. Island Press

American Planning Association (2021) The future of planning is agile, people-centric, and technologically advanced. https://www.planning.org/blog/9211711/the-future-of-planning-is-agile-people-centric-and-technologically-advanced/

Balaji S (2017) India's startups are moving into smaller, second tier cities—why the sudden shift? Forbes. https://www.forbes.com/sites/sindhujabalaji/2017/10/05/thanks-to-regional-rivalries-indias-startup-surge-is-spreading-across-the-country/?sh=4e64a485231b

Basu A, Basu BL, Barbosa V, Dadashpoor H (2022) 'Green placemaking' in Kolkata: Role of urban greens and urban forestry. In: Sahana M, Areendran G, Raj K (eds) Conservation, management and monitoring of forest resources in India. Springer, pp 131–162. https://doi.org/10.1007/978-3-030-98233-1_6

Battistia L, Cuomo F, Manganelli A (2024) Collaborative governance arrangements: what makes nature-based solutions endure? Territory, politics, governance 1–22. https://doi.org/10.1080/21622671.2024.2355317

Beck K, Beedle M, van Bennekum A, Cockburn A, Cunningham W, Fowler M, Grenning J, Highsmith J, Hunt A, Jeffries R, Kern J, Marick B, Martin RC, Mellor S, Schwaber K, Sutherland J, Thomas D (2001) Manifesto for agile software development. http://agilemanifesto.org/

Bhailume SA (2021). An Assessment of land use land cover change and urban growth of Nashik city using geospatial techniques. Appl Ecol Environ Sci 9(2):225–235. https://doi.org/10.12691/aees-9-2-14

Bose S, Mazumdar A, Basu S (2023) Monitoring change in urbanization and green space for eastern Indian cities in 30 years-a comparison between Kolkata and Bhubaneswar. IOP Conf Ser Earth Environ Sci 1164(1). https://doi.org/10.1088/1755-1315/1164/1/012013

Burton I (2004) Climate change and the adaptation deficit. Adaptation and impacts research group (AIRG) meteorological service of Canada, Environment Canada, November. https://publications.gc.ca/collections/collection_2012/ec/En57-41-1-2004-eng.pdf

Chitra V (2022) emembering the river: flood, memory and infrastructural ecologies of stormwater drainage in Mumbai. Urban Studies 59(69):1855–1871. https://doi.org/10.1177/00420980211023381

CPCB (2000) Guidelines for development of greenbelts. Cebtral Pollution Control Board, India

Decker EH, Elliott S, Smith FA, Blake DR, Rowland FS (2000) Energy and material flow through the urban ecosystem. Annu Rev Energy Env 25(1):685–740

Department of Town and Country Planning Government of Kerala (2010) Development plan for Kochi city region 2031 (Draft): Volume 1—study and analysis

Faber A, Hernandez-Mendez A, Rehm SV, Matthes F (2018) An agile framework for modeling smart city business ecosystems. In: Proceedings of the 20th international conference on enterprise information systems 2:39–50. https://doi.org/10.5220/0006696400390050

Farias HTM de, Brasil AB, Barbosa FL (2020) Visualizing under pressure: parametric modeling of urban morphology as an agile and transparent tool for participatory planning in Brazil. Congreso SIGraDi 2020. São Paulo: Blucher 676–682. https://doi.org/10.5151/sigradi2020-93

Franck K, Stevens Q (eds) (2006). Loose space: possibility and diversity in urban life. Routledge.

Frantzeskaki N, Mahmoud IH, Morello E (2022) Nature-based solutions for resilient and thriving cities: opportunities and challenges for planning future cities. In: Mahmoud IH, Morello E, de Oliveira FL, Geneletti D (eds) Nature-based solutions for sustainable urban planning. Springer, pp 3–18. https://doi.org/10.1007/978-3-030-89525-9_3

García Sánchez F, Govindarajulu D (2023) Integrating blue-green infrastructure in urban planning for climate adaptation: lessons from Chennai and Kochi, India. Land Use Policy 124:106455. https://doi.org/10.1016/j.landusepol.2022.106455

Ghosh S (2019) A city growth and land-use/land-cover change: a case study of Bhopal, India. Model Earth Syst Environ 5(4):1569–1578. https://doi.org/10.1007/s40808-019-00605-y

Hämäläinen M (2021) Urban development with dynamic digital twins in Helsinki city. IET Smart Cities 3(4):201–210. https://doi.org/10.1049/smc2.12015

Hohmann RP, Roberts B (2020) Connecting systems of secondary cities: the role of infrastructure and inter-city collaborations. Urbanet. https://www.urbanet.info/systems-of-secondary-cities/

Hölscher K, Frantzeskaki N, Kindlon D, Collier MJ, Dick G, Dziubała A, Lodder M, Osipiuk A, Quartier M, Schepers S, De Sijpe Van K, der Have van C (2024) Embedding co-production of nature-based solutions in urban governance: emerging co-production capacities in three European cities. Environ Sci Policy 152(103652). https://doi.org/10.1016/j.envsci.2023.103652

Kadaverugu A, Nageshwar Rao C, Viswanadh GK (2021) Quantification of flood mitigation services by urban green spaces using InVEST model: a case study of Hyderabad city, India. Model Earth Syst Environ 7(1):589–602. https://doi.org/10.1007/s40808-020-00937-0

Kauark-Fontes B, Ortiz-Guerrero CE, Marchetti L, Hernández-Garcia J, Salbitano F (2023) Towards adaptive governance of urban nature-based solutions in Europe and Latin America—a qualitative exploratory study. Sustainability (Switzerland) 15(5). https://doi.org/10.3390/su15054479

Krämer M, Ludlow D, Khan Z (2013) Domain-specific languages for agile urban policy modelling. In: Rekdalsbakken W, Bye RT, Zhang H (eds) Proceedings—27th European conference on modelling and simulation, ECMS 2013. https://doi.org/10.7148/2013-0673

Law A (2023) New set of tier-2 cities emerging as real estate growth drivers: MD Cushman & Wakefield, India. The Hindu: Businessline. https://www.thehindubusinessline.com/economy/new-set-of-tier-2-cities-emerging-as-real-estate-growth-drivers-md-cushman-wakefield-india/article67389092.ece

Loukaitou-Sideris A (1996) Cracks in the city: addressing the constraints and potentials of urban design. J Urban des 1(1):91–103. https://doi.org/10.1080/13574809608724372

Mergel I, Gong Y, Bertot J (2018) Agile government: systematic literature review and future research. Gov Inf Q 35(2):291–298. https://doi.org/10.1016/j.giq.2018.04.003

Minakshi M, Singh S, Pateriya B (2017) Geospatial analysis of urban sprawl patterns in Amritsar City, Punjab, India. Int J Appl Inf Syst 11(8):8–14. https://doi.org/10.5120/ijais2017451636

Nagarathinam SR (2022) Spatial-temporal analysis of land use: Coimbatore corporation, Tamil Nadu. Int Res J Eng Technol 83–86. www.irjet.net

Nur MI, Juana T, Zain FC, Astuti DR (2021) Study of agile governance on strengthening reforestation program in West Java Indonesia. IOP Conf Ser Earth Environ Sci 905(1). https://doi.org/10.1088/1755-1315/905/1/012009

Pathirana A, Radhakrishnan M, Ashley R, Quan NH, Zevenbergen C (2018) Managing urban water systems with significant adaptation deficits—unified framework for secondary cities: part II—the practice. Clim Change 149(1):57–74. https://doi.org/10.1007/s10584-017-2059-0

Pawar M, Dhote M (2023) The city green landscapes: environmental benefits and typologies of green landscapes in Delhi. In: Chatterjee U, Bandyopadjyay N, Setiawati MD, Sarkar S (eds) Urban commons, future smart cities and sustainability. Springer, pp 645–663

Plaza AB (2016) Blog poll: invest in infrastructure to make Asia's secondary cities 'smart.' Asian Development Blog. https://blogs.adb.org/blog/blog-poll-invest-infrastructure-make-asia-s-secondary-cities-smart

Roy-Basu A, Bharat GK, Chakraborty P, Sarkar SK (2020) Adaptive co-management model for the East Kolkata wetlands: a sustainable solution to manage the rapid ecological transformation of a peri-urban landscape. Sci Total Environ 698:134203. https://doi.org/10.1016/j.scitotenv.2019.134203

Russell JS (2011) The Agile city: building well-being and wealth in an era of climate change. Island Press

Salem F (2016) A smart city for public value: digital transformation through agile governance—the case of "Smart Dubai." World Government Summit Publications. https://ssrn.com/abstract=2733632
Salliou N, Arborino T, Nassauer JI, Salmeron D, Urech P, Vollmer D, Grêt-Regamey A (2023) Science-design loop for the design of resilient urban landscapes. Socio-Environ Syst Model 5(18543). https://doi.org/10.18174/sesmo.18543
Santhanam H, Majumdar R (2022) Quantification of green-blue ratios, impervious surface area and pace of urbanisation for sustainable management of urban lake—land zones in India -a case study from Bengaluru city. J Urban Manag 11(3):310–320. https://doi.org/10.1016/j.jum.2022.03.001
Schwimmer E, Schaufler C (2023) Towards adaptive planning of urban spaces in the context of a new agile urbanism. In: Alberti F, Matamanda AR, He B-J, Galderisi A, Smol M, Gallo P (eds) Urban and Transit Planning City Planning: Urbanization and Circular Development, 3rd ed. Springer Nature, pp 3–10
Shah A, Garg A, Mishra V (2021) Quantifying the local cooling effects of urban green spaces: evidence from Bengaluru, India. Landsc Urban Plan 209(104043). https://doi.org/10.1016/j.landurbplan.2021.104043
Simon D, Vora Y, Sharma T, Smit W (2021) Responding to climate change in small and intermediate cities: comparative policy perspectives from India and South Africa. Sustainability (Switzerland) 13(4):1–16. https://doi.org/10.3390/su13042382
Soe RM, Drechsler W (2018) Agile local governments: experimentation before implementation. Gov Inf Q 35(2):323–335. https://doi.org/10.1016/j.giq.2017.11.010
Taheri R (1990) Urban area as a part of an ecosystem. Energy Build 15–16:129–132
Tajuddin N, Dabrowski M (2021) Enabling socio-ecological resilience in the global south: insights from chennai, india. Sustainability 13(19):10522. https://doi.org/10.3390/su131910522
Tessler ZD, Vorosmarty CJ, Grossberg M, Gladkova I, Aizenmen H, Syvitski JPM, Foufoula-Georgiou E (2015) Profiling risk and sustainability in coastal deltas of the world. Science 349(6248):638–643. https://doi.org/10.1126/science.aab3574
The Economic Times (2023) De-Bengalurued: why India's tech industry is now rushing to 26 smaller cities. ET Online. https://economictimes.indiatimes.com/news/company/corporate-trends/de-bengalurued-why-indias-tech-industry-is-now-rushing-to-26-smaller-cities/articleshow/103245493.cms?from=mdr
Thevapalan T (2021) Karaikudi vina tradition: a comparative study. MRes Thesis
UNEP (2014) The adaption gap report: a preliminary assessment. http://www.unep.org/climatechange/adaptation/gapreport2014/
Valsan S (2017) Livability quotient: a paradigm shift in India's emerging cities
Vögt V, Harrs JA, Reinhart V, Hollenbach P, Bühler MM, Tewes T (2023) Implementing agile data workflows to unlock climate-resilient urban planning. Climate 11(174). https://doi.org/10.3390/cli11090174
Wamsler C, Wickenberg B, Hanson H, Alkan Olsson J, Stålhammar S, Björn H, Falck H, Gerell D, Oskarsson T, Simonsson E, Torffvit F, Zelmerlow F (2020) Environmental and climate policy integration: targeted strategies for overcoming barriers to nature-based solutions and climate change adaptation. J Clean Prod 247:119154. https://doi.org/10.1016/j.jclepro.2019.119154
West D, Jia J (2023) The climate sprint: an agile process for catalytic collaboration towards a just transition. In: Starik M, Rands GP, Deason JP, Kanashiro P (eds) Handbook of multi-level climate actions. Edward Elgar, pp 242–254
World Economic Forum (2018) Agile cities: preparing for the fourth industrial revolution. Global future council on cities and urbanization. http://www3.weforum.org/docs/WP_Global_Future_Council_Cities_Urbanization_report_2018.pdf
Xiao J, Lavia L, Kang J (2017) Towards an agile participatory urban soundscape planning framework. J Environ Planning Manage 61(4):677–698. https://doi.org/10.1080/09640568.2017.1331843
Zhang Y, Yang Z, Yu X (2006) Measurement and evaluation of interactions in complex urban ecosystem. Ecol Model 196(1–2):77–89. https://doi.org/10.1016/j.ecolmodel.2006.02.001

Part II
NBS—Urban Infrastructure Solutions and Development Strategies

Chapter 6
Urban Climate Shelters: A Nature-Based Solution for Urban Resilience

Ombretta Caldarice, Bruna Pincegher, Maria Pizzorni, and Nicola Tollin

Abstract In recent years, European cities have been experimenting with planning and designing solutions to operationalise the Paris Agreement, improving the mitigation of the impacts of Global Greenhouse Gas (GHG) emissions, and supporting adaptation to multiply effect on urban challenges, direct threat slow/rapid on-setting risk. A growing number of local governments have started to contrast the impacts of climate-induced extreme events, including the challenges posed by rising temperatures, thanks to ecosystem-based approaches and Nature-based Solutions (NbS) to make cities more resilient. Among many, Urban Climate Shelter (UCS) is based on transforming critical urban spaces into climate refuges to foster security from heat, flood, and drought events through a multi-hazard approach by using NbS and activating a process of co-production with the wider local community. Recently, this kind of solution has been applied to schools, transforming schoolyards into shelters for the population during extreme temperatures and functioning as a cooling project. The chapter explores the potential of UCS in schoolyards as NbS for urban resilience since they can contribute significantly to integrating climate mitigation and adaptation actions locally. Reviewing case studies in European cities that have successfully carried out climate shelters in schools in the past decade, including Madrid, Barcelona, London, Paris, Amsterdam, and the Hague, the chapter argues that incorporating NbS in school settings seems to be a promising path to creating co-benefits for urban resilience. In the end, the chapter focused on the lessons learned from implementing NbS in schoolyards, including the societal challenges within the co-benefits of NbS to urban resilience and suggesting that UCS in schoolyards can be a codified solution for sustainable urbanisation.

Keywords Nature-based Solutions (NbS) · Urban Climate Shelter (UCS) · Climate change mitigation and adaptation · Urban resilience · Schoolyard regeneration · Co-production

---

O. Caldarice · B. Pincegher (✉)
Politecnico di Torino, Turin, Italy
e-mail: bruna.pincegher@polito.it

M. Pizzorni · N. Tollin
University of Southern Denmark, Odense, Denmark

M. Firoz C et al. (eds.), *Nature-Based Solutions for Urban and Peri-Urban Areas*, Disaster Risk Reduction, https://doi.org/10.1007/978-981-96-5933-3_6

6.1 Introduction. Nature-Based Solutions for Urban Resilience

Despite occupying only 3% of the world's surface, cities contribute over 70% of GHG emissions, making them both vulnerable to and a significant handler of climate change (De Munck et al. 2013). Driven by rapid urbanisation, with estimations indicating two-thirds of the global population living in cities by 2050, a critical need arises for a paradigm shift toward sustainable urban development that seamlessly integrates nature into planning, design and governance (McPhearson et al. 2023). In response to this urgency, cities globally are actively exploring solutions aligned with the Paris Agreement's 1.5–2 °C temperature targets (UNFCCC 2015). These efforts encompass both mitigation and adaptation strategies. Mitigation efforts aim to reduce GHG emissions, while adaptation efforts focus on preparing for extreme weather events like heatwaves, floods, and droughts, collectively contributing to a more sustainable future. Following this path, cities are increasingly experimenting with Nature-based Solutions (NbS) to improve urban ecosystems as alternatives to "hard" and "grey" approaches for mitigating and adapting to the challenges posed by climate change and rapid urbanisation. Despite the spread of several definitions introduced in the last decade in the academic and policy discourses, NbS stand for meeting the targets of climate change and biodiversity agendas since nature could provide central benefits for human health and wellbeing.

In the last few years, the cases for NbS have been expanded by the global research communities, policymakers and practitioners, amplifying the catalogue of possible solutions. In particular, European cities have recently started implementing projects to transform the school environment using NbS to sustain the transition of cities toward a more sustainable and resilient future. Although these solutions are not being disseminated, they can transform schools into climate shelters by mainstreaming nature-based interventions in their schoolyards. Incorporating NbS in school settings has the potential to provide equitable access to safe and inclusive public spaces, especially for children and the most vulnerable members of the population. These initiatives, which can be listed among the Urban Climate Shelter (UCS), aim at transforming schoolyards from traditional methods based on impervious surfaces and low-albedo materials to more sustainable, natural, and resilient approaches while preserving the use of the schoolyard for educational purposes supporting biodiversity. Transforming schoolyards involves integrating nature into the planning and design of climate actions at the local level. This approach emphasises the dual purpose of preserving the schoolyard for educational activities while opening it to the public during non-school hours and summertime, serving as a climate shelter for everyone. This move aims to enhance accessibility to green spaces and foster a more inclusive and environmentally conscious urban environment.

This chapter falls under this heading and aims to frame the potential of NbS for urban resilience and sustainable urbanisation. Moreover, the chapter examines how

Urban Climate Shelters can enhance urban resilience by merging urban climate mitigation with climate adaptation, employing an ecosystem-based strategy, and leveraging NbS. To keep up this perspective, the chapter explores six European pioneer projects—Paris, Barcelona, Madrid, London, Amsterdam, and The Hague—that have successfully transformed schoolyards into climate shelters in the past decade. The comparative case study analysis, developed by desk-based research, identifies the specific types of NbS implemented in each project and the societal challenges they address. For each case study, the climate hazards, the project implementation assessment and plans for upscaling were explored. After this introduction, the conceptual framework on NbS and climate change is presented in Sect. 6.2. Considering the NbS as a broad umbrella concept, the co-benefits raised by NbS solutions for urban resilience are described. Section 6.3 presents Urban Climate Shelters as an NbS and the review of European case studies that used NbS to transform schoolyards. Section 6.4 provides the results and discusses the benefits of transforming schoolyards in UCS using NbS. In Sect. 6.5, conclusions call for implementing Urban Climate Shelters in schoolyards as a codified solution for urban resilience, supporting the upscaling at the European level of these projects, together with future research work.

6.2 Conceptual Framework. Nature-Based Solutions for Addressing Urban Challenges

Defining NbS is challenging, as several definitions have been introduced in the last decade. A good starting point is a widely accepted definition by the IUCN (2017) Report that frames NbS as actions that protect, manage sustainably, and restore natural or modified ecosystems, effectively addressing societal challenges while providing benefits for human well-being and biodiversity. Historically, terms such as urban forestry (UF), green and blue infrastructure (GI, BI), and ecosystem services (ESS) have been utilised to express similar concepts (Escobedo et al. 2019). Subsequently, additional concepts have been introduced, including ecosystem-based adaptation (EbA), ecosystem-based disaster risk reduction (Eco-DRR), blue-green infrastructure (BGI), low-impact development (LID), best management practices (BMPs), water-sensitive urban design (WSUD), sustainable urban drainage systems (SuDs), and ecological engineering (EE) (Ruangpan et al. 2020). As encompassed by the Directorate-General for Research and Innovation of the European Commission (2021), the NbS is a broad umbrella concept to various key existing concepts (Fig. 6.1).

A turning point for NbS happened when many cities strived to respond to climate change pressures in an inclusive and resilient way, reconnecting people with nature and upgrading ecologically urban spaces (Giudice et al. 2023). For example, GI-based planning is recognised as a relevant approach to addressing climate change in the EU strategy on adaptation. On this path, IUCN estimates that one-third of the

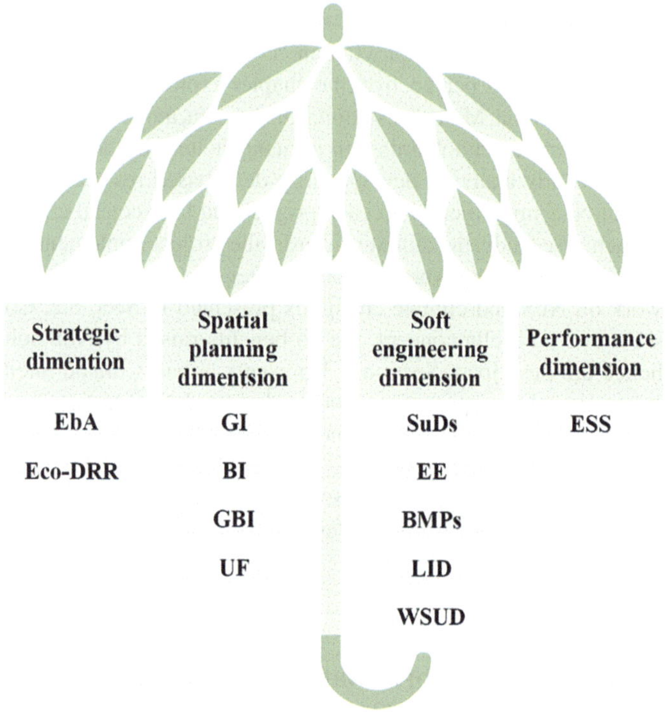

Fig. 6.1 NbS as an umbrella concept and the relation to key existing concepts. *Source* Authors elaboration from EU (2021)

emission reductions needed to meet the Paris Agreement goals will be achievable through NbS. Globally, NbS's focus on climate change is estimated to deliver $170 billion annually in benefits related to ecosystem services (IUCN 2017). Ultimately, NbS can be framed as "*a cornerstone for addressing the twin crises of biodiversity loss and climate change*" (McPhearson et al. 2023, p. 4).

6.2.1 The Role of Nature-Based Solutions in Climate Mitigation and Adaptation

Nowadays, many cities are shifting towards a resilient approach by utilising NbS. This involves mainstreaming nature in cities, upgrading green spaces, and implementing solutions like green roofs, rain gardens, street trees, and restored natural areas (wetlands, forests) to address climate challenges and enhance the quality of life (Kabisch et al. 2017; Frantzeskaki et al. 2019; Keeler et al. 2019). Traditional top-down planning methods seem inadequate for unlocking the full potential of NbS, requiring a paradigm shift.

Despite NbS's undeniable potential in cities, its implementation faces unique challenges (Kabisch et al. 2022). Dense and complex urban environments create fierce competition for space, requiring more meticulous planning and design approaches. Additionally, human-altered environmental conditions necessitate tailored approaches for each city due to their significant impact on biodiversity.

Beyond physical hurdles, successful implementation demands collaborative governance (Russo et al. 2017). This involves navigating diverse stakeholder interests (planners, citizens, NGOs) through collaborative design processes with multi-disciplinary teams (Hofmann et al. 2012; Buchel and Frantzeskaki 2015; De Vreese et al. 2016). These processes ensure inclusivity, respect cultural contexts, and foster a sense of ownership through shared leadership, open communication, and community involvement (Frantzeskaki 2019).

In order to address these challenges, Frantzeskaki (2019) proposed a list of seven lessons related to all stages of proof-of-concept and implementation of NbS in cities:

- Lesson #1 nature-based solutions need to be aesthetically appealing to citizens;
- Lesson #2 nature-based solutions create new green urban commons;
- Lesson #3 experimenting with nature-based solutions requires trust in the local government and in the experimentation process itself;
- Lesson #4 co-creation of nature-based solutions requires diversity and learning from social innovation;
- Lesson #5 nature-based solutions require collaborative governance;
- Lesson #6 is an inclusive narrative of the mission for nature-based solutions that can enable integration into many urban agendas;
- Lesson #7 design nature-based solutions to learn and replicate them in the long term.

Fostering an environment of experimentation allows for testing and refining innovative solutions (Russo et al. 2017). Flexibility and transparency are crucial for both planners and citizens. In the end, broad NbS implementation and biodiversity conservation are considered major strategic aims in urban planning and decision-making, highlighting the diverse benefits of NbS across economic, social, and environmental spheres.

6.2.2 Nature-Based Solutions: Types, Impacts, and Co-benefits

To effectively address the challenge of planning, co-producing, and implementing NbS in an urban context, the framework provided by the EU Report "Evaluating the Impact of Nature-Based Solutions" (2021) is recommended. It evaluates the different types of NbS and lists 12 indicators related to the societal challenge. According to this Report (EU 2021), NbS include three types (Fig. 6.2). Type 1 NbS improves the use or protection of natural ecosystems; Type 2 NbS manages ecosystem suitability

Fig. 6.2 NbS types. *Source* Authors elaboration from EU (2021)

or function; Type 3 NbS designs and manages newly created ecosystems (Eggermont et al. 2015).

To measure the impacts of NbS, a suite of recommended and additional indicators is identified for each of the twelve societal challenge areas (Fig. 6.3). Indicators of NbS impact have been classified as structural, process or outcome-based. Structural indicators support infrastructure and resources in place in NbS to achieve the desired goals; process indicators refer to the NbS efficiency, quality, or consistency to achieve the desired goals; outcome indicators refer to NbS impacts (EU 2021).

In the end, integrating NbS into urban planning, design, and policies is an added value since it provides social and economic benefits alongside environmental ones. They combine technical, financial, and social innovations to create systemic solutions, particularly relevant for cities facing the dual challenges of urbanisation and climate change (Raymond et al. 2017). Frantzeskaki (2019) proposed a list of co-benefits raised by the analysis of some case studies showing how nature-based solutions can contribute to urban resilience, including fifteen aspects:

- Restoration of ecosystems, soil, space;
- Social capital through community support;
- Social capital through stewardship of places;
- Social capital through civic empowerment and actions;
- Environmental education;
- Green jobs;
- Habitat;
- Water retention/flood protection;

Fig. 6.3 Indicators of societal challenge areas. *Source* Authors elaboration from EU (2021)

- Recreation;
- Heat regulation;
- Sense of place;
- Social cohesion;
- Food;
- Water inundation;
- Thermal comfort.

6.3 Urban Climate Shelters as a Nature-Based Solution

In recent years, cities have been using their schools as a mechanism to provide awareness of and adapt to climate change. More specially, in the last decade, European cities have started implementing a new initiative to transform schoolyards into a more sustainable and resilient approach, using NbS. This innovation of transforming schoolyards does not have a standard definition among the projects already created. However, the Urban Climate Shelter (UCS) concept used in this chapter is based on the on-going conceptualisation of the European project MAINCODE - MAINstreaming nature to CO-DEsign urban climate shelters in schoolyards, funded by the DUT Partnership (2025–2027). UCS is based on transforming critical urban spaces to adapt cities to climate change by offering protection from extreme weather events like heat waves, floods, and droughts. They achieve this by transforming the spaces using NbS and designing them collaboratively with local communities, ensuring bottom-up initiatives. Transforming schoolyards into UCS and incorporating NbS is a promising path to boost multiple co-benefits from urban nature and climate change mitigation and adaptation to enhanced health, well-being, and social justice for all citizens, in particular children and those mainly vulnerable to extreme heat (Baró et al. 2022; Flax et al. 2020).

6.3.1 Climate Shelters in Schoolyards: Reviewing European Case Studies

This chapter discusses six European case studies: Barcelona, Paris, Madrid, London, Amsterdam, and the Hague (Fig. 6.4), and evaluates the projects they have implemented to transform their schoolyards. Each project focuses on different climate hazards, such as heatwaves, flooding, and drought. Upscaling is crucial for enhancing a more significant long-term impact and effectiveness at the city scale, focusing on a sustainable future, and fostering urban resilience. Therefore, some cities plan to transform their schoolyards further, considering their individual goals to adapt to their specific hazards. For instance, Barcelona, Paris, and Madrid focus on heat hazards, while Amsterdam focuses on rainwater absorption. London addresses a broader range of hazards, including floods, heat, and droughts, and The Hague tackles flooding and heat. Table 6.1 offers a synthesis presentation of these case studies, while more detailed explanations of each case study can be found in the following sections.

6.3.1.1 Paris

The “School OASIS” project is an initiative by the city of Paris to transform schoolyards into urban cooling islands by 2050. The project was sponsored by Urban Innovation Actions (UIA) and began in 2018 with the aim of transforming ten schoolyards into cooler areas with more greenery and shade. However, the project has been very successful and has transformed 75 schools by the beginning of 2024. The project aims to transform all schoolyards in Paris into urban cooling islands by 2050. The project includes provisions such as permeable grounds, more vegetation, water points, creative games adapted to different needs, quiet spots, and better space distribution (Sitzoglou 2020). The involvement of numerous stakeholders, such as the school, community, and the city of Paris, was crucial to the project’s success. To ensure the Oasis schoolyards’ maintenance a guide has been created that includes

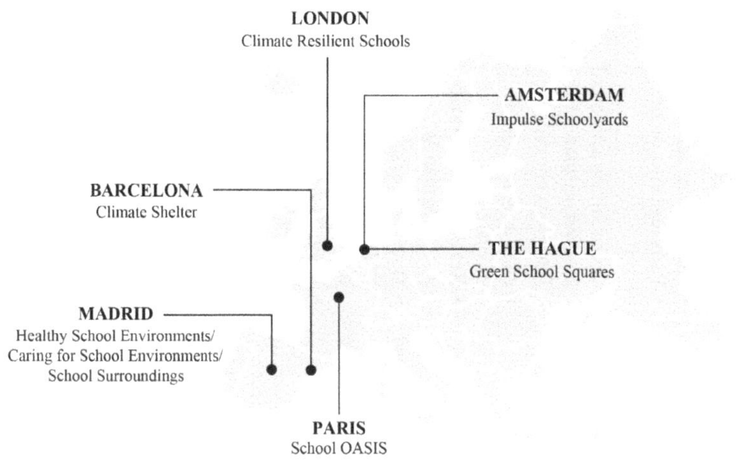

Fig. 6.4 Case studies of cities and projects. *Source* Author's elaboration

four different recommendations covering ground, greenery, and furniture (Sitzoglou 2020; Paris City Council 2018).

6.3.1.2 Barcelona

The city of Barcelona has launched a project called "Climate Shelters" to reduce heat-related deaths during warmer summers. Sponsored by Urban Innovation Actions (UIA), the project began in 2020 by transforming 10 schools. The primary goal is to provide climate shelters within a 10 min walkable distance for the city's residents. As part of the upscaling efforts, the city of Barcelona developed the "Let's Transform the School Yards" project to continue transforming more schoolyards into climate shelters. The goal is to transform a minimum of 10 schoolyards per year, contributing to the increase in green areas in the city. Overall, Barcelona has transformed approximately 29 schools into climate shelters. These shelters are publicly accessible spaces that anyone can use to escape the heat, focusing on protecting the most vulnerable members of the population, such as the elderly and children. Schools were chosen due to their central location and accessibility. The project has been a collaborative effort with many stakeholders involved. It has successfully provided a clear statement and understanding of the process and outcome of the project. This initiative is a step towards creating a safer and more comfortable environment for the residents of Barcelona (Cartalis 2020a, b; Gerència d'Àrea d'Ecologia Urbana, Urban Innovative Actions 2022; Vetter 2020).

Table 6.1 Analysis of European case studies

City	Project name	Project year	Schools transformed	Upscaling	Main climate hazards	Funding agency	Sectorial plans and policies on climate change
Paris	School OASIS	2018	75	760 by 2050	Heat	Urban innovation actions/	• Paris climate action plan • Paris resilience strategy
Barcelona	Climate shelter	2020	29	Ten new schools by year	Heat	Urban innovation actions/	• Green infrastructure and biodiversity plan 2020 • Barcelona climate plan
Madrid	Healthy school environments/ caring for school environments/ school surroundings	2018	3	Not specified	Heat	Public initiative	• Guia diseno entornos escolares • Recovery transformation and resilience plan Madrid
London	Climate resilient schools	2020	Not specified	100	Flood, heat and droughts	Private and public partnership	• How London schools and early years settings can adapt to climate change • Resilience Strategy 2020 • The London plan
Amsterdam	Greening school playgrounds	2016	95	Not specified	Flood-rainwater absorption	Perfect Interreg Europe	• Greening school playgrounds report • Green infrastructure vision 2050 • Strategy for climate adaptation
The Hague	Green school squares	2011	187	Not specified	Heat and flood-rainwater absorption	Fonds 1818 private initiate	• Memorandum on sustainability: clean energy in a green city • Resilience strategy

Source Author's elaboration

6.3.1.3 Madrid

Madrid has undertaken public initiatives and projects aimed at creating healthy and caring school environments by transforming three schools. These initiatives, known as "Healthy School Environments/Caring for School Environments/School Surroundings" projects, have three main objectives: climate adaptation, health and social cohesion, and urban regeneration. The project aims to focus on the immediate surroundings of the schools, including the patio and different zoning areas. The design criteria for climate adaptation include increasing shade, improving soil moisture and infiltration, and designing a water cycle system. However, the future upscaling of the project remains uncertain due to a lack of resources (De Madrid et al. 2016; Serrano et al. 2017; Vetter 2020).

6.3.1.4 London

London has been taking steps to transform their schools to adapt to climate change. By implementing this strategy, the city is coping with the impacts of climate change, including heat risk, flood risk, and water scarcity. The "Climate Resilient Schools" project in London aims to transform 100 schools using public and private funding (Greater London Authority 2020). The document titled "how London schools can adapt to climate change" provides an understanding of the risks and priorities, along with measures that can be taken to protect the schools from climate change. The document covers important topics such as responding and recovering from severe weather events, learning and raising awareness, funding, and monitoring. The project is a collaborative effort between the Greater London Authority and private initiatives (Greater London Authority 2020).

6.3.1.5 Amsterdam

Amsterdam's "Impulse Schoolyards" project, funded by Perfect Interreg Europe, has transformed 85 schools from 2016 to 2019 and aims to transform an additional 60 schools from 2020 to 2024 (Flax et al. 2020). The project was initiated in 2016 with the goal of including 25% more greenery in the schoolyards to promote healthier playtime. Community participation was also included in the project to raise awareness for sustainability. The schools selected for the project are open to the public after school hours, on weekends and holidays, providing extra public space for the urban area (Antoniadis et al. 2020). The collaboration between public bodies has formed a strong team of experts working on the project (Giezen and Pellerey 2021).

6.3.1.6 The Hague

In 2011, the private initiative "Green School Squares" (Fonds 1818) transformed 187 schools in the Hague. The city aims to provide a safer environment for children to play. To achieve this, they use sustainable solutions such as changing impermeable pavement to green, planting trees, and creating cooling areas. These solutions increase water absorption and help the city adapt to climate change. The municipality plans to upscale the projects, but it poses some challenges since it is a private initiative (Giezen and Pellerey 2021).

6.4 Results and Discussion: Nature-Based Solution for Transforming Schoolyards into Urban Climate Shelters

Following the framework developed by the European Commission, Directorate-General for Research and Innovation (2021) and the framework given by Frantzeskaki (2019), presented above, this analysis concentrated on understating NbS application in transforming schoolyards in climate shelters. To evaluate the impact of NbS in sheltering school environments, the European case studies were analysed, focusing on three key components: (1) which type of NbS is implemented in the case studies; (2) which societal challenge area is addressed in the case studies; (3) which co-benefits of urban resilience are implemented in the case studies. Integrating NbS into planning and design to transform schoolyards into UCS can foster urban resilience. Table 6.2 summarises the results of this analysis.

The analysis based on the NbS types, preliminary reveals that the case studies primarily focus on Type 3 NbS, which uses natural elements to design and manage newly created ecosystems. Green spaces (gardens), trees, and shrubs are present in all six studies, reflecting a common adaptation strategy for heat hazards. Furthermore, green building elements like green roofs/walls, infiltration planters, and green furniture are found in Paris, Barcelona, London, and Amsterdam. Finally, water management components appear most frequently in cities like London, Amsterdam, and The Hague, aim at adapting to flooding, utilising infiltration, filtration, biofiltration structures (rain gardens, infiltration basins, filtration systems), and natural or semi-natural water storage and transport structures (ponds).

The societal challenges addressed are mainly related to the increase in the vegetation provided by the regeneration of the schoolyards, especially from the plantation of trees and the addition of gardens. Such societal challenges from this addition are climate resilience, natural and climate hazards, green space management, biodiversity enhancement, air quality, and place regeneration. These elements help to reduce risk levels, deliver social, human and environmental benefits, restore degraded ecosystems, reduce air temperature and enhance green space and people-nature connection (EU 2021). Furthermore, other challenges addressed include knowledge

Table 6.2 NBS type, societal challenge areas and co-benefits in the European case studies

City	NbS type	Societal challenge areas	Co-benefits of NbS to urban resilience
Paris	• Green space • Trees and shrubs • Green built environment • Infiltration, filtration and biofiltration structures	• Climate resilience • Water management • Natural and climate hazards • Green space management • Biodiversity enhancement • Air quality • Place regeneration • Knowledge and social capacity building • Participatory planning and governance • Social justice and social cohesion • Health and wellbeing	• Restoration of ecosystems, soil, space • Social capital through civic empowerment and actions • Environmental education • Water retention/flood protection • Recreation • Heat regulation • Sense of place • Social cohesion • Thermal comfort
Barcelona	• Green space • Trees and shrubs • Green built environment	• Climate resilience • Natural and climate hazards • Green space management • Biodiversity enhancement • Air quality • Place regeneration • Knowledge and social capacity building • Participatory planning and governance • Social justice and social cohesion • Health and wellbeing	• Restoration of ecosystems, soil, space • Social capital through civic empowerment and actions • Environmental education • Recreation • Heat regulation • Sense of place • Social cohesion • Thermal comfort

(continued)

Table 6.2 (continued)

City	NbS type	Societal challenge areas	Co-benefits of NbS to urban resilience
Madrid	• Green space • Trees and shrubs	• Natural and climate hazards • Green space management • Biodiversity enhancement • Air quality • Place regeneration	• Restoration of ecosystems, soil, space • Social capital through civic empowerment and actions • Environmental education • Recreation • Heat regulation • Social cohesion • Thermal comfort
London	• Green space • Trees and shrubs • Green built environment • Natural or semi-natural water storage and transport structures • Infiltration, filtration and biofiltration structures	• Climate resilience • Water management • Natural and climate hazards • Green space management • Biodiversity enhancement • Air quality • Place regeneration • Health and well-being	• Restoration of ecosystems, soil, space • Social capital through civic empowerment and actions • Environmental education • Water retention/flood protection • Recreation • Heat regulation • Water inundation • Thermal comfort

(continued)

Table 6.2 (continued)

City	NbS type	Societal challenge areas	Co-benefits of NbS to urban resilience
Amsterdam	• Green space • Trees and shrubs • Green built environment • Natural or semi-natural water storage and transport structures	• Climate resilience • Water management • Natural and climate hazards • Green space management • Biodiversity enhancement • Air quality • Place regeneration • Health and well-being	• Restoration of ecosystems, soil, space • Social capital through civic empowerment and actions • Environmental education • Water retention/flood protection • Recreation • Social cohesion • Water inundation
The Hague	• Green space • Trees and shrubs • Infiltration, filtration and biofiltration structures	• Climate resilience • Water management • Natural and climate hazards • Green space management • Biodiversity enhancement • Air quality • Place regeneration • Health and well-being	• Restoration of ecosystems, soil, space • Social capital through civic empowerment and actions • Environmental education • Water retention/flood protection • Recreation • Heat regulation • Water inundation • Thermal comfort

and social capacity building, participatory planning and governance, social justice and cohesion, and health and wellbeing. This includes educational initiatives that contribute to sustainable urban places, social transformation, increased social inclusion in cities, and enhanced access to clean air, safe water, and secure shelter (EU 2021). These challenges are especially tackled in the cases of Paris and Barcelona since they enhance awareness and aim to involve many stakeholders from the overall community in all the steps.

From the compilation of fifteen co-benefits outlined by Frantzeskaki (2019), approximately half of these are evident in the six case studies (Fig. 6.5). Among the co-benefits addressed across all cases is the restoration of ecosystems, soil, and space, as all projects endeavour to reduce social sealing by enhancing vegetation and unpaved areas. Furthermore, another common co-benefit is heat regulation and thermal comfort, as many case studies focus on adapting to heat-related hazards. Additionally, some projects include water retention/flood protection and prevention of water inundation co-benefits. Lastly, other notable co-benefits include fostering

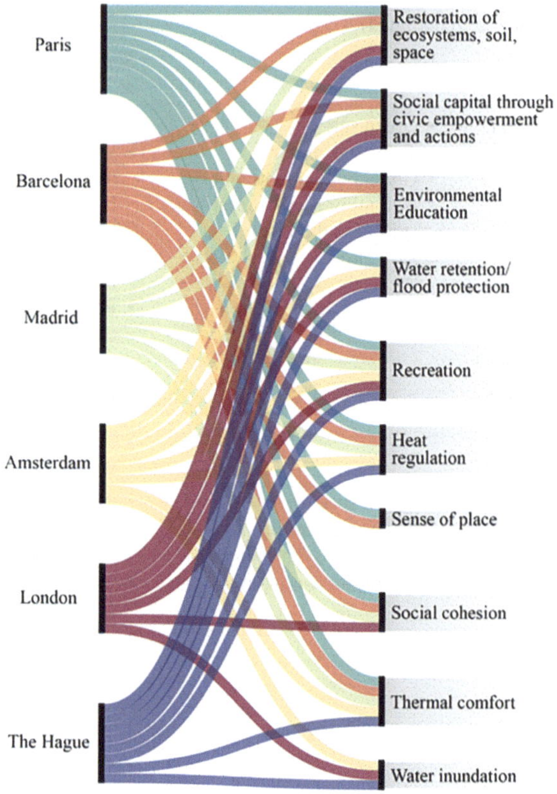

Fig. 6.5 Co-benefits of NbS to urban resilience in the case studies. *Source* Author's elaboration

social capital through civic empowerment and actions, environmental education, recreation opportunities, a sense of place and social cohesion.

6.5 Conclusions. Urban Climate Shelter in Schoolyards as a Codified Solution for Urban Resilience

This study supports that transforming schoolyards through the implementation of NbS can be a codified solution in helping cities face their existing climate challenges, supporting urban resilience and sustainable urbanisation. As a matter of fact, some European cities, including Barcelona and Paris, have begun implementing this process to upscale and transform all their school environment into UCS.

While this chapter has mainly addressed UCS through the lens of NbS, it is essential to recognise that NbS represents just one aspect of the broader UCS initiative. UCS can catalyse regenerative urbanism within cities, facilitate knowledge expansion, and disseminate UCS methodologies to inspire cities globally. UCS integrates NbS into planning and design, fostering resilient and sheltered urban environments.

The impacts of UCS go beyond physical transformations. As outlined by Pincegher et al. (2024), they signify a shift towards fortifying governance structures and raising awareness about how urban spaces can contribute to climate change mitigation and adaptation. This marks a departure from traditional approaches, introducing experimental governance initiatives that embrace new patterns of interaction and allow for some level of conflict. Schoolyards can evolve into spaces for broader community use, and through participatory co-production and co-design involving citizens, solutions are customised to meet their specific needs and preferences. This approach fosters a sense of ownership and active engagement within the community, aligning with the principles of an experimental governance model that encourages diverse perspectives (Brunetta et al. 2024).

UCS also advocates the reinforcement of intersectional social-climate justice (Amorim-Maia et al. 2022). It incorporates social equity into planning strategies; it considers the accessibility of green and cool areas to the most vulnerable populations, including children, the elderly, women, socio-economically disadvantaged groups, and foreign populations. It not only examines the spatial distribution of socio-vulnerable groups but also adopts a co-production approach with them. It tackles both climate and social vulnerabilities, enhancing sustainable urbanisation and urban resilience, together with climate justice (Amorim-Maia et al. 2023).

UCS, by presenting a comprehensive methodology encompassing analysis of climate hazards and social vulnerability, identification of local needs, selection of suitable schoolyard sites, engagement of relevant stakeholders, co-design elaboration, and ensuring a successful upscaling process, allows cities to embrace these principles and bridge the knowledge gap, translating research into actionable policies. UCS aligns with the lessons related to all stages of proof-of-concept and implementation of NbS in cities. It includes NbS with and for the community, enhances urban greenery, enhances bottom-up approaches through co-production and co-design with the engagement of numerous stakeholders, ensures the upscaling of, and finally aims to prove co-benefits using NbS to urban resilience. Through such concerted efforts, cities can unlock the true potential of NbS, fostering sustainable and resilient urban environments in collaboration with their citizens. This way, UCS can effectively address numerous societal challenges, extending beyond environmental concerns to social issues and also serve as co-benefits to reinforce urban resilience towards sustainable urbanisation.

References

Amorim-Maia AT, Anguelovski I, Chu E, Connolly J (2022) Intersectional climate justice: a conceptual pathway for bridging adaptation planning, transformative action, and social equity. Urban Clim. 41:101053. https://doi.org/10.1016/j.uclim.2021.101053

Amorim-Maia AT, Anguelovski I, Connolly J, Chu E (2023) Seeking refuge? The potential of urban climate shelters to address intersecting vulnerabilities. Landsc Urban Plan 238:104836. https://doi.org/10.1016/j.landurbplan.2023.104836

Antoniadis D, Katsoulas N, Papanastasiou D (2020) Thermal environment of urban schoolyards: current and future design with respect to children's thermal comfort. Atmosphere 11(11) (MDPI AG). https://doi.org/10.3390/atmos11111144

Baró F, Camacho DA, Perez del Pulgar C, Ruiz-Mallén I, García-Serrano P (2022) Nature-based climate solutions in European schools: a pioneering co-designed strategy towards urban resilience. In: Ruiz-Mallén I, March H, Satorras M (eds) Urban Resilience to the Climate Emergency. Unravelling the transformative potential of institutional and grassroots initiatives. Urban Book Series. Springer, Cham, pp 25–146

Buchel S, Frantzeskaki (2015) Citizens' voice: a case study about perceived ecosystem services by urban park users in Rotterdam, the Netherlands. Ecosyst. Serv. 12,169–177. https://doi.org/10.1016/j.ecoser.2014.11.014

Brunetta G, Caldarice O, Mohabat Doost D (2024) Multi-level governance: mainstreaming climate commitments. In: Urban climate action: policy brief series. https://gggi.org/report/urban-climate-action-policy-brief-series/. Accessed 15 May 2023

Cartalis C (2020a) The Climate Shelters project Journal N° 1. https://www.uia-initiative.eu/sites/default/files/2020-05/Barcelona_GBGAS2C_Journal.pdf. Accessed 22 May 2023

Cartalis C (2020b) The Climate Shelters project Journal N° 2 climate adaptation. https://www.uia-initiative.eu/sites/default/files/2021-02/GBGAS2C_Barcelona_Journal%202.pdf. Accessed 22 May 2023

De Madrid A, Serrano PG, Leal P, Lucila L, Peña U, Hernández C, Colaborador V, Moreno MB (2016) Proyecto micos. Entornos escolares saludables. https://madridsalud.es/wp-content/uploads/2016/04/MICOS-Regeneracion-urbana.pdf. Accessed 15 May 2023

De Munck C, Pigeon G, Masson V, Meunier F, Bousquet P, Tréméac B, Merchat M, Poeuf P, Marchadier C (2013) How much can air conditioning increase air temperatures for a city like Paris, France? Int J Climatol 33:210–227. https://doi.org/10.1002/joc.3415

De Vreese R, Leys M, Dendoncker N, Van Herzele A, Fontaine CM (2016) Images of nature as a boundary object in social and integrated ecosystem services assessments. Reflections from a Belgian case study. Ecosyst Serv 22:269–279. https://doi.org/10.1016/j.ecoser.2016.06.008

Eggermont H, Balian E, Azevedo JMN, Beumer V, Brodin T, Claudet J, Fady B, Grube M, Keune H, Lamarque P, Reuter K, Smith M, van Ham C, Weisser WW, Le Roux X (2015) Nature-based solutions: new influence for environmental management and research in Europe. Gaia Ecol Perspect Sci Soc 24(4):243–248. https://doi.org/10.14512/gaia.24.4.9

Escobedo FJ, Giannico V, Jim CY, Sanesi G, Lafortezza R (2019) Urban forests, ecosystem services, green infrastructure and nature-based solutions: nexus or evolving metaphors? Urban Forest Urban Green 37:3–12. https://doi.org/10.1016/j.ufug.2018.02.011

European Commission, Directorate-General for Research and Innovation (2021) Evaluating the impact of nature-based solutions: a handbook for practitioners. Publications Office of the European Union. https://doi.org/10.2777/244577

Flax L, Korthals Altes R, Kupers R, Mons B (2020) Greening schoolyards—An urban resilience perspective. Cities 106, 102890. https://doi.org/10.1016/j.cities.2020.102890

Frantzeskaki N (2019) Seven lessons for planning nature-based solutions in cities. Environ Sci Policy 93:101–111. https://doi.org/10.1016/j.envsci.2018.12.033

Giezen M, Pellerey V (2021) Renaturing the city: factors contributing to upscaling green schoolyards in Amsterdam and the Hague. Urban Forest Urban Green 63, 127190. https://doi.org/10.1016/j.ufug.2021.127190

Giudice B, Novarina G, Voghera A (eds) (2023) Green infrastructure planning strategies and environmental design. Springer, Cham

Gerència d'Àrea d'Ecologia Urbana, Urban Innovative Actions (2022). Project Refugis Climàtics. https://bcnroc.ajuntament.barcelona.cat/jspui/bitstream/11703/127541/1/Informe_BonaPractica_RefugisClimatics.pdf. Accessed 18 May 2023

Greater London Authority (2020) How London Schools and early years settings can adapt to climate change. In: Greater London Authority, City Hall, The Queen's Walk, More London, London SE1 2AA. www.london.gov.uk. Accessed 15 May 2023

Hofmann M, Westermann JR, Kowarik I, van der Meer E (2012) Perceptions of parks and urban derelict land by landscape planners and residents. Urban Forest Urban Green 11:303–312. https://doi.org/10.1016/j.ufug.2012.04.001

IUCN (2017) International Union for Conservation of Nature annual report 2017. https://portals.iucn.org/library/sites/library/files/documents/2018-007-En.pdf. Accessed 29 February 2024

Kabisch N, Korn H, Stadler J, Bonn A (2017) Nature-based solutions to climate change adaptation in urban areas—linkages between science, policy and practice. In: Kabisch N, Korn H, Stadler J, Bonn A (eds) Nature-based solutions to climate change adaptation in urban areas. Theory and practice of urban sustainability transitions. Springer, Cham

Kabisch N, Frantzeskaki N, Hansen R (2022) Principles for urban nature-based solutions. Ambio 51:1388–1401. https://doi.org/10.1007/s13280-021-01685-w

Keeler BL et al (2019) Social-ecological and technological factors moderate the value of urban nature. Nat Sustain 2(1):29–38. https://doi.org/10.1038/s41893-018-0202-1

McPhearson T, Kabisch N, Frantzeskaki N (eds) (2023) Nature-based solutions for cities. Edward Elgar Publishing, Cheltenham

Paris City Council (2018) Paris Resilience Strategy. https://www.paris.fr/pages/paris-resiliente-4264/. Accessed 15 May 2023

Paris Oasis Schoolyard Programme, France—English (no date). https://climate-adapt.eea.europa.eu/en/metadata/case-studies/paris-oasis-schoolyard-programme-france. Accessed 15 June 2023

Pincegher B, Caldarice O, Pizzorni M, Tollin N (2024) Co-production as a planning strategy. Insights from Implementing Urban Climate Shelter in Schoolyards. In: AESOP annual congress proceedings, vol 36. GAME CHANGER? Planning for just and sustainable urban regions-book of abstracts, Paris, July 8-12 2024

Raymond CM et al (2017) A framework for assessing and implementing the co-benefits of nature-based solutions in urban areas. Environ Sci Policy 77:15–24. https://doi.org/10.1016/j.envsci.2017.07.008

Ruangpan L, Vojinovic Z, Di Sabatino S, Leo LS, Capobianco V, Oen AMP, McClain ME, Lopez-Gunn E (2020) Nature-based solutions for hydro-meteorological risk reduction: a state-of-the-art review of the research area. Nat Hazard 20:243–270. https://doi.org/10.5194/nhess-20-243-2020

Russo A et al (2017) Edible green infrastructure: an approach and review of provisioning ecosystem services and disservices in urban environments. Agr Ecosyst Environ 242:53–66. https://doi.org/10.1016/j.agee.2017.03.026

Serrano GP, Laredo LP, Pena UL (2017) Guía de diseño de entornos escolares. https://www.madridsalud.es/pdf/guia_diseno_entornos_escolares_opt.pdf. Accessed 22 May 2023

Sitzoglou M (2020) The OASIS Schoolyards project Journal N° 1. https://www.uia-initiative.eu/sites/default/files/2020-06/Paris_OASIS_Journal.pdf. Accessed 22 May 2023

United Nations/Framework Convention on Climate Change (2015) Adoption of the Paris Agreement, 21st Conference of the Parties. United Nations, Paris

Vetter R (2020) Cooling schools experiences from C40's Cool Cities Network, C40 Knowledge. https://www.c40knowledgehub.org/s/guide-home?language=en_US#article1. Accessed 18 April 2023

Chapter 7
Vertical Gardens in Public Open Spaces—Nature-Based Solutions for Sustainable Urbanization in Belgrade City, Serbia

Mina Radić Sibinović, Marta Brković Dodig, and Angela Million

Abstract Nature-based solutions (NbS) are recognized as effective tools for mitigating urbanization's effects, enhancing environmental quality, promoting human well-being, and supporting economic development. When carefully planned, NbS present elements of green infrastructure (GI), essential for the sustainable functioning of cities. Since 2013, the European Union countries have launched a GI Strategy and a Biodiversity Strategy. As part of them, many cities such as Paris, London, Barcelona, Kalisz, etc. have developed specific technical guidelines for implementing NbS, including vertical greenery systems (VGS). In Serbia, urban planning has recently started to focus on GI, but further developments are needed. To contribute to this discourse, we reviewed the scientific literature to present VGS genesis in Belgrade and analyzed how the current "Plan of the General Regulation for the Green Areas System in Belgrade" from 2019 (2019 PGRGASB) presents VGS. Afterward, through field research, we examined six VGS case studies in public open spaces in Belgrade. This was followed by an international expert workshop to critically evaluate the VGS planning and design proposed by the 2019 PGRGASB. Recommendations were formulated for enhancing VGS planning and design, relevant not only to Serbia but also to other South-East European countries.

Keywords Critical evaluation · Green infrastructure planning · Vertical greenery systems · South-East Europe · Belgrade

M. Radić Sibinović (✉) · A. Million
Technische Universität Berlin, Berlin, Germany
e-mail: mina.radic.sibinovic@campus.tu-berlin.de

M. Brković Dodig
Faculty for Media and Communications, Singidunum University, Belgrade, Serbia

M. Firoz C et al. (eds.), *Nature-Based Solutions for Urban and Peri-Urban Areas*, Disaster Risk Reduction, https://doi.org/10.1007/978-981-96-5933-3_7

7.1 Introduction

The exponential increase of inhabitants in cities requires providing sustainable Nature-based solutions (NbS) for urban areas. One way of achieving this is by making green space planning and redevelopment fundamental to sustainable city planning (Mahmoud and El-Sayed 2011). Yet, in the case of Southeast Europe, the availability of green spaces in urban areas (Table 7.1) is much lower than in Scandinavia or Western Europe (Pauleit et al. 2019). Besides monitoring the percentages of green spaces, proper implementation of NbS and green infrastructure (GI) establishment are equally important to strategically tackle sustainable urban development challenges (Szulczewska et al. 2017). Vertical greenery systems (VGS), recognized as NbS by the European Commission (Bauduceau et al. 2015), are beneficial for expanding urban green areas by covering walls with vegetation (Park and Jeong 2020). VGS not only aesthetically enrich, but also play a crucial role in the urban environment'eco-remediation' enabling:

> preservation, purification, and improvement of some or all components of the ecosystem, relying on its natural attributes of self-sustainability through dynamic structure change (Šešlija 2012).

Moreover, unlike other NbS such as green corridors, rain gardens, urban forests, wetlands, etc. (Castellar et al. 2021) citizens themselves can build VGS with little guidance (Million et al. 2014).

In 2013, the European Union (EU) launched a GI strategy, which is also part of the EU Biodiversity Strategy 2020 ("European Commission," n.d.). Therefore, cities such as Paris, London (Giordano et al. 2017), Barcelona ("Barcelona City Council" n.d.), and Kalisz ("Kalisz" n.d. 2021), integrated VGS as a part of their strategies on urban agriculture and combating climate change. London, for example, has developed a technical guide proposing VGS, highlighting their health, climate change adaptation, air quality improvement, and biodiversity conservation (Giordano

Table 7.1 Greenery percentage in European capitals

City, Country	Greenery percentage (%)	Sources
Zagreb, Croatia	75.1	Šiško and Polančec (2022)
Vienna, Austria	More than 50	Wieshofer et al. (2015)
London, Great Britain	48–51	"London datastore" (n.d.)
Warsaw, Poland	47.7	Szulczewska et al. (2017)
Stockholm, Sweden	47	Borges et al. (2019)
Helsinki, Finland	40	Borges et al. (2019)
Berlin, Germany	33.7	"Berlin" (2022)
Tirana, Albania	More than 27	Teqja et al. (2020)
Belgrade, Serbia	12.38	Urban Institute of Belgrade (2019)
Skopje, North Macedonia	12	"City of Skopje" (n.d.)

et al. 2017) benefits. Despite initiatives and projects in the EU member states, it will be difficult to meet environmental sustainability if surrounding countries such as South-East Europe, lag in the development of GI. Conversely, in one of those countries—Serbia spatial policies never or rarely mention VGS. In 2019, for the first time, the Plan of the General Regulation for the Green Areas System in Belgrade (2019 PGRGASB) was adopted, forming a legal basis to increase green areas and establish GI. However, a detailed analysis reveals discrepancies within the plan, as well as between the plan's technical and textual documentation. What is more, a substantial gap persists between research recommendations supporting GI and VGS as NbS solutions, and the contemporary destructive urban planning practice towards GI in Belgrade. To invigorate the debate this chapter takes a comprehensive approach. First, it reviews the literature on the genesis of VGS in Belgrade and VGS inclusion in the 2019 PGRGASB. Second, it explores living wall case studies and presents outcomes from an expert workshop offering suggestions for improving VGS planning in Serbia.

7.2 Literature Review

7.2.1 Desk-Based Research

Search through the planning documents and academic databases, academic search engines, and University repositories was based on the following keywords: 'the concept', 'the genesis', and 'the application of VGS in Belgrade'. The term 'vertical greenery systems' was changed to 'living walls' and 'green facades' in the subsequent search rounds. Discovered findings were grouped into 3 thematic categories further presented in subsections below: 2019 PGRGASB, VGS in public open spaces in Belgrade, and contemporary construction GI practice in Belgrade.

7.2.1.1 2019 Plan of the General Regulation for the Green Areas System (PGRGASB)

The first and current planning document related to the system of green areas in Serbia, 2019 PGRGASB, also declared as an indicative planning document (Stojić 2024), aims for Belgrade's sustainable development. The greenery percentage should increase from 12,38 to 22,74%, to reach the norm of 123 m^2 per capita (Urban Institute of Belgrade 2019). The 2019 PGRGASB envisages the preservation and improvement of the existing and the construction of new green areas. It aims to connect the existing patched system of green areas to one interconnected GI system for Belgrade. While the 2019 PGRGASB textual documentation does not explicitly mention the term NbS, it recognizes VGS as contemporary elements supporting GI development. However, the Plan names VGS only as tools for improving ambient values and enhancing the energy efficiency of public service facilities (Urban Institute of Belgrade 2019, pp. 32, 33, 60). Moreover, the typology of forests and green areas in the 2019 PGRGASB does not include VGS. Instead, it develops 5 main types and 27 sub-types as presented in Table 7.2 (Urban Institute of Belgrade 2019).

Table 7.2 Forests and green areas typology from the 2019 PGRGASB

N	Type	Sub-type
1	Forests and forest land	Forests and forest land
		Wet habitats on forest land
2	Public green areas	Parks
		Squares
		Green areas on the banks of the Danube and Sava rivers
		Green areas in the catchment area of the Danube and Sava rivers
		Green areas in open residential blocks
		Protective green belts
		Green corridors
		Special-purpose green areas
		Wet habitats
		Nurseries
3	Public green areas and forests within areas for public purposes	Public green areas of public services' buildings and complexes
		Forests within public buildings and complexes
		Public green areas within traffic areas
		Avenues of tree rows within traffic areas
		Public green areas within communal areas
		Public green areas within infrastructure facilities and complexes
4	Green areas within public and/or other purpose areas	Green areas of sports facilities and complexes
5	Green areas and forests within the other purpose areas	Green areas of housing (HGA)
		Green areas in mixed urban centers
		Green areas of commercial content
		Green areas of business zones
		Green areas of business parks
		Green areas of religious buildings and complexes
		Other green areas
		Green areas around agriculture

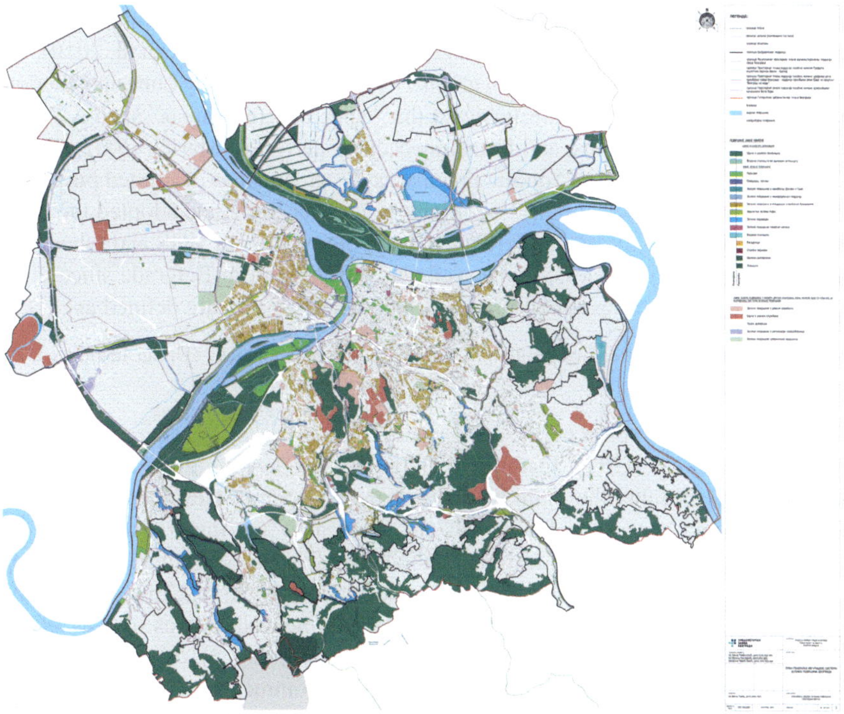

Fig. 7.1 The overview map of the planned green areas system (archive number 350–3052/09, drawing number 3). Available at: https://www.urbel.com/planovi/1547/PGRSZP_plan_namena_LIST_03-pregledna.pdf

The accompanying 2019 PGRGASB technical documentation also does not map existing or planned VGS (Fig. 7.1). Therefore, VGS are poorly conceptualized and integrated into the textual and technical parts of the Plan. This suggests that VGS is not yet represented in Serbia in the ecological, architectural, functional, aesthetic, or cultural sense (Ćerimović 2016).

7.2.1.2 VGS in Public Open Spaces in Belgrade

Scientific literature developed a broad VGS construction typology, clearly differentiating between green facades and living walls (Radić et al. 2019). Living walls present VGS that place growing media with incorporated irrigation systems into a wall. Unlike them, green facades are VGS that contain climbing vegetation with soil at the base of the wall. Most of the literature published in Serbia mentions VGS in the context of climate change, despite the wide range of their benefits such as thermal performances, reduction of air pollution, reduction of noise, positive effect on hydrology, social benefits, visual effects, educational effects, habitat for urban wildlife, economic benefits, and prolonged facade existence (Radić et al. 2019).

In 1987, an extension of the urban greenery typology defined by Lunc and Teodoronski was suggested by Anastasijević so that "specific green areas" such as VGS, green roofs, terraces, ponds, and fountains became part of it (Maričić 2007). Previously, typology consisted of "public green areas", "green areas of limited purpose", and "green areas of special purpose". Many years later, Belgrade adopted the 2019 PGRGASB that lists forests and green areas typology without VGS and other NbS as GI elements. Nevertheless, living walls have been built in Belgrade since 2019 within public open spaces. In Serbian legislation, such spaces are defined as

> areas bounded by sections of regulatory lines of surrounding blocks (Urban Institute of Belgrade 2009).

Notably, certain types of public open spaces, such as parks and squares, inherently serve as green areas (Urban Institute of Belgrade 2009). Vegetation, together with built elements presents the main element of public open spaces and, thus should be attractive and accessible for its citizens. Driven by the imperative to address environmental challenges, the design of public open spaces is evolving, emphasizing the 'change of space types intended for greenery' (Vujičić 2019). This explains the existence of ancillary spaces and the development of new NbS types such as VGS.

Despite the data produced by Serbian researchers (Table 7.3), certainly available to local planning authorities, the 2019 PGRGASB recommends using VGS only for ambiance improvement and increased energy efficiency of public buildings (Urban Institute of Belgrade 2019). The unavailability of green spaces in Southeast European cities may also be explained by the lack of proper green areas management (Pauleit et al. 2019). Moreover, the authority tasked with GI care received the lowest rating from the surveyed landscape architecture experts in the study conducted in Serbia (Vujičić et al. 2019).

7.2.1.3 GI Contemporary Construction Practice in Belgrade

In Belgrade, large green areas are being destroyed by mega-projects such as 'Belgrade Waterfront', being built on an area of almost 90 ha near the confluence of the Sava and the Danube rivers and the protected historical core of the city (Perić 2020). Similarly 'Avala Film Complex' in Košutnjak forest is pushed by private investors to be turned into a residential complex for 8000 people ("eKapija" n.d.). 'Tesla City' is planned for the forest settlement Makiš, with an area of 2.000.000 m^2, where the main natural Belgrade water source lies ("Gradnja" n.d.). Smaller-scale 'Gondola project' was planned to link the oldest Belgrade Fortification 'Kalemegdan' and 'Ušće' park, which could significantly destroy the environmental and cultural properties of this monument complex (City of Belgrade 2018). For this project, the Europa Nostra (2019) issued a critical report with a warning. Although the 2019 PGRGASB proposes the preservation and improvement of existing, and planning new (Urban Institute of Belgrade 2019), construction practices since 2010 have led to the

Table 7.3 VGS genesis concerning GI in Serbia

Year	Aim	VGS type	VGS and GI relation	Source
/	Genesis	Green facades	Assumption that green facades first grew by chance, e.g., wild ivy spread on the facade	Paper authors
1987	Extension of the urban greenery classification	Living walls and green facades	N. Anastasijevic proposed an addition of "specific green areas" to the classification of urban greenery	Maričić (2007)
2012	Ecological benefits: reduction of noise, heat, and air pollution	Living walls	First living wall constructed within bar "Brankov" near Branko's bridge in Belgrade, Serbia, with an area of almost 50m^2	"RTS" (n.d.)
2014	Climate change adaptation	Living walls and green facades	VGS as an integral part of GI strategies is a suggestion for Serbia based on EU and American countries' experience	Cvejić et al. (2014)
2019	Climate change mitigation	Living walls and green facades	Identified new forms of GI such as: linear parks, integrated green streets, green roofs, urban pockets, green dots, green facades, and living walls	Vujičić (2019)
2019	Adoption of the first planning document related to green areas	Living walls and green facades	The 2019 PGRGASB also envisages the conversion of the existing green area system to the city's GI infrastructure	Urban Institute of Belgrade (2009)
2019	Ambiance improvement	Living walls	Since the 2019 PGRGASB adoption living walls are becoming part of various types of public open spaces	Paper authors

loss of of Belgrade's green spaces from 12,38 to 9% (Stojić 2024). Large citizens' protests prevented harmful projects like 'Gondola', 'Tesla City', and 'Avala Film Complex' from being realized, thus further destroying large green areas of Belgrade. Currently, the number of such projects is growing further evidencing Belgrade's urban planning malpractice.

7.3 Materials and Methods

This research uses qualitative methodology with a case study approach applying different research methods to complement findings from various sources. The research was carried out through two phases: desk-based research of planning documents and academic literature, and primary research consisting of fieldwork in six case study locations in Belgrade, followed by an expert workshop. First, document analysis and literature review were employed to review planning documents

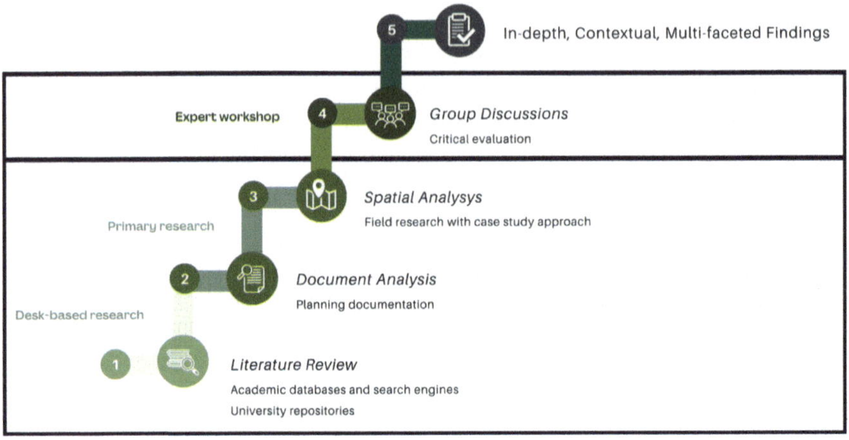

Fig. 7.2 Research methodology

and academic literature, providing a foundational understanding of the scientific evidence—presented above in the literature review. Field research used structured observations in six case study locations in Belgrade, collecting primary data within the public open spaces. Afterwards, an expert workshop was conducted to gather insights from specialists from practice and academia through semi-structured group discussions. This combination of methods—document analysis, field research, and expert workshops—facilitates an in-depth, contextual, and multi-faceted exploration of the research topic (Fig. 7.2).

7.3.1 Field Research

Field research followed a qualitative methodology and an inductive approach, based on case studies (Gerring 2004), and in-situ observations (Ehhalt 1980). Six public open spaces with living walls in Belgrade were observed in June 2021. Firstly, they were analyzed using a framework consisting of location parameters categories and spatial typologies, developed by the Urban Institute of Belgrade (2009). Secondly, technical characteristics were identified using the living walls construction typology developed by Radić et al. (2019). Finally, the existence of ecological, aesthetic, and educational benefits formed a final analytical lens. Due to the lack of public and expert debate on the topic of VGS planning and design in public open spaces in Serbia, an international online expert workshop was organized and moderated by the first author in October 2021. It aimed to steer discussion among experts to critically evaluate VGS planning and design as proposed by the 2019 PGRGASB and to develop recommendations for improvement informed by European best practices. The workshop was chosen as a good way to:

draw on the expertise and professional knowledge of participants (Payne et al. 2019).

Four European experts in urban planning and architecture, affiliated with institutions such as the Technical University Berlin (Germany), the University Union Nikola Tesla in Belgrade (Serbia), the Polytechnic University of Catalonia (Spain), and an architect representing the city of Essen (Germany), European Green Capital in 2017, participated. In the introductory part, VGS as presented in the 2019 PGRGASB were discussed, and the analysis of the case studies was deliberated. Afterward, experts were tasked to:

- Map positive and negative aspects of the 2019 PGRGASB and Living Walls case studies;
- Provide suggestions for enhancing VGS planning in Belgrade;
- Share examples of great living wall projects and NbS practices.

Finally, audio recordings were transcribed and analyzed using thematic analysis (Braun and Clarke 2006). Therefore, Sect. 4.2 Expert workshop data analysis is divided into subsections according to themes.

7.4 Results

7.4.1 Case Study Analysis

With the help of "Mapa," (n.d.) the platform, we have identified case studies (CS) presented in Figs. 7.3 and 7.4.

According to the spatial typology of public open spaces developed by the Urban Institute of Belgrade (2009) CS1 is defined as a square, CS3, CS5, and CS6 are plateaus, CS4 is a mini-park, while CS2 cannot be defined as public open space because the living walls belong to the street. CS4, a mini-park, aligns with the 2019 PGRGASB forest and green area typology. This variety advocates that VGSs can be installed in very different public open spaces. CS3 and CS4 are not connected to the GI system because no other vegetation was identified in their vicinity. This confirms that GI is not established in the city. Moreover, the spatial distribution of living walls is not harmonized across the city.

CS3, CS4, and CS5 are free-standing living walls categorized as trough planters and pocket systems (Radić et al. 2019), installed many years after the public open spaces were constructed. CS1 and CS6 are examples of living walls integrated into building facades, planned during the building's design phase. CS1 and CS6 were integrated into building facades during design, and exhibit plugged-in and geotextile felt systems, both with integrated irrigation. CS2 is a unique green facade and living wall hybrid, including ground-rooted and potted plants using a fence for support.

The impacts of VGS ecological benefits are almost insignificant due to the small size and lack of GI connection. CS2, CS4, and CS5 aimed to reduce traffic noise but have insufficient plant density. Although all observed open spaces became more

Case study	Analysis	
Case study 1 - Square of Sant Sava	Type of open public space: **Square** VGS is part of green infrastructure: **on the local level**	
	VGS description: **Two facade vertical gardens** VGS construction type: **living walls / plug-in system with integrated irrigation system** Identified vegetations: **Ageratumhoustonianum, Galinsogaquadriradiata, Impatienswallerina, Heucherasanguinea, Sedumhispanicum, etc.**	
	Identified benefits: **ecological - yes, on the local level** **aesthetical - yes, on the local level** **educational - no.**	
	VGS condition: **On both vertical gardens vegetation was recently planted, so the garden is not dense. Irrigation system of one garden was showing functioning issue.**	
Case study 2 - Street dividing line in front of shopping mall "Ušće"	Type of open public space: **Street** VGS is part of green infrastructure: **on the local level**	
	VGS description: **Linear vertical garden on the fence of the middle dividing line** VGS construction type: **construction type does not belong to any classification (it consists of plots attached to existing fence)** Identified vegetations: **shurbs, flowers, etc. (garden is not approachable)**	
	Identified benefits: **ecological - yes, on the local level** **aesthetical - yes, on the local level** **educational - no.**	
	VGS condition: **plants are in good condition, although they are placed sparsely. There was identified issue with a water excess from the integrated irrigation system called drop by drop. Streets are very wet after the watering, which makes them dangerous for driving.**	
Case study 3 - Plateau of Djoke Vještice	Type of open public space: **Public space next to a public building** VGS is part of green infrastructure: **no**	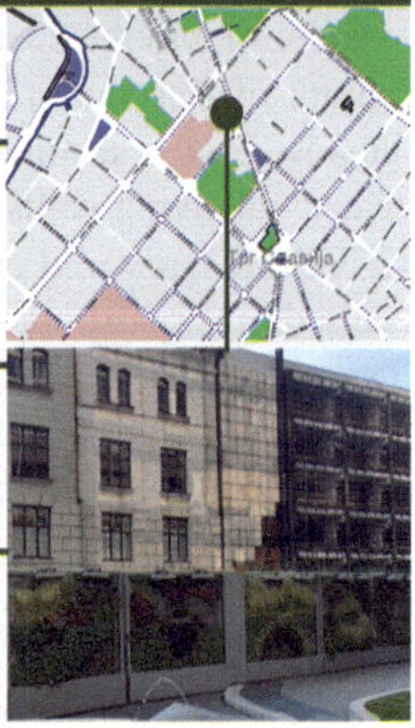
	VGS description: **Seven free-standing vertical gardens** VGS construction type: **living walls / trough planters with integrated irrigation system** Identified vegetations: **Cerastiumtomentosum, Ceratostigmaplumbaginoides, Delospermacooperi, etc.**	
	Identified benefits: **ecological - yes, on the local level** **aesthetical - yes, on the local level** **educational - no.**	
	VGS condition: **three of seven gardens are partly dry. All of them have provided artificial light. The rest of the water is being collected and returned into irrigation system, so the walking area is not wet and slippery.**	

Fig. 7.3 CS1, CS2, CS3 analysis

Case study 4 - Minipark of Djuke Vojvoda Brana

Type of open public space: Minipark
VGS is part of green infrastructure: no

VGS description: Two free-standing corner-vertical gardens
VGS construction type: living wall / pocket system with integrated irrigation system
Identified vegetations: Ageratumhoustonianum, Galinsogaquadriradiata, Impatienswallerina, Heucherasanguinea, Sedumhispanicum, etc.

Identified benefits:
ecological - yes, on the local level
aesthetical - yes, on the local level
educational - no.

VGS condition: VGS plants are in good condition, although there was identified issue with a water excess from the irrigation system.

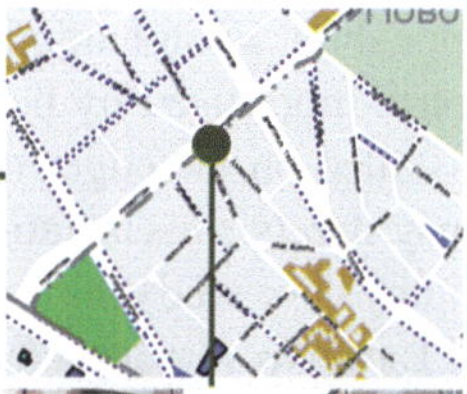

Case study 5 - Plateau of Slobodan Jovanović

Type of open public space: Public space next to a public building
VGS is part of green infrastructure: on the local level

VGS description: Three free-standing vertical gardens
VGS construction type: living walls / two are trough planters, one has pocket system, all three with integrated irrigation system
Identified vegetations: Cerastiumtomentosum, Ceratostigmaplumbaginoides, Delospermacooperi

Identified benefits:
ecological - yes, on the local level
aesthetical - yes, on the local level
educational - no.

VGS condition: Some plants are not in good condition on the living wall with trough planters, while plants on the living wall with the pocket system are completely dry.

Case study 6 - Entrance plateau of the shopping center "Ada Mall"

Type of open public space: Public space next to a public building
VGS is part of green infrastructure: on the local level

VGS description: Two facade vertical gardens
VGS construction type: living walls / geotextile felt system with integrated irrigation system
Identified vegetations: Iberissempervirens, Euonymusfortunei, Heucherasanguinea, Tiarellacordifolia, Pachysandraterminalis, Ajugareptans, etc.

Identified benefits:
ecological - yes, on the local level
aesthetical - yes, on the local level
educational - no.

VGS condition: Some of the VGS plants are partly dry, although both gardens are maintained often. Plateau is clean despite the high frequency.

Fig. 7.4 CS4, CS5, CS6 analysis

appealing, visitor numbers only increased in CS4. This mini-park became a safer space more used by both children and adults. CS3, CS4, and CS5 are living walls installed on the edges of plateaus. They present a visual and physical barrier to the busy streets, thus isolating these open spaces. Educational benefits were not identified in any case study, although VGS in CS5 stands on a plateau in front of the Faculty of Law. No student interest was observed in VGS use or maintenance.

All living walls, except CS6, are maintained by the municipal utility "Greenery Belgrade". During observation, it was noticed that CS1, the last one to be installed, contains a modern irrigation system with built-in screens, indicating maintenance problems. Problems with water leakage and small pond formation in front of the VGS were noted in CS2 and CS4. Water leakage makes the surrounding pavements very wet and slippery, which can be especially dangerous in CS2. Conversely, CS5 reveals a lack of proper watering with many plants being partly or completely dry.

7.4.2 Expert Workshop Data Analysis

The workshop template is presented in Fig. 7.5, while the further text describes the main findings thematically.

7.4.2.1 VGS Not Listed in Planning Documentation Typologies

Experts commented positively on the 2019 PGRGASB for mentioning VGS for the first time in a public document in Serbia as a GI element that can improve ambient values and the energy efficiency of the buildings. Despite such description, the typology of forests and green areas included in the Plan contains no VGS or any other new forms of green urban structures. The Overview Map (Fig. 7.1) from the technical part of the Plan does not include VGS either. This indicates that the plan does not develop multiple VGS classification types that should be the starting point for implementing VGS as GI elements.

7.4.2.2 Public Administration Engagement in VGS Without Public Participation

Neither the 2019 PGRGASB nor the VGS in the case studies were designed through broad participatory processes involving citizens. The plan was entirely prepared by the experts of the Belgrade Urban Planning Institute, while the maintenance is being carried out by the municipal utility "Greenery Belgrade". The lack of public engagement was expectedly criticized, but the leadership and the initiative of the city of Belgrade through engaging the municipal utility were evaluated as positive. For example, Berlin has the opposite scenario. VGS projects are driven by citizens, social initiatives, community gardens, or private operators, unlike in Belgrade,

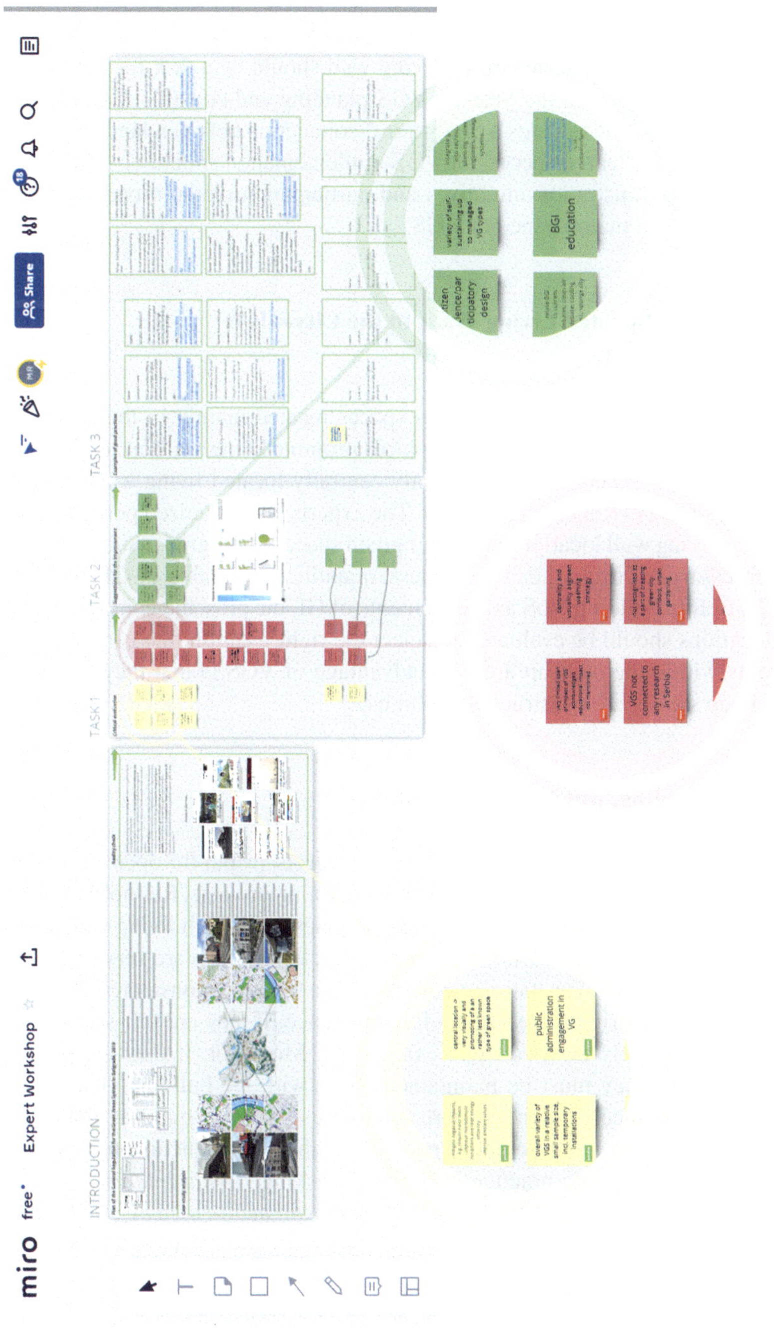

Fig. 7.5 Workshop template

where the city administration itself takes the initiative. In Essen, winner of the award 'European Green Capital' 2017, small VGS projects were also initiated by the city administration. Nevertheless, Essen's green management is not yet sure whether the construction and maintenance of the living wall should be a task for the public or private sector. Contrary to the Serbian VGS planning and construction practice, public engagement at NbS ensures a 'long-term' connection between landscape and citizens (Mell 2017). Clearly, successful VGS implementation should be a balanced interplay of city administration initiatives and participation of the broader public, both in the design and maintenance of VGS.

7.4.2.3 Centrally Located Living Walls in the City—Is It Greenwashing?

Living walls were installed in public open spaces located mainly in the center of Belgrade. Such busy and visually exposed places promote living walls as less known NbS. One workshop expert wondered if only centrally located living walls were a greenwashing strategy (Rinaldi et al. 2021). The experts also criticized poor connectivity between living wall locations and other green spaces in Belgrade. Such connections are necessary to ensure the effectiveness, quantity, and quality of GI (Mertens et al. 2022). Before building NbS as the elements of GI, the pros and cons of several possible locations should be evaluated to select the right location that can optimize their benefits within the built-up area. An advantage of VGS is that they could be built after open spaces are constructed and in use.

7.4.2.4 Self-standing, But Not Self-sustaining

The experts described the scale of living wall projects in public open spaces in Belgrade as very small and characterized them as learning projects. Resembling pilot studies, they could be experimental sites used for assessing which living wall types are suitable for which location. Unlike green facades, living walls are very technical and technology-dependent vegetation elements. Some types are considered the most complicated NbS, requiring a great deal of maintenance. If they are self-sustaining, living walls would likely be used more extensively. Moreover, if they are part of the building facade, they must be maintained along with the building. Therefore, such a high-maintenance element is more optimal as part of a building than as a free-standing element in a public open space. Regarding this conclusion, Expert 3 wondered about the opposite practice in Belgrade, asking, "*What interest does the municipality have in investing in free-standing living wall maintenance*?" According to Mertens et al. (2022), the public budget for GI is always lower than for other forms of infrastructure, which is the reason for:

> insufficient care of green and open spaces and their very often underdeveloped plant growth; neglected green spaces; and losses in biodiversity and the ecosystem services that these deliver.

This criticism depicts the situation in Belgrade as well.

7.4.2.5 VGS Great Benefits on Different Levels—Have They Been Considered?

One workshop expert stated, "*If we have green in cities today, we also have to discuss why we need this green and what benefits it has for the city*". Naming the project Grüne Hülle: Kö-Bogen II in Düsseldorf, Germany ("Handwerk + Bau" n.d.), experts implied that proper GI planning has a great impact at different levels. Indeed, besides effects on climate change mitigation, VGS also brings benefits for urban sustainability, such as food production (Utami et al. 2012), citizen health (Ghazalli et al. 2019), education, sustainable water management (Sheweka and Magdy 2011), urban wildlife (Mayrand and Clergeau 2018), etc. Another expert suggested that urban planners, architects, and landscape architects should discuss VGS implementation to answer some of the cities' contemporary issues, thus acting as a temperature booster, sustainable shading element for the summer city concerts, water absorption element in combination with photovoltaic panels to be used for irrigation of another urban greenery, etc. Hence, a starting point in any VGS design should be a discussion on how VGS influence cities' sustainable development and what kind of benefits they should bring.

7.4.2.6 VGS as a Part of Urban Technical Infrastructure

Workshop experts explained that living walls, unlike green facades, are very sophisticated and technically demanding. They agreed that VGS should be planned in synchronicity with other elements of cities' technical infrastructure. For this reason, technical and maintenance expertise is required. Moreover, collaboration among diverse professionals like engineers, architects, and urban planners is crucial for the technical development of urban infrastructure, addressing the diverse demands of VGS in urban conditions. This suggestion goes along with the statement that there is

> the need to integrate GI planning with planning other urban infrastructure (e.g. transport, water management) and structures (e.g. built-up areas) (Szulczewska et al. 2017).

Workshop experts recommend engineers initiate collaboration by clearly classifying VGS as appropriate GI elements for specific urban climates. Such early classification aids in developing efficient action plans. Moreover, broader discussions on integrating VGS as well as other NbS into city technical infrastructure enhance the understanding that various GI elements require diverse technical knowledge, maintenance, and intensity.

7.4.2.7 VGS as Education Tools

According to one expert, "*Blue-Green Infrastructure (BGI) is not currently widespread in our cities. There are no cities or world regions where BGI is a common practice*". Both, experts and the public need education on BGI. In Belgrade, the implementation of living walls in public spaces lacks ties to SDGs, BGI education, and research making their educational potential waisted. Both in Serbia and in Western European countries, participatory co-design projects in schools can be a way forward (Brković and Parnell 2017). They can be educational when knowledgeable people are involved in the planning and design stages, as users have an understanding and responsibility for the green element they choose, build, and maintain. Another expert cited Barcelona VGS co-design projects in schools, that included pupils in design and installation. As seen in a similar green roof project (Nadal et al. 2018) while children eagerly partake, the lack of elders' support due to incomprehension of the importance of the topic, can result in a missed learning opportunity for all.

Based on the interconnection between GI elements and the waters, BGI is recognized as a sustainable strategy to mitigate the negative aspects that urbanization brings to ecosystems (Din Dar et al. 2021). VGS as facets of our built environment could serve as *par excellence* built environment education tools, where architects, urbanists, children, and teachers could create meaningful learning experiences that are

> personally relevant, self-directed, with self-initiated topics; allowing a multitude of learning styles, personal interest, individual capabilities, and creative impulses to be enacted,

potentially transforming pupils

> into the cognizant future citizens, able to identify locally important BE issues, and take knowledgeable action towards them (Brković Dodig 2013).

While the involvement of architectural offices and academic researchers is a prerequisite, the involvement of community members across generations has a huge bonus through their informed engagement.

7.4.2.8 The Discrepancy Between the Planning Documentation and the Reality

While Serbia ventured into translating relevant GI planning policies from Europe, constructions in parks, on cultural heritage sites, investors' urbanism, and city megaprojects, demonstrate that the reality contradicts the 2019 PGRGASB. Workshop experts unanimously agreed that the gap between policy and practice occurs everywhere. In Belgrade, as in other capitals, but probably to a lesser extent:

> the development and expansion of cities have traditionally been guided by the criterion of maximizing the constructed area to increase profitability (Pérez-Urrestarazu et al. 2015).

Berlin, as an example, has many public policies that look promising (Bürgow et al. 2017), but in reality, green spaces have not increased with population growth in the last decade (Senate for Urban Development and Housing 2017). Unexpectedly, Belgrade's candidacy for the European Green Capital award in 2023 was considered positive by experts, although the discrepancy between the plan and reality seems absurd. One commented: "*Sometimes it needs these big illusions to be somehow reached or never reached. If you are there you can always see the places that cannot be considered sustainable*". This suggests that large and, at first glance, unrealistic goals may be the catalyst for raising benchmarks to the next level where achieving them may be possible.

7.5 Discussion and Recommendations

Existing European VGS planning policies recognize and build upon existing empirical evidence on VGS benefits (Giordano et al. 2017). Conversely, the 2019 PGRGASB does not exploit enough empirical evidence available in Serbia, mentioning only two VGS benefits—raising ambiance values and improving the energy efficiency of public buildings. More detailed and elaborated VGS inclusion in planning documentation is a prerequisite for successful implementation. Therefore, the Plan should be complemented by an implementation task force to discuss the varied VGS classifications and make them operative for planning and implementation practices.

Living walls are mostly built as free-standing structures in central locations in Belgrade, which can be interpreted in two ways: either as a greenwashing strategy or raising awareness of decision-makers of the VGS potential as future GI elements; thus, erecting them in prominent locations and presenting them as innovative NbS. While marketing such living walls in the city center can be beneficial, city administration needs to complement VGS installations with appropriate implementation strategies.

To educate urban planners and architects on the VGS's full potential, it is crucial to link VGS with their benefits in ongoing discussions on mitigating high air pollution and urban heat islands. VGS location selection should be based on a comprehensive interdisciplinary spatial analysis. Moreover, preparing urban infrastructure for city-wide living wall integration, alongside other NbS requires collaboration among various experts, including road engineers, sewage and stormwater specialists, landscape architects, urban planners, and architects.

For educating children and young, it is recommended to organize design and maintenance workshops as part of curricula or extracurricular activities in public spaces featuring VGS. Encouraging intergenerational participation in these activities can provide a valuable educational experience. Additionally, the installation of urban furniture, such as interactive information boards featuring a list of plants, urban fauna, and an explanation of their benefits, is suggested for open public places hosting living walls.

To fully realize the social benefits of VGS, surveys measuring tenant satisfaction with perceived positive impacts should be carried out. Citizens should be aware of their right to participate in activities when VGS are in public spaces and be informed about their positive aspects, design options, installation, and maintenance processes. In Berlin, unlike in Belgrade, the construction of VGS was driven by the public. Therefore, it is not surprising that a study by Nikolić et al. (2020) found a lack of environmental awareness among Serbian youth. Public participation in general and concerning VGS in Serbia is urgently needed, and Citizen Science projects (Peter et al. 2019) could be a complementary measure. Also, organizing various small-scale activities in public spaces to highlight the benefits of VGS can significantly enhance the overall quality of life in cities. Supporting local citizens in building and maintaining living walls, either individually or through weekend actions facilitated by municipal utilities like "Greenery Belgrade," can foster community engagement. A great example of how to encourage citizens to build VGS on their facilities' facades is the Polish city, of Kalisz. Since January 2022, all residents who install VGS or green roofs can be exempted from property tax ("Kalisz" n.d. 2021), which is exactly what Serbian researchers recommend (Vujičić et al. 2019). To further support the VGS application, an online building cadastre could be developed to inform interested citizens whether their building is suitable for a certain VGS construction type.

7.6 Conclusion

Summarized data show that VGS are a new concept in GI planning in Belgrade, mostly mentioned in the context of climate changes caused by the effects of globalization and urbanization. No domestic or international papers discuss VGS planning policies and practices in public open spaces in Belgrade. Moreover, no VGS implementation guides or strategies exist. This study contributes to fulfilling the mentioned gap, by critically evaluating VGS planning recommendations in the 2019 PGRGASB.

Although a key limitation of this study is a lack of comprehensive empirical measurements confirming ecological benefits such as thermal comfort, reduced air pollution and noise, positive effects on hydrology, and increased urban wildlife, it can be concluded that VGS in Belgrade do not have a significant environmental impact at the city level due to the living walls' small sizes and the lack of connection to a GI network in Belgrade. This underscores the need for an implementation strategy that integrates VGS into cities' GI. Planning should rely on precise scientific research results and standards, moving beyond the descriptive approach seen in the 2019 PGRGASB. To achieve widespread implementation, living walls require an adopted typology and developed planning guidelines for specific climate and urban conditions, as in developed European countries.

Developing recommendations for improving VGS planning policies, their implementation, and maintenance, as suggestions for further research, could drive the discussion in Serbia and regional South-East European countries. Since these research findings showed that the public usually pushes VGS construction in Western

European cities, prompt educational actions about NbS importance are needed in Serbia. Therefore, we see developing a design method in a participatory fashion for a low-cost VGS as an educational tool for tackling local sustainability issues as our next research endeavor.

Acknowledgements The work of Mina Radić Sibinović was funded through the doctoral scholarship awarded by the *Global Center of Spatial Methods for Urban Sustainability* SMUS funded by the *German Academic Exchange Service—DAAD* with funds from the *German Federal Ministry for Economic Cooperation (BMZ),* Project-ID 57526630. The work of Marta Brković Dodig has been supported through a research project RELATE by The Science Fund of the Republic of Serbia, grant no: 7726555. The authors are grateful for the participation of all the experts in the online workshop that supported the development of the present research. The work of Angela Million is also financed by SMUS funded by the *German Academic Exchange Service—DAAD* with funds from the *German Federal Ministry for Economic Cooperation (BMZ),* Project-ID 57526630.

References

Barcelona City Council (n.d.). https://www.barcelona.cat/en. Accessed 4 July 2024

Bauduceau N, Berry P, Cecchi C, Elmqvist T, Fernandez M, Hartig T, Tack J (2015) Towards an EU Research and Innovation Policy Agenda for Nature-based Solutions & re-naturing cities: Final Report of the Horizon 2020 Expert Group on "Nature-based Solutions and Re-naturing Cities." Publications Office of the European Union. https://doi.org/10.2777/765301

Berlin (2022). https://www.berlin.de/sen/uvk/natur-und-gruen/stadtgruen/daten-und-fakten/gruenflaechen-gruenanlagen/. Accessed 6 Sep 2022

Borges L, Kjell B, Moa N, Asli T, Dis T, Perjo L, Weber R (2019) White paper on Nordic sustainable cities. https://doi.org/10.30689/R2017:3.1403-2503

Braun V, Clarke V (2006) Using thematic analysis in psychology. Qual Res Psychol 3:77–101. https://doi.org/10.1191/1478088706qp063oa

Brković M, Parnell R (2017) Schools as 3D textbooks for sustainability education. In: Million A, Heinrich AJ, Coelen T (eds) Education, space and urban planning: education as a component of the city. Springer International Publishing, Cham, pp 79–90. https://doi.org/10.1007/978-3-319-38999-8_7

Brković Dodig M (2013) Sustainable schools as the 'third teacher': creating a design framework for sustainable schools in Serbia, learning from practices in England, Germany, and Spain

Bürgow G, Franck V, Höfler J, Million A, Nebert T, Steglich A (2017) Roof water-farm—A building block of climate-sensitive and circular-oriented urban development. In: Kost S, Kölking C (eds) Transitory urban landscapes: what kind of agriculture does the city need?. Springer pp 115–134. https://doi.org/10.1007/978-3-658-13726-7_8

Castellar JAC, Popartan LA, Pueyo-Ros J, Atanasova N, Langergraber G, Säumel I, Acuña V (2021) Nature-based solutions in the urban context: terminology, classification and scoring for urban challenges and ecosystem services. Sci Total Environ 779:146237. https://doi.org/10.1016/j.scitotenv.2021.146237

Ćerimović V (2016) Landscape architecture—3D vision, knowledge and learning. Akademska Misao, Belgrade

Cvejić J, Tutundžić A, Bobić A, Radulović S (2014) Cities adaptation to climate change: guidelines and recommendations from the aspect of Belgrade Green Infrastructure Planning. https://doi.org/10.13140/RG.2.1.4371.2163

City of Belgrade (2018) Detailed Regulation Plan for Kalemegdan Gondola-Ušće, City Municipality Old Town and New Belgrade. Official Gazette of the City of Belgrade, No.65/2018. Retrieved fromhttps://www.sllistbeograd.rs/pdf/2018/65-2018.pdf#view=Fit&page=1

City of Skopje (n.d.) https://skopje.gov.mk/mk/gup-na-gradot/?fbclid=IwAR1YQySvCi-P09uDYK0kvWFgwZyvsO62xDzjEesKqM6iTL1LJ15sAMVH4FU. Accessed 8 Oct 2022

Din Dar MU, Shah AI, Bhat SA, Kumar R, Huisingh D, Kaur R (2021) Blue Green infrastructure as a tool for sustainable urban development. J Clean Prod 318:128474. https://doi.org/10.1016/j.jclepro.2021.128474

Ehhalt DH (1980) In situ observations. Philos Trans R Soc Lond Ser A Math Phys Sci 296(1418):175–189

eKapija (n.d.) https://www.ekapija.com/news/2936196/na-kosutnjaku-u-planu-izgradnja-stambeno-poslovnog-kompleksa-sta-ce-biti-sa. Accessed 1 May 2022

Europa Nostra (2019) https://www.europanostra.org/stop-harmful-cable-car-project-on-the-belgrade-fortress/. Accessed 10 Oct 2022

Green Infrastructure—Environment—European Commission (n.d.) https://ec.europa.eu/environment/nature/ecosystems/strategy/index_en.htm. Accessed 5 Dec 2022

London Green and Blue Cover—London Datastore (n.d.) https://data.london.gov.uk/dataset/green-and-blue-cover. Accessed 6 Sep 2022

Gerring J (2004) What is a case study and what is it good for? Am Polit Sci Rev 98(2):341–354. https://doi.org/10.1017/S0003055404001182

Ghazalli AJ, Brack C, Bai X, Said I (2019) Physical and non-physical benefits of vertical greenery systems: a review. J Urban Technol 26(4):53–78. https://doi.org/10.1080/10630732.2019.1637694

Giordano R, Montacchini E, Tedesco S, Perone A (2017) Living wall systems: a technical standard proposal. Energy Procedia 111:298–307. https://doi.org/10.1016/j.egypro.2017.03.093

Grüne Hülle: Kö-Bogen II (n.d.) https://www.handwerkundbau.at/fassade/gruene-huelle-koe-bogen-ii-13873. Accessed 13 Jan 2022

Kalisz: Properties with green roofs free of tax? (2021) https://constructionmarketexperts.com/en/news/kalisz-properties-with-green-roofs-free-of-tax/. Accessed 13 Jan 2022

Mahmoud AHA, El-Sayed MA (2011) Development of sustainable urban green areas in Egyptian new cities: the case of El-Sadat City. Landsc Urban Plan 101(2):157–170. https://doi.org/10.1016/j.landurbplan.2011.02.008

Mapa (n.d.) http://gispublic.zelenilo.rs/giszppublic/Map. Accessed 12 Jan 2022

Maričić T (2007) System of green areas in big cities—Examples of Berlin and Belgrade. 57:145

Mayrand F, Clergeau P (2018) Green roofs and green walls for biodiversity conservation: a contribution to urban connectivity? Sustainability 10(4):985. https://doi.org/10.3390/su10040985

Mell IC (2017) Green infrastructure: reflections on past, present and future praxis. Landsc Res 42(2):135–145. https://doi.org/10.1080/01426397.2016.1250875

Mertens E, Stiles R, Karadeniz N (2022) Green may be nice, but infrastructure is necessary. Land 11(1):89. https://doi.org/10.3390/land11010089

Million A, Bürgow G, Steglich A, Raber W (2014) Participatory and multifunctional infrastructures for urban neighborhoods. In: Finding Spaces for Productive Cities, from Rob Roggema und Greg Keeffe. VHL University of Applied Sciences, pp 659–678

Nadal A, Pons O, Cuerva E, Rieradevall J, Josa A (2018) Rooftop greenhouses in educational centers: a sustainability assessment of urban agriculture in compact cities. Sci Total Environ 626:1319–1331. https://doi.org/10.1016/j.scitotenv.2018.01.191

Nikolić V, Vukić T, Maletaski T, Andevski M (2020) Students' attitudes towards sustainable development in Serbia. Int J Sustain High Educ 21(4):733–755. https://doi.org/10.1108/IJSHE-11-2019-0336

Park J, Jeong M (2020) Analysis on the types of benefits of gardens in urban areas: comparison of Korea and overseas cases. J People Plants Environ 23(6):667–681. https://doi.org/10.11628/ksppe.2020.23.6.667

Pauleit S, Ambrose-Oji B, Andersson E, Anton B, Buijs A, Haase D, Konijnendijk van den Bosch C (2019) Advancing urban green infrastructure in Europe: outcomes and reflections from the GREEN SURGE project. Urban Forest Urban Green 40:4–16. https://doi.org/10.1016/j.ufug.2018.10.006

Payne S, Hughes S, Wilkinson J, Hasselaar J, Preston N (2019) Recommendations on priorities for integrated palliative care: transparent expert consultation with international leaders for the InSuP-C project. BMC Palliat Care 18(1):32. https://doi.org/10.1186/s12904-019-0418-5

Pérez-Urrestarazu L, Fernández-Cañero R, Franco-Salas A, Egea G (2015) Vertical greening systems and sustainable cities. J Urban Technol 22(4):65–85. https://doi.org/10.1080/10630732.2015.1073900

Perić A (2020) Public engagement under authoritarian entrepreneurialism: the Belgrade waterfront project. Urban Res Pract 13(2):213–227. https://doi.org/10.1080/17535069.2019.1670469

Peter M, Diekötter T, Kremer K (2019) Participant outcomes of biodiversity citizen science projects: a systematic literature review. Sustainability 11(10):2780. https://doi.org/10.3390/su11102780

Radić M, Brković Dodig M, Auer T (2019) Green facades and living walls—A review establishing the classification of construction types and mapping the benefits. Sustainability 11(17):4579. https://doi.org/10.3390/su11174579

Rinaldi C, Giovanardi M, Lucarelli A (2021) Keeping a foot in both camps: sustainability, city branding and boundary spanners. Cities 115:103236. https://doi.org/10.1016/j.cities.2021.103236

RTS (n.d.) https://www.rts.rs/lat/vesti/drustvo/1118024/prvi-zeleni-zidovi-u-beogradu.html. Accessed 27 Feb 2024

Senate for Urban Development and Housing (2017) Versorgung mit öffentlichen, wohnungsnahen Grünanlagen/Provision of public, residential green spaces. https://www.www.berlin.de›_assets ›nutzung›de-texte

Šešlija D (2012) Green roofs as a method of ecoremediation of urban ecosystems (S. Spasić, Ed.). Civil Society Organization "Ecoist.", Belgrade. 47–56

Sheweka S, Magdy A (2011) The living walls as an approach for a healthy urban environment. Energy Procedia 6:592–599. https://doi.org/10.1016/j.egypro.2011.05.068

Šiško D, Polančec V (eds) (2022) Zagreb in Figures 2021. City of Zagreb, City Office for Economy, Environmental Sustainability and Strategic Planning Sector for Strategic Information and Research Department of Statistical and Analytical Affairs https://www.zagreb.hr/UserDocsImages/1/Zagreb%20in%20figures%202021%20-%20web.pdf

Stojić B (2024) Why do we talk about green infrastructure—introduction. In Stojić B, Andrić O, Čukić I (eds) Green infrastructure in urban planning. Institute for Urban Policies/Ministry of Space, Belgrade, pp 11–16

Szulczewska B, Giedych R, Maksymiuk G (2017) Can we face the challenge: how to implement a theoretical concept of green infrastructure into planning practice? Warsaw case study. Landsc Res 42(2):176–194. https://doi.org/10.1080/01426397.2016.1240764

Teqja Z, Lekaj E, Libohova Z, Bardhi N (2020) Assessment of the green space of Tirana city and its impact on some public health indicators. 19:1–13

Tesla city (2016) https://www.gradnja.rs/tesla-grad-istorijat-jedne-ideje/. Accessed 1 May 2022

Urban Institute of Belgrade (2009) Study of public spaces in Belgrade—Phase I analysis of public spaces in the old town, pp 1–206. http://urbel.com/studija-javnih-prostora-prva-faza/?script=lat

Urban Institute of Belgrade (2019) Plan of the general regulation for the green areas system in Belgrade. Official Gazette of the City of Belgrade. http://sllistbeograd.rs/pdf/2019/110-2019.pdf#view=Fit&page=1

Utami SNH, Darmanto, Jayadi R (2012) Vertical gardening for vegetables. Acta Horticult (958):195–202. https://doi.org/10.17660/ActaHortic.2012.958.23

Vujičić D, Tutundžić A, Radić B, Čepić S, Stefanović M, Tubić L, Jevtović A (2019) Legislative framework in the function of green infrastructure development. Association of Landscape Architects of Serbia, Belgrade. https://docplayer.rs/202291279-законодавни-оквир-зелене-у-функцији-развоја-инфраструктуре.html

Vujičić D (2019) Tendencies in design of urban open spaces in the context of ecological changes. Kraljevo

Wieshofer I, Prochazka E, Knoll T, Cserny A, Magistrat der Stadt Wien (eds) (2015) Green and open spaces: sharing the outdoors; thematic concept. Vienna City Administration, Municipal Dep. 18 (MA 18), Urban Development and Planning, Wien

Chapter 8
Planning Nature-Based-Solutions (NbS) Through Rainfall-Runoff Simulations for Slums of Developing Countries: A Case of Nagpur, India

Namrata T. Gaurkhede and Vinayak S. Adane

Abstract The chapter introduces the critical issue of floods in India, particularly affecting vulnerable populations in slum areas, exacerbated by unplanned urbanization and climate change. Despite the proven effectiveness of NbS in flood mitigation, their implementation in developing countries, including India, remains limited. The chapter emphasizes the need to understand the unique challenges of urban areas and slums in developing countries before planning NbS interventions. It highlights the paradigm shift towards sustainable stormwater management systems and emphasizes the multifunctional benefits of NbS in creating resilient communities. Drawing on a case study of Nagpur, the chapter illustrates the urgency of NbS interventions in flood-prone slum areas. Challenges hindering NbS implementation in slums, such as locational settings, population and built density, socio-economics, housing quality, physical infrastructure availability, literacy rate, poor and unhygienic living conditions, and governance issues, are described. Furthermore, criteria for selecting NbS components are outlined, guiding their simulation in Nagpur's slum areas using SWMM software. Simulation results demonstrate the effectiveness of certain NbS techniques, such as, rooftop disconnection, infiltration trench/basin, and bioretention cell/area, highlighting the importance of context-specific solutions. The chapter also provides initial stage research and simulations needed for Indian cities. It advocates integrating NbS into urban planning strategies to enhance urban resilience and promote sustainable growth in flood-prone slum areas.

Keywords Sustainable urban development · Urban resilience · Community engagement · Socio-economic factors · Governance issues · Multifunctional benefits

---

N. T. Gaurkhede (✉) · V. S. Adane
Department of Architecture and Planning, Visvesvaraya National Institute of Technology, Nagpur, India
e-mail: namrata2707@gmail.com

V. S. Adane
e-mail: vsadane@arc.vnit.ac.in

M. Firoz C et al. (eds.), *Nature-Based Solutions for Urban and Peri-Urban Areas*, Disaster Risk Reduction, https://doi.org/10.1007/978-981-96-5933-3_8

8.1 Introduction

Floods are the most prevalent natural disaster globally, with India experiencing thousands of fatalities and significant economic losses (Singkran and Kandasamy 2016; CRED 2021). In 2020, the world witnessed 26% more storm events, 23% more floods than the yearly average of 163, and 18% more flood deaths than the annual average of 5,233. India saw the most significant adverse human effect, with 19.6 million people impacted by floods (CRED 2021). Floods are not limited to coastal, low-lying areas (Haji 2021), or cities along major rivers or valleys (CPHEEO 2019). In recent years, due to a combination of unplanned urbanization (Mguni et al. 2016) and climate change (Haji 2021), flood occurrences have increased in urban inland areas, otherwise disaster-free (Gaurkhede and Adane 2023).

Slums, the informal settlements (Wolff et al. 2023) that start in the peri-urban areas and then integrate with the urban area with the city's expansion (Parkinson et al. 2007), are victims of flood hazards (Sagala et al. 2022). Developing countries are characterized by a higher number of slums, for example, over one billion people reside in slums, with the largest proportion concentrated in Eastern and Southeast Asia (Wolff et al. 2023). Slums are characterized by highly dense settlements with poor socioeconomic and infrastructure conditions.

Nature-based Solutions (NbS) are ideal for rainfall-runoff control, improved water quality, and the possibility of using stormwater as a resource and an element of urban amenity (Tucci 2001). NbS and its synonymous concepts have been tested in developed countries for over 50 years (Fletcher et al. 2015). Studies have been conducted to understand its multifunctional benefits (Fletcher et al. 2015; Lähde et al. 2019), performance (Nie and Jia 2018), cost-effectiveness (Webber et al. 2019), implementation (Charlesworth and Mezue 2017), operation, maintenance (Ortega et al. 2023), adaptability (Haji 2021), framing standards and guidelines and testing them (Lashford et al. 2019), and others. Thus, developed countries have come a long way in NbS research (Lashford et al. 2019). However, developing countries started such studies only in the last decade (Vasconcelos and Barbassa 2021). NbS integration into urban planning is increasingly important, especially in low and middle-income countries (McClymont et al. 2020). Therefore, developing countries must also conduct rigorous studies and experiments with socio-economic, developmental, and political limitations (Lashford et al. 2019) to set indigenous standards (Vasconcelos and Barbassa 2021) and development practices.

According to the United Nations (Tucci 2001), a fascinating fact is that most developing countries have tropical/subtropical climates. The UN also states that planning for NbS in developing countries requires consideration of special factors (Tucci 2001; Gaurkhede and Adane 2024), including climate, land use pattern, socio-economic conditions, and governance structures. Developing countries have higher intensity and annual average rainfall (Lashford et al. 2019) compared to the temperate climate in developed countries. Slums in developing countries are the most vulnerable pockets, causing challenges and barriers (Vasconcelos et al. 2022) in planning and implementing NbS, which should be identified and considered.

In 2003, when the world's slum population was 32% (Charlesworth and Mezue 2017), it was 43% in developing countries (Arimah 2011). The slum population will be two billion by 2030 with current trends (Riley et al. 2007; Jorgenson and Rice 2016). This unplanned growth will further affect urbanization and increase informal settlements (Mguni et al. 2016). Cities in developing Asian and African countries will observe maximum urbanization (Mguni et al. 2016) due to natural increase and migration.

According to the 2011 Census of India, the slum population was 65.49 million, 17.45% of the urban population (Sinha et al. 2011). The average slum population in many Indian cities ranges from 30 to 40% (Deshkar 2019). In the 2001 census, Nagpur stood fourth in the country, with 35.9% of the slum population (Sawhney 2013). In 2008, 36% of Nagpur's population lived in slums (NMC 2008), with 28% of slum dwelling unit properties (MoUD 2015). Nagpur, a sub-tropical inland city, has been facing flooding for the past decade (Gaurkhede and Adane 2023), affecting several slums (DMC-NMC 2022). On 23rd September 2023, Nagpur received a peak rainfall of 250 mm in 24 h, leaving the city devastated. Thus, a slum sub-catchment in Nagpur city has been selected for rainfall-runoff simulations.

The multiple-layered benefits of NbS in slums (Wolff et al. 2023) need to be identified. Currently, in India, the guidelines for disaster management (NDMA 2010), the drainage design manual (CPHEEO 2019), and the spatial planning guidelines (URDPFI) (TCPO-MoUD 2015) fail to address NbS, the non-structural flood mitigation measures. Thus, to effectively plan and implement NbS in India, national guidelines must be formulated (Lashford et al. 2019), and aligned with other relevant guidelines. This can be achieved through studies in various parts of the country and sub-climatic zones (Lashford et al. 2019; Li et al. 2019). Thus, this study focuses on initiating the process of setting up standards for designing NbS in slums of Indian cities through available literature and standards. The sub-tropical regions inland cities are chronically getting flooded due to extreme weather conditions increasing the rainfall intensity. Thus, this chapter intends to present the retrofitting of suitable NbS for flood mitigation in slum areas of Nagpur, India.

The study's methodology (Fig. 8.1) involves a literature review, identification and consideration of challenges in adopting NbS in slums of developing countries; selection of suitable NbS techniques; SWMM set-up for rainfall-runoff simulations; presenting results.

8.2 Literature Review

8.2.1 NbS in Slums of Developing Countries

Urban informal settlements face significant challenges and constraints in managing natural ecosystems and integrating green infrastructure. NbS initiatives in slum areas of developing countries have limited examples, some of which have been studied.

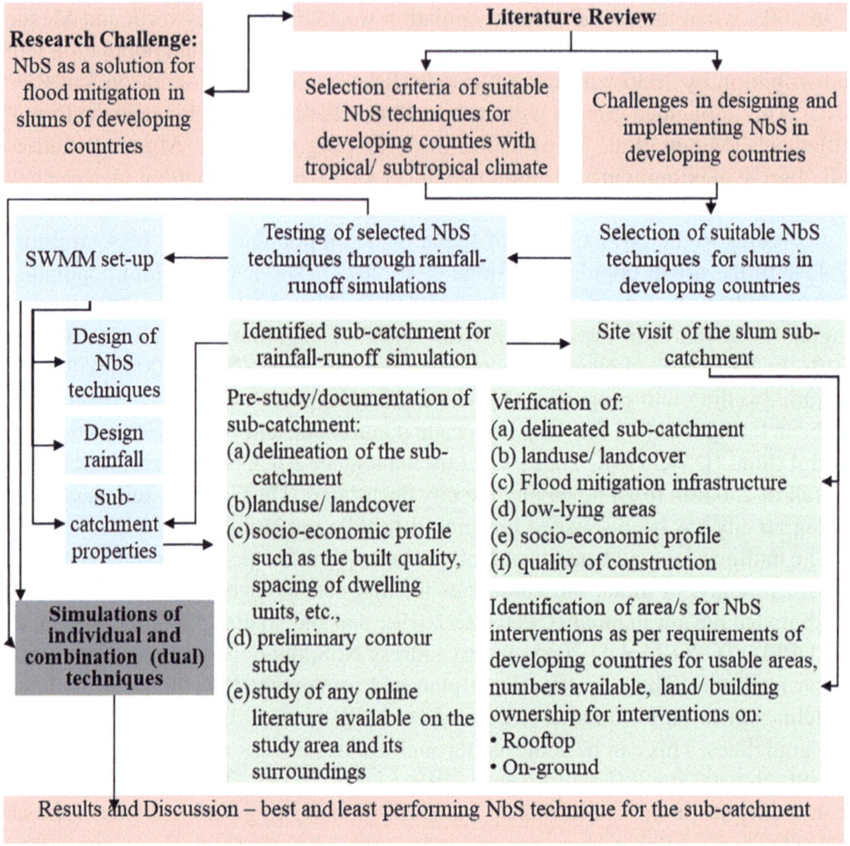

Fig. 8.1 Methodology for study

Adegun in (2014, 2019) explored two slums in Johannesburg, South Africa: Slovo Park and Kya Sands. In Slovo Park, the residents employ building-related practices such as adding sand/gravel, paving ways, and deepening water channels. Despite technical limitations, these initiatives demonstrate the potential for community-led efforts in addressing stormwater issues. However, the need for state and non-state agencies' support to enhance their effectiveness is emphasized. In Kya Sands, the slum community's efforts to green Johannesburg face limitations. The study suggests that local realities must be considered in greening projects to avoid disrupting residents' connections to their environment. The reconstruction of wetlands is proposed as a solution to reduce flooding issues, highlighting the need for context-specific interventions. In a third study, Adegun (2018) examines residents' negative experiences with natural ecosystems and green spaces in informal settlements in Kya Sands and Ruimsig, Johannesburg. Issues with domestic gardens and vegetation due to inadequate services and infrastructure are taken up, again highlighting the importance of the needs of residents and challenges.

Kharisma and Kaswanto (2021) discuss the need for ecological approaches to prevent the negative impacts of land use conversions on the Ciliwung River, Bogor City. Their research on home gardens in formal and informal settlements is an insight into the community perceptions, and their engagement and provides recommendations for sustainable garden management. The impact of climate change on urban green structures in informal settlements is investigated by Roy et al. (2018) in Dares Salaam, Tanzania. The study underscores the importance of cultural ecosystem services and recommends that city authorities focus on creative urban planning for greening informal settlements; enhancing resilience and improving the quality of life for residents.

Mulligan et al. (2020) examine the benefits and limitations of green, blue, and grey infrastructure solutions in informal settings in Kibera, Nairobi. The study finds that co-developing small-scale green infrastructure changes people's valuation, perception, and stewardship of NbS. This has implications for larger-scale infrastructure adoption and suggests that hybrid infrastructure systems may be most effective. In line with these findings, UN-Habitat (2023) identifies best practices for NbS in informal areas, emphasizing community involvement, knowledge and information dissemination, and governance. Coleman et al.'s (2018) study focuses on the adoption of Green Stormwater Infrastructure (GSI) by Vermont residents, giving an insight into the residential adoption of GSI in informal settlements. It reveals that private landowners may prefer practices like infiltration trenches, while roof runoff diversion is more likely part of larger green behavior initiatives.

Santos et al. (2021) analyze the environmental impacts and hydrological effects of NbS in a highly urbanized area in Brazil. Their study simulates piped drainage, NbS, and combination scenarios to conclude that NbS performs best in flood mitigation with a 30-year infrastructure service life. This research supports the adoption of NbS for sustainable urban water management. Similarly, McClymont developed a resilience-driven multi-objective optimization model for incorporating NbS into urban planning. The model identifies suitable options such as rain barrels, green roofs, bio-retention tanks, vegetation, grass swales, and permeable pavements, optimizing water quality and quantity; highlighting the potential for integrated urban planning.

Finally, Toit et al. (2018) review consolidates research on GSI and ecosystem services in sub-Saharan African cities. Despite reviewing 68 papers from 20 countries, only 38% had any conducted research, indicating significant gaps in the literature. It suggests that future research should focus on expanding studies in underrepresented areas and exploring innovative solutions for integrating green infrastructure in informal settlements.

In conclusion, the reviewed studies highlight the potential and challenges of greening and natural ecosystem management in informal settlements. Community-based strategies, NbS, and creative urban planning are essential for enhancing resilience and improving environmental conditions. However, these initiatives require adequate support, context-specific interventions, and active community involvement for success. Although these studies document the community initiatives well and give an idea of possible spaces and interventions in the dense fabric of slums, the aspect of planning and designing NbS interventions for their best possible performance is absent, except in (McClymont et al. 2020; Santos et al. 2021) studies. Addressing the identified research gaps will further inform and refine approaches to sustainable urban development in informal settings.

8.2.2 *Challenges in the Design and Implementation of NbS in Slums of Developing Countries*

There is a much-needed paradigm shift from traditional to modern drainage systems (Nie and Jia 2018), to treat stormwater as a resource for integrated water management (CPHEEO 2019). In addition to flood mitigation, the spaces generated in a city using NbS result in more recreational areas, healthier ecosystems, a cooler city, and better city aesthetics (Sowińska-Świerkosz and García 2022). It can potentially minimize the cost of drainage and water delivery infrastructure while also addressing droughts (Rohilla et al. 2017). These solutions can also increase the land value (Mguni et al. 2016) of slums, alongside improving the quality of life.

Developing countries like China have made regulations for the country through its Sponge City program and are testing them in multiple cities (Lashford et al. 2019). Some African cities have started experimenting with rainwater harvesting through NbS technologies and can provide 2–12 months of drinking water supply (Mguni et al. 2016) and flood mitigation. Other developing countries working on the NbS are Malaysia, Hong Kong, Vietnam, Bangladesh, Indonesia, Philippines, and India, but in the affluent parts of the city.

The designing of NbS in slums (Sagala et al. 2022) of developing countries needs special considerations (Charlesworth and Mezue 2017) in terms of data availability (Tucci 2001), land availability (Mguni et al. 2016), their acceptance by the residents, its operation and maintenance (O&M) (Sagala et al. 2022). Based on the literature, the study has identified certain considerations for designing NbS in informal settlements of developing countries, including their locational setting (Wolff et al. 2023), population and built density (Sagala et al. 2022), socioeconomic status (Sawhney 2013), quality of housing (Arimah 2011), availability of infrastructure (Arimah 2011), education status (Tucci 2001), living conditions (Jorgenson and Rice 2016), and governance and political scenario (Ortega et al. 2023).

8.2.2.1 Locational Setting

The slum areas are mainly located along water bodies (Haji 2021), which are low-lying areas (Chimankar 2016). The squatters informally occupy these areas (Parkinson et al. 2007) and develop into full-scale slums. These low-lying areas always receive runoff, along with solid and liquid waste from surrounding areas. Slums also occupy land parcels reserved by the government (Chimankar 2016) for future development of recreational spaces and public amenities.

8.2.2.2 Population and Built-Density

The population and built density in developing countries are higher than in developed ones (Radcliffe 2018) and even higher in slums (Sagala et al. 2022). Due to higher built density and low ground permeability (Sagala et al. 2022), there are fewer open spaces (Wolff et al. 2023) for planning infrastructure.

8.2.2.3 Socio-economic Status

The slum dwellers, generally migrated population to urban areas (Chimankar 2016) and economically weaker sections were involved in low-income jobs (Sinha et al. 2011; Diep et al. 2022). Thus, they have low resilience to floods and cannot afford to contribute towards building infrastructures. Although community participation (Ballard et al. 2015) is highly recommended in disaster planning and implementation, the weak socio-economic status (Vasconcelos and Barbassa 2021) demotivates the residents to participate.

8.2.2.4 Housing Quality

The slums have residents living in temporary sheds (Arimah 2011) with no/low plinths, making them vulnerable to calamities (Wolff et al. 2023). The roofs are often made of thatch/cloth/plastic or tin/G.I. sheets where joinery may be exposed, letting rainwater into the houses. The temporary or semi-permanent condition of the houses (Chimankar 2016) allows for limited possibility for plot-level NbS interventions (Stovin and Swan 2007).

8.2.2.5 Physical Infrastructure Availability

These unauthorized settlements unsurprisingly lack basic infrastructure facilities (Deshkar 2019) such as piped water supply, metered electrical supply, sewer and stormwater drainage connections, and solid waste management (Mguni et al. 2016; Diep et al. 2022). These inhuman living conditions make these settlements polluted zones (Vasconcelos and Barbassa 2021) and more vulnerable to the primary and secondary effects of flooding (Wolff et al. 2023), such as epidemics (Riley et al. 2007), loss of income sources, and many more. The limited infrastructure is ill-maintained and encroached upon by the residents and street vendors, further hindering the drainage (Gaurkhede et al. 2021).

8.2.2.6 Literacy Rate

The percentage of the literate population in developing countries is 73.5%, which is lower than that of the developed countries (99.2%) (CIA 2019), with worse figures for slums. Poor social infrastructure in slums further affects educational opportunities (Chimankar 2016), prolonging the cycle of poverty (Diep et al. 2022) and vulnerability. Thus, they are unaware of the disaster-related ill effects of and for their settlement. The level of education also affects the community's acceptance (Vasconcelos and Barbassa 2021) of new infrastructure and O&M.

8.2.2.7 Poor and Unhygienic Living Conditions

Due to the lack of infrastructure facilities, housing conditions, and location, the slum dwellers have unhygienic living conditions (Parkinson et al. 2007; Deshkar 2019). After flood events, insect breeding in stagnant waters gives rise to epidemics (Riley et al. 2007; Mguni et al. 2016). This can also affect the functioning of the NbS due to clogging of the porous areas.

8.2.2.8 Governance and Political Scenario

Slum areas are located and mapped under the jurisdiction of the urban local bodies (ULBs) (TCPO-MoUD 2015). Low-income residents are prone to more anti-social or illegal activities (Sawhney 2013), which benefit the political parties. Due to political interests, the slums are notified and are provided with various physical and social infrastructures (Sawhney 2013). Ideally, they should be relocated to safer localities with built-in infrastructure. The created vacant land after relocation should be taken up for construction of the site and regional scale NbS (Chimankar 2016). The availability of land (Mguni et al. 2016; Jayasooriya et al. 2020) for NbS development can be a significant problem without slum relocation.

8.2.3 Selection Criteria for Components and Techniques of Nature Based Solutions

NbS can be considered to have seven components (Ballard et al. 2015) at three scales of control: source, site, and regional. Some of these components are used in cities worldwide without fully understanding their potential, needs, and suitable locations (Martin-Mikle et al. 2015; Kazak et al. 2018). Consequently, this chapter also elaborates on the eight criteria for selecting relevant NbS components for installation, considering the capability of the techniques and the challenges of retrofitting them in slums of developing countries.

1. The primary objective of this study is flood mitigation, making it the first criterion for selecting the NbS techniques (Fletcher et al. 2015). Certain NbS techniques, such as runoff conveyances and pollution controls, contribute indirectly to flood mitigation. However, this study only considers the techniques that contribute directly to flood mitigation (Ballard et al. 2015).
2. Urban areas with flooding issues have an existing partial stormwater drainage network. Some countries are adopting NbS in new developments (Lashford et al. 2019) but not in the existing/older locations that continue to flood. Retrofitting NbS in existing settlements has been underexploited (Stovin and Swan 2007) and is crucial in slums of developing countries' densely built, highly vulnerable urban fabric. Thus, the retrofitting capability of NbS techniques is a criterion (Lashford et al. 2019; Webber et al. 2019).
3. NbS is a land-intensive infrastructure solution, and the high-density slums have hardly any open spaces (Stovin and Swan 2007; Radcliffe 2018). Land procurement, even for public purposes and infrastructure, is tedious, expensive, and impossible for infrastructure development of these unauthorized settlements. Thus, NbS techniques requiring less land area (Chen et al. 2021) can be given preference to avoid implementation barriers (Melville-Shreeve et al. 2017; Jayasooriya et al. 2020).
4. The developing countries struggle to intervene only in crucial infrastructure developments, like water and electrical supply, and wastewater management. Soft loans from world organizations often fund infrastructure systems (MoUD 2015; Vasconcelos and Barbassa 2021). Thus, they commonly have a combined sewer system for wastewater and stormwater (Parkinson et al. 2007). However, this leads to backflushes during heavy rains, increasing the water volume and creating unsanitary conditions (Gaurkhede and Adane 2024). Therefore evaluating the cost-effectiveness (Webber et al. 2019) of NbS to align with governance requirements is essential.
5. Infrastructure developments entail O&M costs other than capital costs (Webber et al. 2019). The ULBs managing the infrastructure must minimize O&M costs (Jayasooriya et al. 2020). To promote good governance, it is crucial to plan for NbS techniques with simple operations (Ortega et al. 2023) and demanding minimal maintenance (Jayasooriya et al. 2020).
6. Land use influences various aspects of the built environment (Kazak et al. 2018), including pollution (Jayasooriya et al. 2020) and porosity levels, and consequently runoff volume (Gaurkhede et al. 2021). The configuration of NbS infrastructure, such as its type, size, and purpose, depends on the compatibility with surrounding land use within the watershed (Martin-Mikle et al. 2015). This can also ensure better stakeholder acceptance.
7. Compatibility of NbS techniques in the highly dense fabric (Diep et al. 2022) of slums (Parkinson et al. 2007) in developing countries is an important criterion. Thus, despite challenges such as a complex mix of built-unbuilt spaces, infrastructure conditions, and socioeconomic and governance systems, it is crucial to integrate NbS into the dense urban fabric (Webber et al. 2019).

8. NbS offers localities additional socioeconomic advantages (Ortega et al. 2023). However, stakeholder acceptance of NbS has been challenging in various countries, including developing nations of China and Malaysia, and developed European countries (Vasconcelos and Barbassa 2021). UNESCO considers socioeconomic aspects more important than climate change in the design of drainage systems in developing countries (Tucci 2001). Better community participation and adaptability can be ensured with the multifunctionality (Lähde et al. 2019) of NbS.

Thus, the selection of these components primarily depends on their being flood mitigation capability (Lashford et al. 2019), spatial factors, retrofitting potential (Rohilla et al. 2017), and land area requirement (Melville-Shreeve et al. 2017); two governance factors, infrastructure development cost (Rohilla et al. 2017; Thorne et al. 2018) and O&M cost (Diep et al. 2022); socio-physical factor, compatible land use (Andjelkovic 2001), high-density areas and social adaptability (Thorne et al. 2018) dependent on its multifunctionality (Lähde et al. 2019; Ortega et al. 2023). Based on these criteria, a literature study from (Ballard et al. 2015) and (Rohilla et al. 2017) is conducted to evaluate NbS technique suitability for slums in developing countries; the selection matrix is summarized in Table 8.1.

As per observations from Table 8.1, all source control measures, including Green Roofs (GR), Rain Barrels (RB), and Roof Disconnection (RD), and all permeable surfaces, such as large open areas and Permeable Pavement (PP), are suitable for slum areas in developing countries. Additionally, harvesting structures are beneficial for storing water for the summers, and green roofs reduce the heat island effect.

The swales, conveyance channels, and filtration systems don't directly mitigate floods; thus, they have not been assessed for other criteria except bioretention (BR). Although BR is unsuitable for land requirement, density, and O&M cost, it has been selected for simulation to check its flood mitigation capability.

Infiltration systems such as Soakaways, Infiltration Trenches (IT)/Infiltration Basins (IB), and Rain Gardens (RG) are suitable for retrofitting in neighbourhoods in developing countries, while infiltration basins are not suitable. Although soakaways are best suited for slum areas, the selected software, SWMM (Rossman 2015), cannot currently simulate them. Thus, only IT and RG have been simulated in the study area.

Retention ponds and wetlands require large land parcels, while the detention and geocellular systems require moderate land area. Although, these are suitable for developing countries but not for slums; thus, they have not been selected for this study. Finally, seven NbS components, RB, RD, GR, PP, BR, IT, and RG, have been simulated in selected slums of Nagpur.

Table 8.1 NbS techniques selection matrix

Component	Techniques	Criteria											
		Flood Mitigation Capability	Retrofitting capability	Land Requirement	Capital Cost	O&M Cost	Site Land use					Suitable in high-density areas	Multi-functionality
							R	C/I	PSP	Rec	T		
Source control	Green Roofs	Good	Good	Good	Good	Good	Good	Good	Good	Good	Bad	Good	Good
	Rain Barrels	Good	Good	Good	Good	Good	Good	Good	Good	Good	Bad	Good	Good
	Rooftop Disconnection	Good	Good	Good	Good	Good	Good	Good	Good	Good	Bad	Good	Good
	Permeable Pavements	Good	Good	Average	Good	Good	Good	Good	Good	Good	Good	Good	Good
	Other Permeable Surfaces	Good	Good	Good	Good	Good	Good	Good	Good	Good	Good	Good	Good
Swales and conveyance channels	Swales	Bad	Since these are not directly useful for flood mitigation, the rest of the criteria were not assessed										
	Canals, rills & other channels	Bad											
Filtration	Filter Strips	Bad											
	Filter Trenches	Bad											
	Bioretention Area	Good	Good	Bad	Good	Average	Good	Good	Good	Good	Good	Bad	Good
Infiltration	Soakaways	Good	Good	Good	Good	Good	Good	Good	Good	Good	Bad	Good	Bad
	Infiltration Trenches	Good	Good	Average	Good	Average	Good	Good	Good	Good	Good	Good	Bad
	Infiltration Basins	Good	Bad	Bad	Good	Bad	Good	Good	Good	Good	Bad	Bad	Good
	Rain Gardens	Good	Good	Average	Good	Average	Good	Good	Good	Good	Bad	Good	Good
Retention and detention	Detention Basins	Good	Good	Average	Good	Good	Good	Good	Good	Good	Bad	Good	Good
	Retention Ponds	Good	Bad	Bad	Good	Good	Good	Good	Good	Good	Bad	Average	Bad
	Geocellular Systems	Good	Good	Average	Good	Average	Good	Good	Good	Good	Bad	Good	Bad
Wet-lands	Wetlands	Good	Bad	Bad	Bad	Good	Bad	Bad	Bad	Good	Bad	Bad	Good

Legend: Good (green), Average (yellow), Bad (red)

8.3 Nagpur and Its Slums

Nagpur is at 319 m of average elevation with a flat terrain and an annual rainfall of 1205 mm. It has 40% of stormwater drainage coverage, which is highly insufficient (MoUD 2015), and 90% coverage of wastewater network (MoUD 2015). Some households are still discharging their waste into the nearby waterbodies, causing pollution.

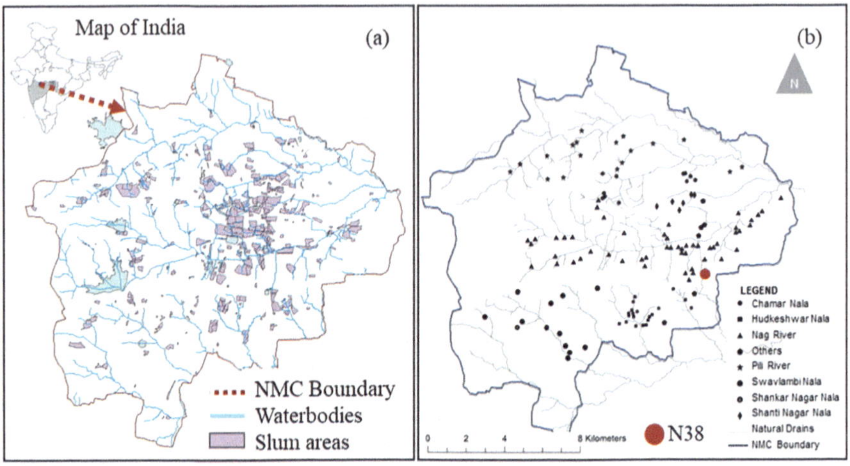

Fig. 8.2 **a** Slums, Nagpur. **b** Points of submergence, based on NMC data

The number of slums in Nagpur increased dramatically between 1971 and 91 (85%) and 1991–2001 (25%). Nagpur contains 446 slums, of which 287 are registered, 137 are not notified, and 22 are newly designated slum areas (NMC 2008). Some informal communities still use lake water for washing, bathing, and household purposes (Deshkar 2019). Slums are concentrated in the center and east of Nagpur (Fig. 8.2a). Nagpur slopes from west to east; most of the slums are located along the waterbodies and are thus located in flood-prone, low-lying areas. NMC has listed some slums as chronic submergence points (Fig. 8.2b) (DMC-NMC 2022).

8.4 Model Setup for Simulations in SWMM Software

The rainfall-runoff simulations for NbS interventions are carried out with SWMM 5.1, an open-source software (Rossman 2015). SWMM requires specific data for the catchment properties, such as geospatial, hydrological, infrastructure, and time series information. These include the land cover and soil properties, contour levels, and the piped drainage systems, if any. It also requires design specifications of various NbS infrastructure components.

8.4.1 Rainfall and Time Series Data

The time series data, i.e., rainfall intensity for 12 h for each Return Period (RP): 2, 5, 10, 25, and 100 years (Table 8.2) is adapted from a draft stormwater project report by the NMC (2008). Based on the historical data, hourly precipitation depth analysis

Table 8.2 Rainfall intensity (mm/h)

Duration (h)	RP (Years)				
	2	5	10	25	100
1	38.66	47.03	55.85	67	83.47
2	24.36	29.63	35.19	42.21	52.58
3	18.59	22.61	26.85	32.21	40.13
4	15.34	18.67	22.17	26.59	33.13
5	13.22	16.09	19.1	22.91	28.55
6	11.71	14.24	16.92	20.29	25.28
7	10.57	12.85	15.26	18.31	22.81
8	9.67	11.76	13.96	16.75	20.87
9	8.94	10.87	12.91	15.48	19.29
10	8.33	10.13	12.03	14.43	17.98
11	7.82	9.51	11.29	13.55	16.88
12	7.38	8.97	10.66	12.78	15.93

for various RPs is considered with probability distribution function and conversion into rainfall intensity data.

8.4.2 Design Specifications for Various NbS Techniques

NbS components comprise typical layers: surface, pavement, soil, storage, drain, and drainage mat (Ballard et al. 2015). Each of these layers has its purpose, and their applicability varies with the type of NbS component (PUB 2018). Considering the design standards of SWMM Manual 5.1 (Rossman 2015), suitable cross-sections of NbS for Nagpur have been designed (Table 8.3).

8.4.3 The Sub-Catchment

N38 is one of the chronic flood hotspots out of the 127 points of submergence (Fig. 8.2b) in Nagpur (DMC-NMC 2022). It is located in *Shivankar Nagar*, a notified slum in *Kharbi* (Fig. 8.3a), near the *Bhandewadi* landfill facility has 600 households and 3000 residents (NMC 2008). The sub-catchment properties, as input for SWMM are presented in Table 8.4.

The site was visited to locate the hotspot and check its socio-economic status through housing quality, open spaces, its built character, surroundings, topography, and infrastructure status. The residents confirmed the presence of the hotspot where

Table 8.3 SWMM input for NbS techniques

Layer	Parameter	RB	RD	GR	PP	BR	RG	IT
Surface	Berm height (mm)			50		200		200
	Vegetation volume fraction			0.15		0.15	0.15	
	Surface roughness (Manning's *n*)		0.012	0.13	0.13	0.13	0.13	0.033
	Surface slope (%)		2	1	2			
Pavement	Thickness (mm)				100			
	Void ratio				0.165			
	Permeability (mm/h)				700			
Soil	Thickness (mm)			100	100	450	450	
	Porosity (volume fraction)			0.464	0.464	0.464	0.464	
	Field capacity (volume fraction)			0.31	0.31	0.31	0.31	
	Wilting point (volume fraction)			0.187	0.187	0.187	0.187	
	Conductivity (mm/h)			0.04	0.04	0.04	0.04	
	Conductivity slope			30	30	30	30	
	Suction head (mm)			8.27	8.27	8.27	8.27	
Storage	Barrel height (mm)	900	900 (depth)					
	Thickness (mm)				300	300		300
	Void ratio				0.625	0.625		0.625
	Seepage rate (mm/h)				10	10	10	10
	Open/covered	Covered						
Drain	Flow coefficient	0.165						

(continued)

Table 8.3 (continued)

Layer	Parameter	RB	RD	GR	PP	BR	RG	IT
	Flow exponent (n)	0.5						
	Drain delay (h)	72						
	Open level (mm)	850						
	Close level (mm)	800						

illegal solid waste is being traded and dumped (Fig. 8.3). Located near a concretized nullah the hotspot is low-lying with water logging up to one-meter height.

The slum has some household workshops and trade in solid waste. The roads surrounding the slum were recently constructed and thus installed with a piped drainage system, making them the extent of the sub-catchment. The slum has newly constructed and unmaintained sewer and water supply lines; however, a stormwater line is not installed. According to the residents, the unmaintained sewer line is polluting the water supply. This process is frequent during the rainy season, and the residents mentioned diseases and deaths yearly due to polluted water supply.

Before the construction of the stormwater pipelines, the cement roads also used to be heavily flooded. The plinth of the slum residences ranged from 7 to 45 cm, and the housing is a mix of temporary, semi-permanent, and permanent structures. The sub-catchment is densely packed with no open spaces except a few empty plots. For recreation, two community halls and a dargah are present in the slum and a temple is in the sub-catchment.

8.5 Proposed NbS Design in the Sub-Catchment

As the study focuses on source and site control techniques for the sub-catchments, a total of twenty rainfall-runoff simulations are carried out: first, without NbS; second, with selected seven individual techniques; third, twelve possible (dual) combinations of the source control to the site control techniques. The simulation results without NbS intervention are displayed in Table 8.5, and the size and number of proposed interventions are in Table 8.6.

The slum in the sub-catchment occupies an area of 33,638 m^2 (38.18%) and is characterized by temporary structures and no open spaces. Thus, only RDs can be installed in 50% of the slum households to divert the stormwater to the nearby open spaces/nullahs, but up to 85% in other households. RBs and GRs were assumed possible in 50% of the outside slum households. Certain identified unused/poorly used land parcels distributed on-site have been proposed with three options: PP, BR,

Fig. 8.3 Site photographs; **a** sub-catchment boundary with slum, **b** landcover, **c** contour

Table 8.4 Sub-catchment properties

Area	M^2	%
Built footprint (IP)	53670	69
Un-built plots (50%P + 50%IP)	8610	11
Road (IP)	15573	20
Total sub-catchment	77854	100
Impervious (IP) area	73548	94.47
Pervious (P) area	4305	5.53
Width	417 m	
Slope	0.010	
Manning's *n* (Impervious)	0.012	
Manning's *n* (Pervious)	0.13	
Impervious storage depth (Dstore)	0.05	
Pervious storage depth (Dstore)	0.3	
Zero-impervious	25%	
Subarea routing	100% to Outlet	
Infiltration data	Horton	

Table 8.5 Rainfall-runoff simulation results without NbS

RP (years)	Total		Runoff				
	Precipitation (mm)	Infiltration (mm)	Impervious (mm)	Pervious (mm)	Total (mm)	Peak (CMS)	Coefficient
100	237.86	0.21	201.24	14.52	215.76	1.83	0.907
25	190.92		159.85	11.52	171.52	1.42	0.898
10	159.23		132.03	9.43	141.52	1.15	0.889
5	134.03		110.06	7.90	117.96	0.93	0.880
2	110.17		89.37	6.39	95.76	0.73	0.869

and IT (Fig. 8.4a, Table 8.5), especially in the government-owned vacant areas. RGs have been proposed land parcels as an amenity (Fig. 8.4b).

8.6 Simulation Results and Discussion

RDs are the best-performing NbS for flood mitigation in the sub-catchment (Fig. 8.5). It reduces the total runoff by 57.1% and peak runoff by 46.6%-49.7%, across all RPs. It is due to their broad applicability to most residential areas' rooftops even if they have semi-permanent and temporary construction. RDs reduce the infiltration rate by 19.35%. This technique involves redirecting stormwater from rooftops to

Table 8.6 Proposed interventions

Component	RB	RD (average rooftop area)		GR (covering 50% of rooftop area)	PPR/BR/IT	RG
Area of each unit (m^2)	0.2917 (others)	45 (slum)	100 (others)	50 (others)	4453	300
No. of units	70	260	120	70	1	6
	50% rooftops	50% rooftops	85% rooftops	50% rooftops	Open spaces/ vacant land	Open spaces
% of impervious area treated	8.52	16.43	17.16	4.26	2.86	1.16
% of pervious areas treated					3.03	1.22

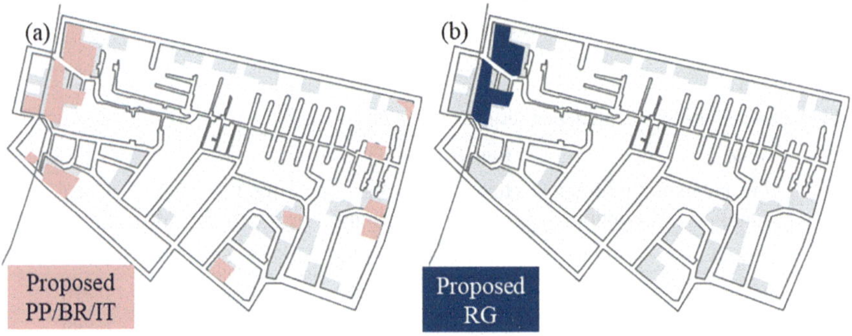

Fig. 8.4 Proposal for: **a** PP/BR/IT. **b** RG

designated discharge or collection areas via trunk lines, requiring open spaces or specially designed infiltration zones for effective water management (Liu et al. 2021).

RDs can be installed quickly and inexpensively (Liu et al. 2021) on permanent and semi-permanent roofs. RDs only need a downpipe with a strainer for stormwater that connects directly to a trunk line and discharges into the open spaces (Liu et al. 2021). To implement the installations on site, NMC can provide material cost while the slum dwellers (Parkinson et al. 2007) do guided labour work (Haji 2021). This will ensure community participation (Singkran and Kandasamy 2016) and ownership of the new infrastructure (Diep et al. 2022). Similarly, a permanent roof construction scheme can be proposed for residences with temporary roofs, ultimately increasing the number of RDs. The reduced infiltration with RDs is due to stormwater collection in the community/open spaces. If the collected water is stored in a retention tank/ pond, it can reduce non-potable water demand (Ahiablame et al. 2013). A detailed

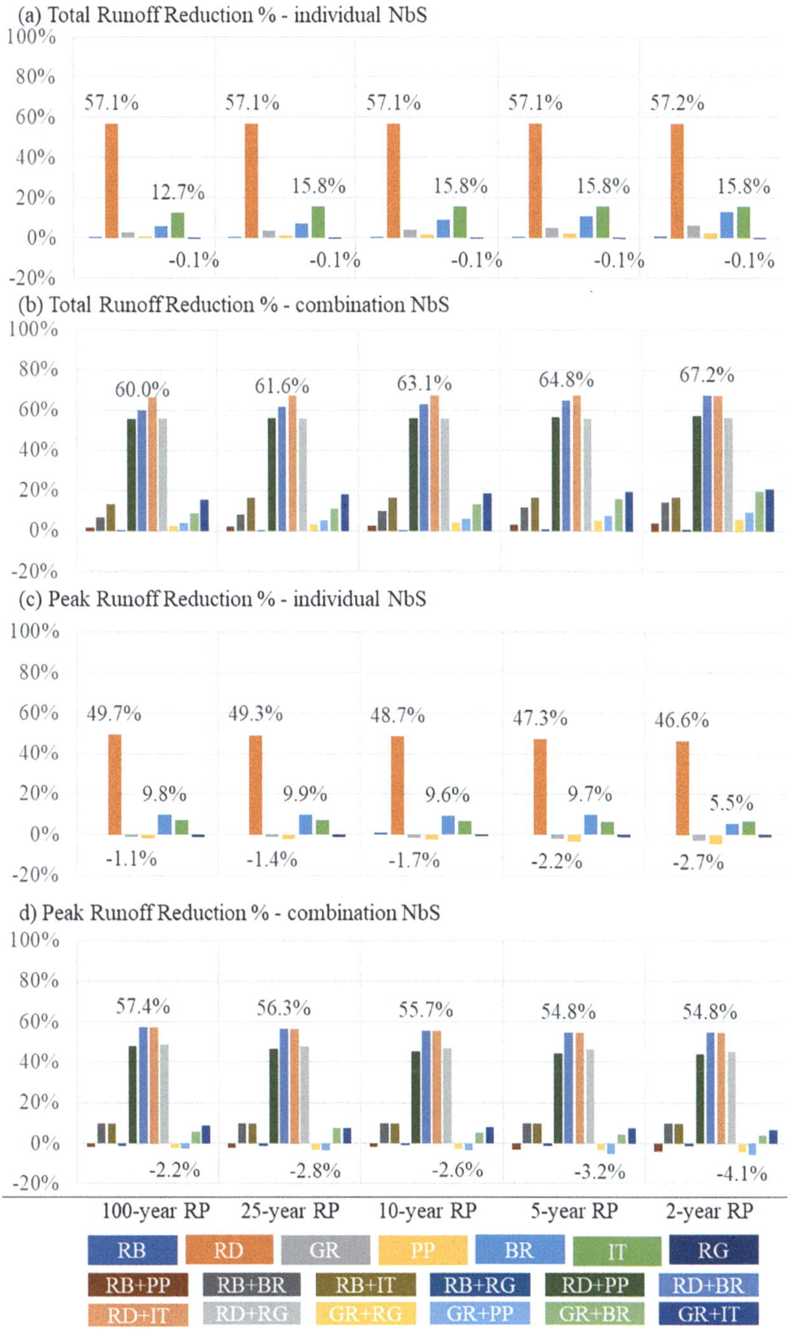

Fig. 8.5 Runoff simulation results

cost–benefit analysis (Liu et al. 2021) and engineering design can support this study. As RDs are installed on impervious rooftop areas, they are the best option for slums.

Infiltration Trenches (Basins) reduce the total runoff by 12.7–15.8% and peak runoff by 5.5–9.9% for all RPs, making it the second-best performing NbS. BRs are the next best performing NbS for total runoff reduction of 5.9–13.1% but are unsuitable for flood mitigation as they increase peak runoff by 5.5–9.9%. The IT and BR increase the infiltration rate by 94.17% and 93% respectively.

IBs can be implemented in large spaces, while smaller areas, such as roadsides, can accommodate Its (CIRIA 2015a). The additional benefits of ITs include reduced runoff volume, improved water filtration, and enhanced porosity of vacant lands and open spaces. A high groundwater table and low hydraulic conductivity in the soil primarily impact the effectiveness of the Its (Liu et al. 2021). In several other studies conducted on the suitability of SuDS for pluvial flooding, IT has performed well (Guptha et al. 2022). The cost of ITs is 40% higher than conventional drainages (Liu et al. 2021); however, they are easy to construct and require low maintenance (CIRIA 2015a).

BRs operate by capturing stormwater and retaining it for a period using filter media, effectively reducing runoff volumes (CIRIA 2015b). They offer additional benefits, including reduced runoff volumes, improved water quality, reduced air pollution, and increased biodiversity (Liu et al. 2021). Although bioretention cells are more expensive to install, they provide the highest return on investment (Liu et al. 2021). The greenery in BRs helps cool the environment and increase humidity during hot days, mitigating urban heating (Liu et al. 2021). A study in Maryland showed that BRs achieved significant reductions in flow peaks, ranging from 44 to 63% (Liu et al. 2021), unlike the N38 sub-catchment.

These solutions also filter water before recharging the groundwater. A combination of IT and BR can also be worked out for a combination of hardscapes and softscapes in open spaces of the sub-catchment, simulated, and the best results can be adopted. This will enhance the quality of open spaces in the locality (Jayasooriya et al. 2020).

RB, GR, and PP have deficient performance with 0.2–0.6% total runoff. While the peak runoff reduction for RB is 0.2%, interestingly, peak runoff has increased by 1.8% for GRs and 2.7% for PP. Thus, RB, GR, and PP may not be suitable for flood mitigation in Nagpur. However, variations in the cross-section design of NbS may result in different results (Kazak et al. 2018). The RBs with water storage reduce dependency on piped water supply. GRs also benefit from home gardens for residences with little/no open space (Webber et al. 2019) and cool roofs (PUB 2018). PPs can be provided in planned open/semi-open recreational and parking spaces (Rohilla et al. 2017; Webber et al. 2019). RBs, GRs, and PPs are the most popular NbS in temperate climates and seem unsuitable for sub-tropical regions.

RGs increase the total and peak runoff by 0.5 and 1.2%. This could be due to their design cross-section or location in pervious areas only for N38. Further simulations will be needed to check the effectiveness of the non-performing NbS. Though functionally similar to BRs, RGs have performed less due to the size and location of the intervention in pervious areas only in the sub-catchment. Considering the recreation and biodiversity benefits of RGs, this study has planned RGs in a much smaller way than the other site control measures.

Due to the performance of RDs, its combinations, i.e., RD + PP, RD + BR, RD + IT, and RD + RG, have runoff reductions of 55.7%, 58.8%, and 50.3%, respectively. Other than RD + IT, IT, RB + IT, and GR + IT combinations reduce total runoff by 10.2% and 10.4%, respectively. The remaining combinations of BR, RB + BR, and GR + BR reduce runoff by 5.9% and 6.3%, respectively. The combinations of RB + PP, GR + RG, and GR + PP have a negligible reduction in the total runoff, and RB + RG increases the total runoff, like other RGs. RDs can discharge the collected water in the large land parcels near nallah, which are currently used for solid waste trading and can be converted into a planned recreational open space or retention pond.

The community should be involved in planning and designing the system (Mguni et al. 2016) to ensure the best results in the O&M stages. Local NGOs and NMC can jointly take the NbS project in these areas, and their involvement can lead to political support for the NbS project. Piped drainage will not be possible in the narrow roads of the slum; thus, the roads can act as conveyors for sustainable nature-based drainage solutions. NbS will also improve the condition of the water supply, sewage, and solid waste management infrastructure in the slum.

8.7 Conclusion

In conclusion, the chapter underscores the urgent need for implementing NbS in flood-prone slum areas of developing countries like India, focusing on the case study of Nagpur. With floods becoming increasingly frequent and severe due to unplanned urbanization and climate change, slum populations are disproportionately affected, facing heightened risks and vulnerabilities.

It highlights the need for NbS interventions tailored to the unique challenges of developing countries. While NbS has been extensively studied and applied in developed nations, its implementation in developing countries, particularly in slum areas, remains relatively nascent. It explores the challenges inherent in designing and implementing NbS in slums, ranging from locational settings and population density to socio-economic factors and governance issues highlighting the importance of context-specific solutions.

The hotspots in Nagpur illustrate the pressing need and suitability of NbS, given its history of flooding and proliferation of slum settlements. By analyzing selection criteria for NbS components and conducting simulations using the SWMM, the chapter identifies feasible NbS solutions for slum areas, including rooftop disconnection, infiltration trenches, and bio-retentions for flood mitigation.

Simulation results demonstrate the effectiveness of specific NbS components, particularly rooftop disconnection, a source control measure, in reducing total and peak runoff by 57.1% and 46.6% respectively re-emphasizing the importance of NbS in slums. Other performing NbS are infiltration trenches (basins) which can be installed in the open spaces and along roads. Bioretention cells also be installed in open spaces but need more road widths for installation, they can be installed in the peripheral roads of the sub-catchment. A combination of RDs and ITs in any slum area can prove beneficial. They can be maintained with the help of communities, similar to the case of Slovo Park slum, South Africa. Although retention and detention areas are not simulated in this study, their importance cannot be neglected, especially for a holistic planning of sustainable water management solutions.

The design of NbS can be modified and tested at other slums of Nagpur, and cities of similar geo-climatic zones in India, such as Indore, Kolhapur, Akola, Amravati, Raipur, etc., and in other developing countries. Implementing NbS in slums has many challenges, and the best possible solutions must be tested on the ground and in other case studies with similar contexts. Overall, the imperative of adopting NbS is a holistic approach to urban water management in flood-prone slum areas of developing countries. It advocates for context-specific solutions that address such settings' socio-economic, environmental, and governance challenges. By integrating NbS into urban planning and development strategies, flood resilience cities can be enhanced while promoting sustainable and inclusive growth.

References

Adegun OB (2014) Coping with stormwater in a Johannesburg, South Africa informal settlement. Proc Inst Civ Eng Munic Eng 167:89–98. https://doi.org/10.1680/muen.13.00030

Adegun OB (2018) When green is grievous: downsides in human-nature interactions in informal urban settlements. J Urban 11:347–361. https://doi.org/10.1080/17549175.2018.1470102

Adegun OB (2019) Green infrastructure in informal unplanned settlements: the case of Kya Sands, Johannesburg. Int J Urban Sustain Dev 11:68–80. https://doi.org/10.1080/19463138.2019.1565412

Ahiablame LM, Engel BA, Chaubey I (2013) Effectiveness of low impact development practices in two urbanized watersheds: retrofitting with rain barrel/cistern and porous pavement. J Environ Manag 119:151–161. https://doi.org/10.1016/j.jenvman.2013.01.019

Andjelkovic I (2001) Guidelines non-structural measures in urban flood management. UNESCO

Arimah BC (2011) The face of urban poverty: explaining the prevalence of slums in developing countries. Urban Dev Multidiscip Perspect. https://doi.org/10.1093/acprof:oso/9780199590148.003.0008

Ballard BW, Wilson S, Udale-Clarke H et al (2015) The SuDS manual. CIRIA, London, UK

Charlesworth S, Mezue M (2017) In: Charlesworth S, Booth C (eds) Sustainable surface water management: a handbook for SUDS, Wiley, pp 301–314

Chen SS, Tsang DCW, He M et al (2021) Designing sustainable drainage systems in subtropical cities: challenges and opportunities. J Clean Prod 280:124418. https://doi.org/10.1016/j.jclepro.2020.124418

Chimankar DA (2016) Urbanization and condition of Urban Slums in India. Indones J Geogr 48:28–38. https://doi.org/10.22146/ijg.12466

CIA (2019) The World Factbook, *Central Intelligence Agency (CIA)*, Field Listing - Literacy

CIRIA (2015a) Infiltration systems. In: CIRIA SuDS manual 2015. CIRIA, London, UK, pp 256–269

CIRIA (2015b) Bioretention systems. In: CIRIA SuDS manual 2015. London, UK, pp 332–359

Coleman S, Hurley S, Rizzo D et al (2018) From the household to watershed: a cross-scale analysis of residential intention to adopt green stormwater infrastructure. Landsc Urban Plan 180:195–206. https://doi.org/10.1016/j.landurbplan.2018.09.005

CPHEEO (2019) Manual on storm water drainage systems. Volume I. Part A: engineering Design. New Delhi

CRED (2021) Disaster year in review 2020 global trends and perspectives

Deshkar S (2019) Resilience perspective for planning urban water infrastructures: a case of Nagpur City. In: Ray B, Shaw R (eds) Urban drought, disaster risk reduction. Springer Nature Singapore Pte Ltd, Singapore, pp 131–154

Diep L, Mulligan J, Oloo MA et al (2022) Co-building trust in urban nature: learning from participatory design and construction of nature-based solutions in informal settlements in East Africa. Front Sustain Cities 4. https://doi.org/10.3389/frsc.2022.927723

DMC-NMC (2022) Disaster managment plan 2022–23. Nagpur Municipal Corporation, Nagpur

dos Santos MFN, Barbassa AP, Vasconcelos AF, Ometto AR (2021) Stormwater management for highly urbanized areas in the tropics: life cycle assessment of low impact development practices. J Hydrol 598:1–11. https://doi.org/10.1016/j.jhydrol.2021.126409

du Toit MJ, Cilliers SS, Dallimer M et al (2018) Urban green infrastructure and ecosystem services in sub-Saharan Africa. Landsc Urban Plan 180:249–261. https://doi.org/10.1016/j.landurbplan.2018.06.001

Fletcher TD, Shuster W, Hunt WF et al (2015) SUDS, LID, BMPs, WSUD and more—The evolution and application of terminology surrounding urban drainage. Urban Water J 12:525–542. https://doi.org/10.1080/1573062X.2014.916314

Gaurkhede N, Adane VS, Khonde S (2021) Identification of interruptions in urban drainage systems and their sustainable solutions for alleviating flood risk in Mumbai, an Indian Megacity. J Integr Disaster Risk Manag 11:108–130. https://doi.org/10.5595/001c.30705

Gaurkhede N, Adane VS (2023) Flood susceptibility analysis using freely available data, gis, and frequency ratio model for Nagpur, India. Appl Ecol Environ Res 21:2341–2361. https://doi.org/10.15666/aeer/2103_23412361

Gaurkhede N, Adane VS (2024) Sustainable drainage solutions for flood mitigation in developing tropical/sub-tropical regions through rainfall-runoff modelling: a case of Nagpur, India. Appl Ecol Environ Res 22:2721–2748

Guptha GC, Swain S, Al-Ansari N et al (2022) Assessing the role of SuDS in resilience enhancement of urban drainage system: a case study of Gurugram City. India. Urban Clim 41:101075. https://doi.org/10.1016/j.uclim.2021.101075

Haji HKS (2021) Nature-based solutions for climate change adaptation and sustainable urban drainage in tropical coastal cities: a case FOR Mombasa City. Graduate School of Engineering and Sciences of İzmir Institute of Technology

Jayasooriya VM, Ng AWM, Muthukumaran S, Perera CBJ (2020) Optimization of green infrastructure practices in industrial areas for runoff management: a review on issues, challenges and opportunities. Water (Switzerland) 12:1–21. https://doi.org/10.3390/W12041024

Jorgenson AK, Rice J (2016) Slum prevalence and health in developing countries: sustainable development challenges in the urban context. Sustain Dev 24:53–63. https://doi.org/10.1002/sd.1606

Kazak JK, Chruściński J, Szewrański S (2018) The development of a novel decision support system for the location of green infrastructure for stormwater management. Sustainability 10. https://doi.org/10.3390/su10124388

Kharisma D, Kaswanto (2021) Management of home garden in formal and informal settlement along riparian Ciliwung river in Bogor. IOP Conf Ser Earth Environ Sci 694. https://doi.org/10.1088/1755-1315/694/1/012021

Lähde E, Khadka A, Tahvonen O, Kokkonen T (2019) Can we really have it all? Designing multi-functionality with sustainable urban drainage system elements. Sustainability 11. https://doi.org/10.3390/su11071854

Lashford C, Rubinato M, Cai Y et al (2019) SuDS & Sponge cities: a comparative analysis of the implementation of Pluvial Flood Management in the UK and China. Sustainability 11:1–14. https://doi.org/10.3390/su11010213

Li C, Peng C, Chiang PC et al (2019) Mechanisms and applications of green infrastructure practices for stormwater control: a review. J Hydrol 568:626–637. https://doi.org/10.1016/j.jhydrol.2018.10.074

Liu T, Lawluvy Y, Shi Y, Yap PS (2021) low impact development (LID) practices: a review on recent developments, challenges and prospects. Springer International Publishing

Martin-Mikle CJ, de Beurs KM, Julian JP, Mayer PM (2015) Identifying priority sites for low impact development (LID) in a mixed-use watershed. Landsc Urban Plan 140:29–41. https://doi.org/10.1016/j.landurbplan.2015.04.002

McClymont K, Fernandes Cunha DG, Maidment C et al (2020) Towards urban resilience through sustainable drainage systems: a multi-objective optimisation problem. J Environ Manag 275. https://doi.org/10.1016/j.jenvman.2020.111173

Melville-Shreeve P, Cotterill S, Grant L et al (2017) State of SuDS delivery in the United Kingdom. Water Environ J 00:1–8. https://doi.org/10.1111/wej.12283

Mguni P, Herslund L, Jensen MB (2016) Sustainable urban drainage systems: examining the potential for green infrastructure-based stormwater management for sub-Saharan cities. Nat Hazards 82:241–257. https://doi.org/10.1007/s11069-016-2309-x

MoUD (2015) City Development Plan for Nagpur, 2041 Ministry of Urban Development, Government of India, Capacity Building for Urban Development. Nagpur

Mulligan J, Bukachi V, Clause JC et al (2020) Hybrid infrastructures, hybrid governance: new evidence from Nairobi (Kenya) on green-blue-grey infrastructure in informal settlements: "Urban hydroclimatic risks in the 21st century: integrating engineering, natural, physical and social sciences to build. Anthropocene 29:100227. https://doi.org/10.1016/j.ancene.2019.100227

NDMA (2010) National Disaster Management Guidelines: Management of Urban Flooding. New Delhi

Nie L, Jia H (2018) China's Sponge city construction: ambition and challenges—Vannforeningen. Vann 2:159–167

NMC (2008) Slum Atlas of Nagpur 2008. Nag

Ortega AD, Rodríguez JP, Bharati L (2023) Building flood-resilient cities by promoting SUDS adoption: a multi-sector analysis of barriers and benefits in Bogotá, Colombia. Int J Disaster Risk Reduct 88:103621. https://doi.org/10.1016/j.ijdrr.2023.103621

Parkinson J, Tayler K, Mark O (2007) Planning and design of urban drainage systems in informal settlements in developing countries. Urban Water J 4:137–149. https://doi.org/10.1080/15730620701464224

PUB (2018) ABC waters programme, Singapore. Singapore

Radcliffe JC (2018) Australia's water sensitive urban design. In: The 2018 international Sponge City conference on 15

Riley LW, Ko AI, Unger A, Reis MG (2007) Slum health: diseases of neglected populations. BMC Int Health Hum Rights 7:1–6. https://doi.org/10.1186/1472-698X-7-2

Rohilla SK, Matto M, Jainer S, Sharda C (2017) Water-sensitive urban design and planning: a practitioner's guide. New Delhi

Rossman LA (2015) storm water management model user's manual version 5.1. Cincinnati

Roy M, Shemdoe R, Hulme D et al (2018) Climate change and declining levels of green structures: life in informal settlements of Dar es Salaam, Tanzania. Landsc Urban Plan 180:282–293. https://doi.org/10.1016/j.landurbplan.2017.11.011

Sagala S, Murwindarti A, Avila BE et al (2022) Sustainable urban drainage system (SUDS) as nature based solutions approach for flood risk management in high-density urban settlement. IOP Conf Ser Earth Environ Sci 986. https://doi.org/10.1088/1755-1315/986/1/012055

Sawhney U (2013) Slum population in India: extent and policy response. Int J Res Bus Soc Sci 2:47. https://doi.org/10.20525/ijrbs.v2i1.62

Singkran N, Kandasamy J (2016) Developing a strategic flood risk management framework for Bangkok, Thailand. Nat Hazards 84:933–957. https://doi.org/10.1007/s11069-016-2467-x
Sinha BRK, Nishad P, Nath B (2011) Characteristics of slum population in India. Natl Inst Heal 7:384–392
Sowińska-Świerkosz B, García J (2022) What are nature-based solutions (NBS)? Setting core ideas for concept clarification. Nat-Based Solut 2:100009. https://doi.org/10.1016/j.nbsj.2022.100009
Stovin VR, Swan AD (2007) Retrofit SuDS-cost estimates and decision-support tools. Proc Inst Civ Eng Water Manag 160:207–214. https://doi.org/10.1680/wama.2007.160.4.207
TCPO-MoUD (2015) Urban and regional development plans forumlation and implemnetation (URDPFI) guidelines volume I, New Delhi
Thorne CR, Lawson EC, Ozawa C et al (2018) Overcoming uncertainty and barriers to adoption of Blue-Green infrastructure for urban flood risk management. J Flood Risk Manag 11:S960–S972. https://doi.org/10.1111/jfr3.12218
Tucci CEM (2001) Urban drainage issues in developing countries. Urban Drain Humid Trop 40:23–40
UN-Habitat (2023) Climate resilience nature-based solutions to build climate resilience in informal areas 70
Vasconcelos AF, Barbassa AP (2021) Sustainable urban stormwater management in developing countries: integrating strategies to overcome Brazilian barriers. Urban Water J 00:1–16. https://doi.org/10.1080/1573062X.2021.1969415
Vasconcelos AF, Barbassa AP, Santos MFN dos, Imani2d M (2022) Barriers to sustainable urban stormwater management in developing countries: the case of Brazil. Land Policy 112. https://doi.org/10.31224/osf.io/9e5sy
Webber JL, Fu G, Butler D (2019) Comparing cost-effectiveness of surface water flood management interventions in a UK catchment. J Flood Risk Manag 12:1–12. https://doi.org/10.1111/jfr3.12523
Wolff E, Rauf HA, Hamel P (2023) Nature-based solutions in informal settlements: a systematic review of projects in Southeast Asian and Pacific countries. Environ Sci Policy 145:275–285. https://doi.org/10.1016/j.envsci.2023.04.014

Chapter 9
Quantifying the Cooling Potential of Nature-Based Solutions to Mitigate the Urban Heat Island: A Case Study of the City of Bhopal

Aishwarya Dwivedi and Rajat Soni

Abstract The Urban Heat Island (UHI) phenomenon is one of the most prominent climatic characteristics of cities around the world. This affects the public health and well-being, socio-economic activities, and ecological balance of the urban area. Nature-based solutions (NbS) are key to reducing these harmful impacts of the UHI. NbS help in limiting the thermal gains of built surfaces and reducing the surrounding temperature by creating a 'cooling effect'. The present study focuses on quantifying this cooling effect of NbS and its mitigation potential. Urban Green Areas (UGA) are taken as the NbS measure under study. 84 UGA patches are identified for this, taking the urban development region of Bhopal city as a case study. The cooling effect for the UGA is calculated by identifying the Cooling Effect Distance (CED) and Cooling Effect Intensity (CEI). Further, the CED and CEI calculated are compared for different landscape characteristics of UGA patches i.e. size, shape, and vegetation features. This gives the influence of landscape characteristics on the cooling effect of UGA. The relationship between variables is obtained using correlation analysis and scatter plots. The study identifies the perimeter of UGA as the most significant factor influencing the cooling effect. The present study, in this way, forms a quantitative basis for the application of NbS for reducing the UHI effect. The outcome of the study helps in identifying the optimum size, shape, vegetation and location of respective NbS in an urban area. Such quantitative evaluation also helps in assessing the preparedness of an urban area concerning increased temperature and gives way to developing a more resilient urban form.

Keywords Nature-based solutions · Urban Heat Island · Urban Green Areas · Cooling effect · Landscape characteristics

A. Dwivedi (✉) · R. Soni
Maulana Azad National Institute of Technology, Bhopal, India
e-mail: aishwarya.d1808@gmail.com

R. Soni
e-mail: sonirajat@gmail.com

M. Firoz C et al. (eds.), *Nature-Based Solutions for Urban and Peri-Urban Areas*, Disaster Risk Reduction, https://doi.org/10.1007/978-981-96-5933-3_9

9.1 Introduction

The urbanization level across the world has increased significantly in the past 2–3 decades and is expected to be 68% by 2050 (United Nations 2019). The urbanization process considerably transforms the land cover composition and natural environment of an area. These transformations alter the land-atmosphere interactions and form an urban microclimate distinct from its rural surroundings (Voogt and Oke 2003; Manoli et al. 2019). One of the most significant features of this urban microclimate is higher temperature as compared to non-urban areas, termed Urban Heat Island (UHI).

9.1.1 Urban Heat Island

The UHI is defined as the urban 'islands' of higher temperatures amidst cooler rural areas (Oke 1982). The UHI effect is experienced by cities across the world, making it one of the key characteristics of urban areas (Peng et al. 2021). Shepherd (2013) described UHI as the 'conclusive example of human alteration to the Earth's climate system'. The UHI is primarily caused by higher heat gain and storage capacity of urban areas (Li et al. 2013a). This increased capacity has four major factors:

a. Built surface covers absorb and emit more solar radiation as compared to natural surfaces. Because of this, a lower amount of solar radiation is reflected into the atmosphere and a higher amount is dissipated in the form of heat.
b. Vegetated surfaces regulate temperature by providing shade and evapotranspiration, which gets severely reduced in urban areas.
c. Urban morphology disrupts the natural airflow which reduces the heat convection to the atmosphere and increases urban temperature.
d. Anthropogenic activities such as industries, transportation, and air conditioning generate additional heat which adds to the atmospheric temperature intensifying UHI.

In addition to this, regional geographic features, climate zones, and weather events also act as moderating or catalytic factors for the UHI effect. The UHI effect is experienced as an atmospheric dome over an urban area (Zhang et al. 2023). The size and intensity of this dome vary with different cities. The UHI dome comprises three layers over the urban areas. The layer closest to the ground is termed Surface Urban Heat Island (SUHI) and represents the land surface temperature of urban areas (Yang et al. 2020). The SUHI is most used as the indicator to measure UHI. The layer between the surface and uppermost layer of urban climate is termed as Canopy layer UHI (TERI 2017). The canopy layer depicts the ambient air temperature closer to the ground and building. The third layer of the UHI dome is termed as Boundary layer UHI and represents the air temperature between rooftops to the height where the urban climate does not affect the atmosphere. These three layers collectively form the UHI dome and depict a quantitative value of urban thermal climate.

9.1.2 UHI Impacts

The UHI has severe impacts on the performance and efficiency of urban functions and built structures. The increase in temperature reduces the thermal comfort of the urban population and affects their health and well-being (Lowe 2016). UHI also decreases the service lifetime efficiency of built infrastructure and results in higher energy consumption and pollution levels (Magli et al. 2015; Roxon et al. 2020). As urban infrastructure elements have significant economic, social, and physical influence on their vicinity, any impact on them has a long-term consequence for urban growth potential. UHI also impacts the physical assets used within cities for economic production, the costs of raw materials and inputs to economic production, the subsequent costs to businesses, and thus output and competitiveness (Sridhar 2010). These impacts become more problematic for developing countries such as India, where rapid development and high density of population have resulted in over-stressed urban infrastructure and already degraded environmental conditions (TERI 2017). The severity of UHI becomes even more prominent considering the vulnerability of Indian cities to extreme climatic events, especially extreme heat (Mohanty and Wadhawan 2021). The vulnerability of Indian cities to extreme heat events is also a factor in consistently increasing urban population. With a higher degree of urbanization, the intensity of UHI increases along with the population's exposure to extreme temperatures. This, combined with the current global warming scenario, creates a highly vulnerable urban future for Indian cities (Tuholske et al. 2021). Such impacts of UHI on physical infrastructure, health and environmental quality of urban areas demands urgency of reducing these harmful impacts. The mitigation of the UHI effect thus becomes an integral part of future urban development. The need to mitigate UHI can be summarized into the following points:

i. *Exposure to extreme heat events*: Consistently increasing global temperature and events of extreme heat waves lead to an increase in UHI intensity and further degradation of the urban environment. Figure 9.6 shows the temperature prediction for 2100 according to different Share Socioeconomic Pathways (SSP) as shown in the Sixth Assessment Report of IPCC (2023). The predicted increase in global surface temperature ranges from 1.40 to 4.50 C, meaning a significant increase in global temperature even if the most sustainable SSP is considered.

ii. *Social and Economic Risks*: UHI leads to inefficiency and damages to city infrastructure to function, resulting in huge economic and social damages. For example, UHI over Delhi causes an increased Net Energy Consumption of 2600 (11.4%) GWh annually, as compared to the scenario is there was no LST present (Kumari et al. 2021).

iii. *A perpetual cycle of rising temperatures*: The impacts of UHI, such as increased energy consumption, air and water pollution, and reduced vegetation lead to further increases in anthropogenic heat and result in a perpetual cycle of more and more UHI effects.

iv. *Window of Opportunity*: Urban areas are projected to grow exponentially in the next 2–3 decades, creating a window of opportunity to build more resilient urban infrastructure.

9.1.3 UHI in Indian Cities

The severity of UHI becomes even more prominent considering the vulnerability of Indian cities to extreme climatic events, especially extreme heat, and drought. According to Mohanty and Wadhawan (2021), nearly 74% of all Indian districts are vulnerable to extreme climate events, representing a composite measure of exposure, sensitivity, and adaptive capacity. Figure 9.3 shows the district-level vulnerability across India, showing a composite vulnerability towards multiple hazards. The vulnerability of Indian cities to extreme heat events is also a factor in consistently increasing urban population. With a higher degree of urbanization, the intensity of UHI increases along with the population being exposed to extreme temperatures. This, combined with the current global warming scenario, creates a highly vulnerable urban future for Indian cities (Tuholske et al. 2021). The vulnerability of Indian cities towards UHI is a factor of the following points:

- A higher rate of urbanization leads to higher impervious surface and reduced natural cover, consequently, resulting in a higher degree of UHI intensity (Yao et al. 2022).
- The rate of urbanization is quite high in Indian cities, which already include 31.3% of the total country's population as of 2011.
- Given that the most valued infrastructure is usually located in cities, the economic and social costs of increased temperature are much higher in urban areas.

Indian cities experience an already overstressed infrastructure scenario, because of rapid and unplanned urbanization. This makes cities especially susceptible to extreme weather events. The overall risk in Indian cities typically is associated more with vulnerability than hazard exposure (Revi 2008). UHI impacts the physical assets used within cities for economic production, the costs of raw materials and inputs to economic production, the subsequent costs to businesses, and thus output and competitiveness (Sridhar 2010). UHI in this way becomes a major climatic stress in Indian cities and needs to be mitigated for sustainable urban growth. The present study focuses on this important aspect of UHI mitigation and presents Nature-based Solutions (NbS) as a key mitigation strategy for combating the UHI.

9.2 UHI Mitigation Through Nature-Based Solutions

NbS play a significant role in reducing the UHI intensity in an urban area. NbS enhance the natural ecosystem of surrounding areas and helps in eliminating the root cause of rising temperature in the built environment. This puts a break in the perpetual cycle of rising temperatures and uplifts the thermal comfort level of an urban area (Dwivedi and Soni 2024). Developing NbS also provides other benefits such as pollution reduction, water filtration, rainwater discharge, energy conservation, and socio-economic welfare of respective communities (Kabisch and Haase 2014). Cities with a higher proportion of green areas and natural land cover tend to be cooler as compared to cities with less green proportion (Akbari and Taha 1992). Within a city, NbS provide a cooling effect to their surroundings, creating 'cool islands' of lower temperatures amidst warmer urban areas. The intensity and range of this cooling effect varies with the types of NbS and their characteristics. The mitigation potential of different NbS interventions is given in the following section.

9.2.1 NbS Types and Mitigation Potential

The NbS interventions are majorly categorized into Increased Surface Greenery (ISG) and Increased Surface Reflectivity (ISR) for reducing UHI impact (Hayes et al. 2022). These measures include developing green areas, shading, and increasing surface albedo (Battista et al. 2021; Cortes et al. 2022; Zheng et al. 2022; Rahman et al. 2023). Several of these measures have been studied in past literature works, which give both quantitative and qualitative accounts of the extent of their mitigation potential. A summary of these measures is provided in Table 9.1. Among different NbS interventions available, 'Urban Green Areas' (UGA) provide a large temperature reduction range and other ecological benefits.

In the present study, UGA is selected as the NbS measure to mitigate UHI for the study area. UGA is defined as the network of natural and designed landscape components, including parks, gardens, urban forests, and marshlands (Ghofrani et al. 2017). UGA directly reduce the thermal gain capacity of urban areas (Aram et al. 2019; Jane Wesley and Brunsell 2019; Xi et al. 2023). UGA stops the cycle of higher thermal mass, increasing temperature, and increasing energy consumption. The cooling effect of BGI has the following major components:

i. *Evapotranspiration*: The evapotranspiration process absorbs heat from the atmosphere to convert water into water vapours and in turn cools the surrounding air (Gunawardena et al. 2017).
ii. *Reflectivity*: Vegetated areas also help in reflecting and absorbing radiant energy, limiting radiation gain by built surfaces. Part of the radiation is utilized by vegetation for photosynthesis and part is reflected into the atmosphere (Gunawardena et al., 2017).

Table 9.1 UHI mitigation measures comparison

S. No	Measure	Type	Reduction range (°C)	Other benefits	References
1	Urban green areas	ISG	0.02–5.94	Ecosystem benefits, flood relief, pollution, water recharge	Tan et al. (2021), Battista et al. (2021), Zheng et al. (2022)
2	Cool roof/facade (high albedo/ reflective coating)	ISR	0.39–3.6	Reduced energy conservation	Battista et al. (2021), Wang et al. (2022), Zheng et al. (2022)
3	Green roof/façade	ISG	0.67–2.76	Reduced energy conservation	Diren-Üstün et al. (2024), Ornam et al. (2024)
4	High albedo pavement	ISR	0.77–1.6		Calis et al. (2022)
5	Grass pavement	ISG	1.35	Reduced impervious surfaces, flood relief	Battista et al. (2021)

iii. *Surface Roughness*: Roughness in urban areas estimates the arrangement of buildings and open spaces. The presence of Blue-green areas increases the surface roughness of urban areas and increases the convection rate of heat into the atmosphere, as compared to a 'smooth' urban morphology (Ferranti et al. 2022).

9.2.2 *Quantifying the Cooling Effect of NbS*

UGA as a NbS measure, thus provide a cooling effect to its surrounding area and plays an important role in mitigating the UHI effect (Bartesaghi Koc et al. 2018). UGA help in reducing the surrounding temperature by 1–4 °C and has a cooling effect up to 300–400 m (Aram et al. 2019). However, this cooling effect is not uniform and varies greatly for different UGA elements. (Xiao et al. 2018). The major factors that influence this cooling effect are the landscape characteristics of a UGA element. These characteristics include the size, shape, and vegetation features of UGA (Zhou et al. 2011; Cheng et al. 2015; Jane Wesley and Brunsell 2019; Varentsov et al. 2023). Understanding the magnitude and reach of cooling effect for different landscape characteristics thus becomes an important step in incorporating UGA for UHI mitigation. This study aims to quantify the cooling effect of UGA in their surrounding areas and to understand its variation patterns across different landscape characteristics.

Quantification of cooling effect involves Cooling Effect Intensity (CEI) and Cooling Effect Distance (CED) for a particular UGA (Jaganmohan et al. 2016; Peng et al. 2021; Xu and Zhao 2023). The CEI is defined as the mean difference between the UGA and surrounding temperature. The CED is defined as the maximum distance

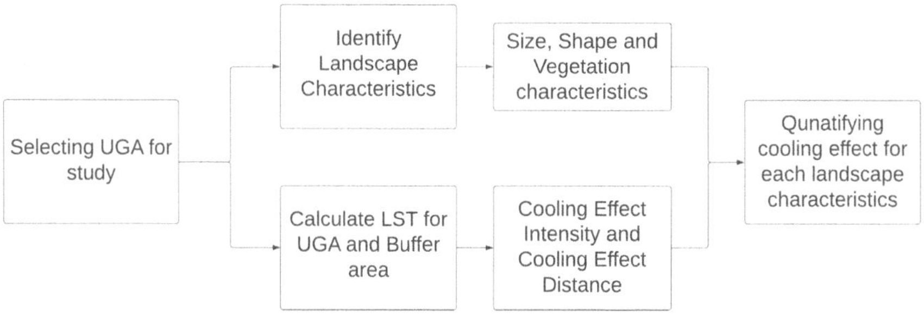

Fig. 9.1 Conceptual framework

up to which the cooling effect is observed for the UGA. Land Surface Temperature (LST) is taken as the measurement of temperature to calculate the CEI and CED. The two indicators thus give the quantitative value of the cooling effect and help in comparing different UGA. The conceptual framework for the study is provided in Fig. 9.1.

9.3 Materials and Methods

The quantification of the cooling effect is undertaken considering the city of Bhopal as a case study. The study incorporates remote sensing and GIS tools to calculate the LST and landscape metrics for UGA. The steps involved in this are given in the points below.

9.3.1 Delineating UGA and Landscape Metrics

The UGA are delineated using Google Earth Engine (Imagery dated: 15/06/2023) within the Bhopal urban planning boundary (Directorate of Town and Country Planning 2023). The UGA are selected based on the following criteria:

a. Only detached patches of green areas enclosed by built surfaces are chosen to reduce the miscalculations in the cooling effect.
b. The minimum area of the green patch: 900 m^2.
c. Minimum distance between two patches: 600 m.
d. Green areas were chosen to incorporate diverse patch sizes, shapes, locations, and surrounding built typologies.

84 distinct UGA are selected in this way, as shown in Fig. 9.2. The landscape characteristics for each UGA are calculated using ArcGIS software. Four indicators are chosen to represent the shape, size, and vegetation parameters of the UGA as shown in Table 9.2.

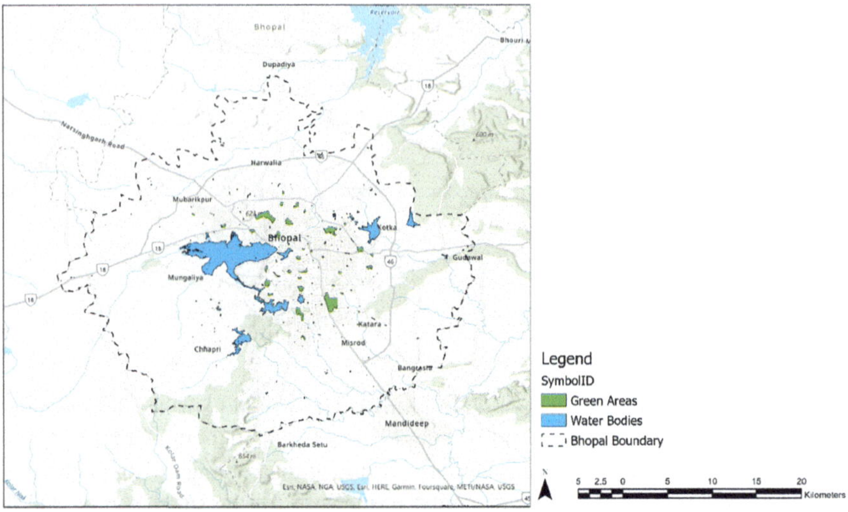

Fig. 9.2 UGA mapping for Bhopal city

Table 9.2 Landscape characteristics indicators

S. No	Landscape characteristic	Indicator	Definition/range	Unit/ formula
1	Size	Area (Xu and Zhao 2023)	Extent of UGA patch	Area of UGA in m^2
2		Perimeter (Geng et al. 2022)	Edge length and complexity of UGA patch shape	Perimeter of UGA in m
3	Shape	Landscape Shape Index (Xu and Zhao 2023)	Compactness of UGA patch (=1 for circle, >1 for other shapes)	$\frac{\text{Perimeter}}{2\sqrt{\pi}\times\text{Area}}$
4	Vegetation	Normalized difference vegetation index (NDVI) (Xu and Zhao 2023)	Type of vegetation (−1 to 1, higher values depicting higher dense green leaves)	$\frac{\text{NIR}-R}{\text{NIR}+R}$

9.3.2 Calculating and Comparing the Cooling Effect for UGA

The LST for the study area is calculated through LANDSAT 8 TIR bands (10 and 11) using the Radiative Transfer Equation (RTE) algorithm (Li et al. 2013b). The LANDSAT imagery used for preparing the LST map is procured for 09/10/2023 considering a non-cloudy sky and post-monsoon condition. The necessary atmospheric corrections are applied using ArcGIS software. The cooling effect is then calculated using the raster calculator in ArcGIS software. CED is calculated first by delineating the buffer distance at which the temperature stops increasing from the

UGA edge (Eq. 9.1). The area up to this distance is taken as the effective impact zone of UGA. The CEI is then calculated by taking the difference between the mean temperatures of the inside and buffer zone of the UGA (Eq. 9.2).

$$\text{CED} = D \text{ at which } \Delta T \text{ is negative for the first time} \tag{9.1}$$

$$\Delta T = T_{\text{Outside}} - T_{\text{Inside}}$$

$$D = \text{Distance from the edge of the park}$$

$$\text{CEI} = T_D - T_{\text{Inside}} \tag{9.2}$$

The CED and CEI thus calculated are then compared for different landscape characteristics of the UGA. A correlation matrix is performed to understand the relationship between each indicator. The CED and CEI represent the cooling effect of each UGA and are taken as the dependent variables. The area, perimeter, LSI, and NDVI represent the landscape characteristics for each UGA and are taken as the dependent variables. Further, scatter plots are created for each of the pairs of dependent and independent variables, showing the goodness of fit for the respective relationship.

9.4 Results and Discussion

The analysis performed based on the above methodology reveals the LST pattern and UGA composition of Bhopal city. The LST map obtained from the analysis is shown in Fig. 9.3. The map shows a temperature range of 19.5–43.8 °C for the urban region. Trees, forests, and waterbodies have lower LST values as compared to the areas with built surfaces, roads, barren land, and agriculture as land cover. LST map for the study area indicates a positive relationship between natural land cover and lower heat island effect, promising the significance of NbS interventions.

9.4.1 Cooling Effect of UGA

The cooling potential of NbS measures is quantified by calculating the cooling effect indicators for the identified UGA entities. The CED and CEI for the selected UGA are calculated using the above LST map. Out of 84 UGA studied, 76% have an effective cooling effect on their surroundings. The remaining 24% does not show a cooling effect and is not included in further calculations. The CED is calculated in the form of consecutive ring buffers of 30 m around the respective UGA. The calculated CED

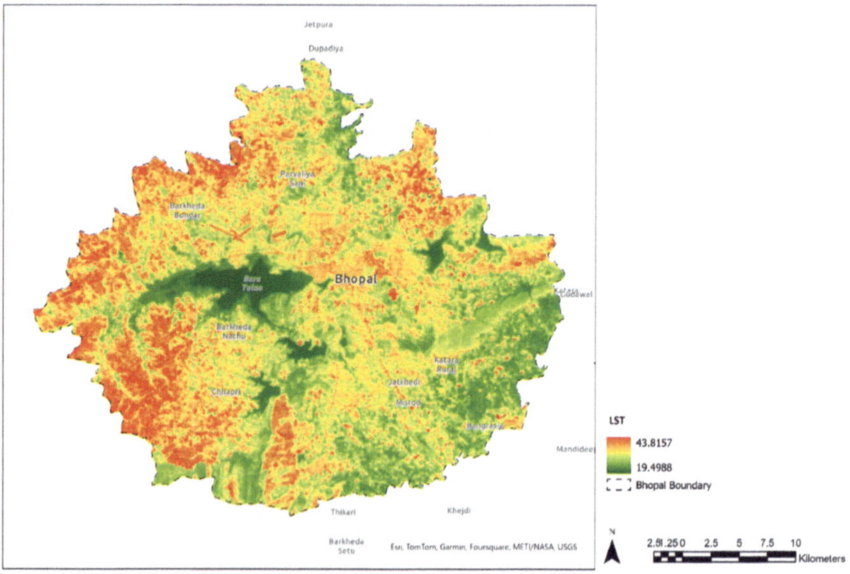

Fig. 9.3 LST map for Bhopal city

is shown in Fig. 9.4a. The CEI is calculated using the CED as the effective cooling distance of each UGA and shown in Fig. 9.4b.

The average CED for Bhopal city is 118 m from the edge of the UGA. 94% of the UGA show a CED less than 180 m and only 4 show a CED more than 180 m. Similarly, the average CEI for the UGA studied is 0.68 °C. 75% of the UGA has a CEI less than 1 °C and the number of UGA having larger CEI sharply declines further. Only 6 UGA have a CEI of more than 2 °C and one with a CEI of more than 2.5 °C. These average values indicate an average cooling potential of such NbS and provide a quantitative measure in this regard.

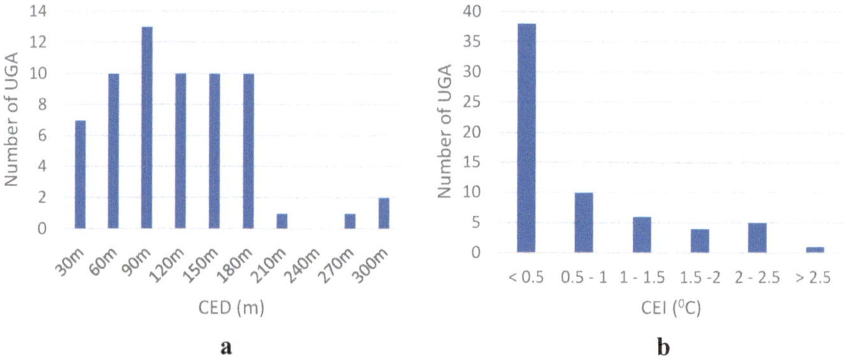

Fig. 9.4 **a** CED calculated for the selected UGA, **b** CEI calculated for the selected UGA

9.4.2 Landscape Dependence of the Cooling Effect

The landscape characteristics of the selected UGA are calculated showing the shape, size, and vegetation features of the respective areas. The average area and perimeter of the UGA are 79,289 m^2 and 957 m, respectively. 75% of the UGA has an area of less than 50,000 m^2, which corresponds to a community-level park in an urban area (Ministry of Urban Development 2015). The mean LSI for the selected UGA is 1.4, which shows that on average UGA patches have a regular and compact shape profile. The UGA has a moderate mean NDVI value of 0.28, which depicts shrubs and grasslands and corresponds to the vegetation most found in UGA. The landscape characteristics thus calculated are used to prepare a correlation matrix with the CED and CEI of each UGA (Table 9.3).

All the values of the correlation matrix show a positive correlation with respective indicators. CEI shows higher correlation values with all the landscape characteristics studied. The highest correlation is obtained for CEI and the perimeter of UGA. Area and NDVI also show a moderate positive correlation with CEI. The CED has a weak to moderate positive correlation with the landscape characteristics. CED also shows the highest correlation with perimeter, as compared with other landscape characteristics. LSI shows a very weak correlation with both CEI and CED among all landscape characteristics and depicts a negligible influence over the cooling effect. To further understand the influence of each of the landscape characteristics on the cooling effect, scatter plots for each pair of variables are shown in Figs. 9.5 and 9.6.

CEI has the highest r^2 (coefficient of determination) value of 0.6 with perimeter, which shows a statistically significant relationship. Similarly, r^2 for CEI with area and NDVI also have a moderate goodness of fit. On the contrary, CED in general has a very low r^2 value for all the indicators and shows an insignificant statistical relationship. The low r^2 shows the influence of some extraneous variables on the relationship and further study is needed to explore these variables. The multiple adjusted r^2 for CED and CEI with landscape characteristics is 0.13 and 0.71 respectively, again showing a similar statistical relationship.

Table 9.3 Correlation matrix for landscape characteristics, CED, and CEI

	Area	Perimeter	LSI	NDVI	CED	CEI
Area	1					
perimeter	0.92	1				
LSI	0.17	0.34	1			
NDVI	0.34	0.48	0.12	1		
CED	*0.37*	*0.41*	*0.06*	*0.27*	1	
CEI	*0.64*	*0.77*	*0.07*	*0.61*	0.50	1

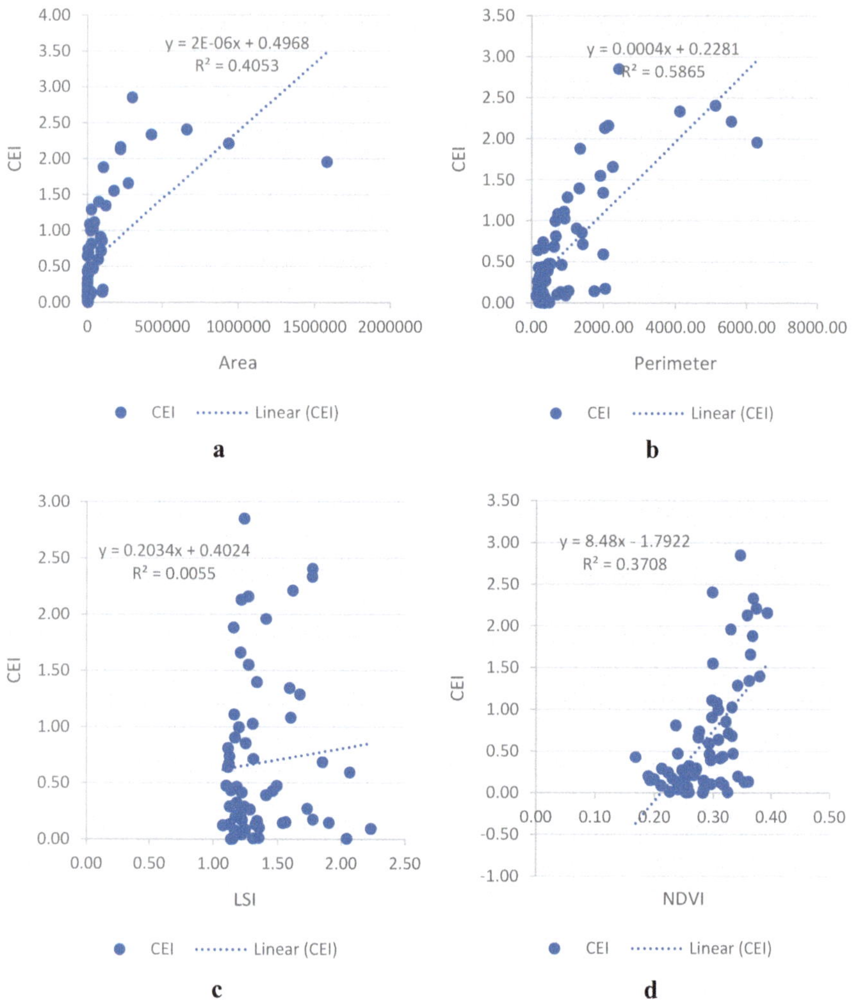

Fig. 9.5 **a** Scatter plots for CEI and area, **b** scatter plots for CEI and perimeter, **c** scatter plots for CEI and LSI, **d** scatter plots for CEI and NDVI

9.4.3 Discussions

The analysis performed thus gives a quantitative aspect to the cooling potential of NbS. The UGA studied for this shows a variety of cooling effect intensities. The majority of UGA show lower cooling effect measures than the mean values for the whole city. This is evident by the accumulation of CED and CEI values near the lowest quarter of the obtained values. The smaller cooling effect is because of the small size of the majority of UGA as compared to the urban scale. This gives a preliminary relationship between the size of the UGA and their corresponding cooling effect.

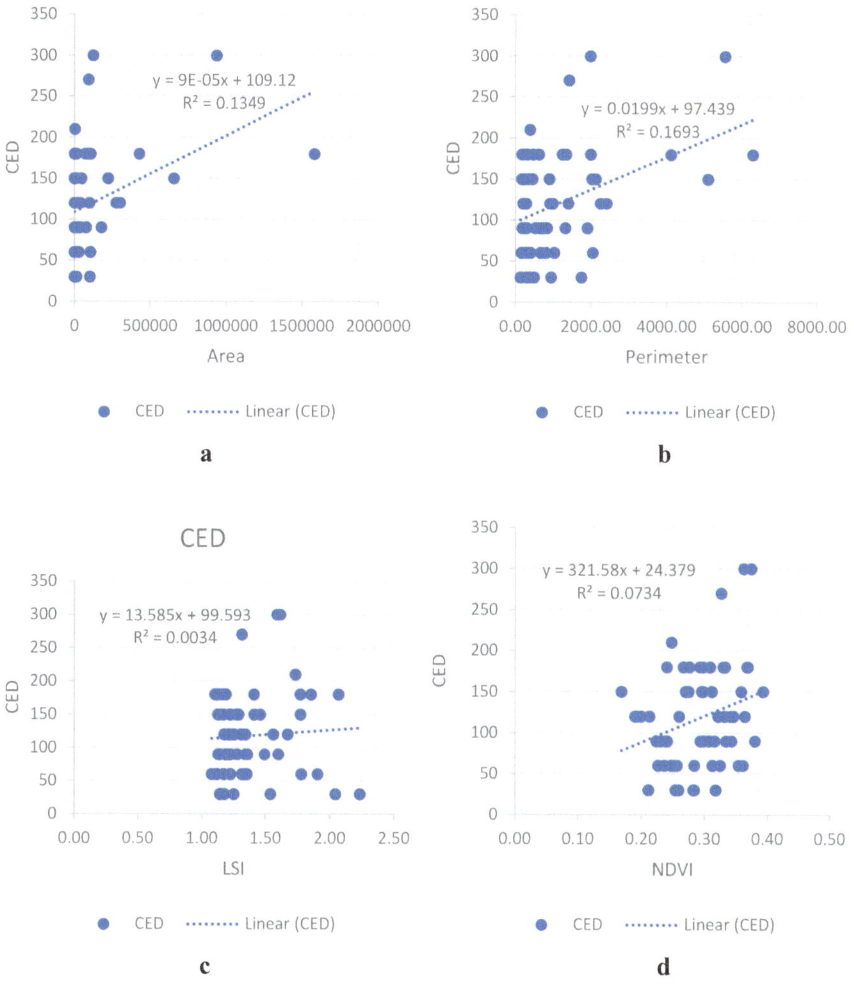

Fig. 9.6 **a** Scatter plots for CED and area, **b** scatter plots for CED and perimeter, **c** scatter plots for CED and LSI, **d** scatter plots for CED and NDVI

Further, the highest values of CEI are obtained for UGA with large patch sizes (4,63,175 m^2 average area) and large forest cover. This shows the importance spatial scale of NbS interventions for effectively combating the UHI effect.

The CED shows that the average cooling range of UGA is 150–300 m from the edge of the patch. Thus, to achieve optimum cooling effect in a city, the maximum distance between UGA should not be more than 300–600 m. Similarly, dense vegetation is required to reach an optimum cooling intensity of UGA. The comparison between CED, CEI, and the size of the park shows a need to incorporate larger UGA within the urban area to compensate for the rising UHI effect. The correlation analysis also shows a similar positive relationship between the size, shape, and vegetation of

UGA and the cooling effect. The perimeter has a larger influence on the cooling effect of UGA, which shows that the more complex shaped patches have a more intense and outreaching cooling effect. Area and NDVI of the UGA also have a moderate positive correlation with the cooling effect, with larger areas and vegetation showing larger CED and CEI values. The LSI has a very weak relationship with the cooling effect, which shows the compactness of a UGA patch has little to no effect on its cooling effect. The relationship between landscape characteristics and CEI is much more prominent than the CED. This shows that the landscape characteristics have a very limited effect on the extent to which the cooling effect is observed in the urban surroundings. Factors such as urban morphology, surrounding climate, and urban activity pattern can further explain the variation in CED for different UGA. This is further evident through the scatter plots prepared for the variables. The CEI shows 71% variations explained by the given data, while CED shows only 13% variations explained by the given data. The low r^2 for CED demonstrates a need to incorporate more variables and datasets to comprehensively explain the observed variations.

The above quantification thus illustrates how the cooling effect varies across an urban area. This helps in benchmarking the necessary amount of NbS intervention required to mitigate the UHI effect and in comparing different cities or periods. The calculated cooling effect hence becomes an important part of delineating NbS in an urban planning framework. This is especially crucial for residential areas where the impact of UHI is mostly felt. Further studies are needed to explore the available and required UGA for mitigating UHI based on the quantified cooling effect values. The development of NbS becomes even more important for developing cities such as Bhopal, where high ambient temperature has become a major climatic hazard in recent years. The rising temperature suggests the UGA existing in the city is not adequate to maintain the thermal balance required by the current urban setting. The study thus validates the significance of NbS for UHI mitigation and becomes an analytical base for including NbS in the future urban development scenario.

9.5 Conclusions

Urban Heat Island (UHI) has become a major climatic feature of modern cities that severely impacts the health and well-being of the urban population. Indian cities are especially susceptible to the UHI due to high exposure to global warming impacts and underdeveloped urban areas. One of the most significant measures to mitigate UHI is to develop a Nature-based Solution (NbS). The NbS exhibit a cooling effect on their surroundings and effectively reduce the land surface temperature. NbS provide a cooling effect through increasing land surface greening and reflectance. Surface greening has a much higher potential for UHI mitigation due to higher cooling intensity and range. The mitigation potential of Urban Green Areas (UGA) is quantified in terms of Cooling Effect Distance (CED) and Cooling Effect Intensity (CEI) for the city of Bhopal in the present study. CED and CEI are calculated using Land Surface Temperature (LST) identified through remote sensing and GIS. The CED values vary from 30 to 300 m with an average value of 118 m. Similarly, the CEI values vary from 0.003 to 2.8 °C and have an average value of 0.68 °C. The dependence of these cooling effect indicators on the size and vegetation characteristics of UGA is analysed further. The perimeter of UGA patches has the most prominent influence over the cooling effect with a correlation coefficient of 0.77 and 0.41 for CEI and CED respectively. The CEI shows a much more statistically significant relationship with the landscape characteristics and has a multiple-adjusted r^2 value of 0.71. CED has a multiple adjusted r^2 value of 0.13 and shows a major influence of extraneous variables on the relationship. The present analyses thus help in identifying the appropriate shape, size, and location of UGA for effectively mitigating the UHA effect. In general, the UGA patches with complex shapes and dense vegetation have a larger cooling effect over their surroundings. It is recommended that the distance between UGA patches should be 300–600 m to effectively reduce the UHI effect. Moreover, one large UGA is preferable to multiple small green areas to maximize cooling intensity. The present study in this way mobilizes the adoption of NbS for an urban area and recertifies the significance of NbS in sustainable development.

References

Akbari H, Taha H (1992) The impact of trees and white surfaces on residential heating and cooling energy use in four Canadian cities. Energy 17:141–149. https://doi.org/10.1016/0360-5442(92)90063-6

Aram F, Higueras García E, Solgi E, Mansournia S (2019) Urban green space cooling effect in cities. Heliyon 5:e01339. https://doi.org/10.1016/j.heliyon.2019.e01339

Bartesaghi Koc C, Osmond P, Peters A (2018) Evaluating the cooling effects of green infrastructure: a systematic review of methods, indicators and data sources. Sol Energy 166:486–508. https://doi.org/10.1016/j.solener.2018.03.008

Battista G, Roncone M, de Lieto VE (2021) Urban overheating impact: a case study on building energy performance. Appl Sci 11:8327. https://doi.org/10.3390/app11188327

Calis G, Yildizel SA, Keskin US (2022) Investigation of color pigment incorporated roller compacted high performance concrete as a mitigation tool against urban heat island. Case Stud Construct Mater 17:e01479. https://doi.org/10.1016/j.cscm.2022.e01479

Cheng X, Wei B, Chen G et al (2015) Influence of park size and its surrounding urban landscape patterns on the park cooling effect. J Urban Plan Dev 141. https://doi.org/10.1061/(ASCE)UP.1943-5444.0000256

Cortes A, Rejuso AJ, Santos JA, Blanco A (2022) Evaluating mitigation strategies for urban heat island in Mandaue City using ENVI-met. J Urban Manag 11:97–106. https://doi.org/10.1016/j.jum.2022.01.002

Directorate of Town and Country Planning G of MP (2023) Draft Bhopal Development Plan 2031: Volume 1. Bhopal

Diren-Üstün DH, Ünal YS, Bilgen Sİ et al (2024) Effects of land-use mitigation scenarios on urban heat island intensity in Istanbul. Atmos Res 297:107083. https://doi.org/10.1016/j.atmosres.2023.107083

Dwivedi A, Soni R (2024) Impacts of urban heat island effect on critical urban infrastructure: a review of studies published between 2012 and 2022. Environ Rev. https://doi.org/10.1139/er-2023-0108

Ferranti EJS, Fontana Oberling D, David Quinn A (2022) Transport resilience to weather and climate: an interdisciplinary view from Rio de Janeiro. Proc Inst Civ Eng Urban des Plan 175:103–121. https://doi.org/10.1680/jurdp.21.00006a

Geng X, Yu Z, Zhang D et al (2022) The influence of local background climate on the dominant factors and threshold-size of the cooling effect of urban parks. Sci Total Environ 823:153806. https://doi.org/10.1016/j.scitotenv.2022.153806

Ghofrani Z, Sposito V, Faggian R (2017) A comprehensive review of blue-green infrastructure concepts. Int J Environ Sustain 6. https://doi.org/10.24102/ijes.v6i1.728

Gunawardena KR, Wells MJ, Kershaw T (2017) Utilising green and bluespace to mitigate urban heat island intensity. Sci Total Environ 584–585:1040–1055. https://doi.org/10.1016/j.scitotenv.2017.01.158

Hayes A, Jandaghian Z, Lacasse M et al (2022) Nature-based solutions (NBSs) to mitigate Urban Heat Island (UHI) effects in Canadian cities. Buildings 12:925. https://doi.org/10.3390/buildings12070925

IPCC (2023) Summary for policymakers. In: Climate change 2022—Impacts, adaptation and vulnerability. Cambridge University Press, pp 3–34

Jaganmohan M, Knapp S, Buchmann CM, Schwarz N (2016) The bigger, the better? The influence of urban green space design on cooling effects for residential areas. J Environ Qual 45:134–145. https://doi.org/10.2134/jeq2015.01.0062

Jane Wesley EJ, Brunsell NA (2019) Greenspace pattern and the surface Urban Heat Island: a biophysically-based approach to investigating the effects of urban landscape configuration. Remote Sens (Basel) 11:2322. https://doi.org/10.3390/rs11192322

Kabisch N, Haase D (2014) Green justice or just green? Provision of urban green spaces in Berlin, Germany. Landsc Urban Plan 122:129–139. https://doi.org/10.1016/j.landurbplan.2013.11.016

Kumari P, Garg V, Kumar R, Kumar K (2021) Impact of urban heat island formation on energy consumption in Delhi. Urban Clim 36. https://doi.org/10.1016/j.uclim.2020.100763

Li Y, Zhu L, Zhao X et al (2013a) Urbanization impact on temperature change in china with emphasis on land cover change and human activity. J Clim 26:8765–8780. https://doi.org/10.1175/JCLI-D-12-00698.1

Li Z-L, Tang B-H, Wu H et al (2013b) Satellite-derived land surface temperature: current status and perspectives. Remote Sens Environ 131:14–37. https://doi.org/10.1016/j.rse.2012.12.008

Lowe SA (2016) An energy and mortality impact assessment of the urban heat island in the US. Environ Impact Assess Rev 56:139–144. https://doi.org/10.1016/j.eiar.2015.10.004

Magli S, Lodi C, Lombroso L et al (2015) Analysis of the urban heat island effects on building energy consumption. Int J Energy Environ Eng 6:91–99. https://doi.org/10.1007/s40095-014-0154-9

Manoli G, Fatichi S, Schläpfer M et al (2019) Magnitude of urban heat islands largely explained by climate and population. Nature 573:55–60. https://doi.org/10.1038/s41586-019-1512-9

Ministry of Urban Development (2015) Urban and Regional Development plans formulation and implementation (URDPFI) guidelines volume 1

Mohanty A, Wadhawan S (2021) A district level assessment mapping India's climate vulnerability

Oke TR (1982) The energetic basis of the urban heat island. Q J R Meteorol Soc 108:1–24. https://doi.org/10.1002/qj.49710845502

Ornam K, Wonorahardjo S, Triyadi S (2024) Several façade types for mitigating urban heat island intensity. Build Environ 248:111031. https://doi.org/10.1016/j.buildenv.2023.111031

Peng J, Dan Y, Qiao R et al (2021) How to quantify the cooling effect of urban parks? Linking maximum and accumulation perspectives. Remote Sens Environ 252:112135. https://doi.org/10.1016/j.rse.2020.112135

Rahman T, Zudhy Irawan M, Noor Tajudin A et al (2023) Knowledge mapping of cool pavement technologies for urban heat island mitigation: a systematic bibliometric analysis. Energy Build 291:113133. https://doi.org/10.1016/j.enbuild.2023.113133

Revi A (2008) Climate change risk: an adaptation and mitigation agenda for Indian cities. Environ Urban 20:207–229. https://doi.org/10.1177/0956247808089157

Roxon J, Ulm FJ, Pellenq RJM (2020) Urban heat island impact on state residential energy cost and CO2 emissions in the United States. Urban Clim 31. https://doi.org/10.1016/j.uclim.2019.100546

Shepherd JM (2013) Impacts of urbanization on precipitation and storms: physical insights and vulnerabilities. In: Climate vulnerability. Elsevier, pp 109–125

Sridhar KS (2010) Carbon emissions, climate change, and impacts in India's cities. In: India infrastructure report

Tan X, Sun X, Huang C et al (2021) Comparison of cooling effect between green space and water body. Sustain Cities Soc 67:102711. https://doi.org/10.1016/j.scs.2021.102711

TERI (2017) Final report on urban planning characteristics to mitigate climate change in context of urban heat island effect. Bangalore

Tuholske C, Caylor K, Funk C et al (2021) Global urban population exposure to extreme heat. Proc Natl Acad Sci 118. https://doi.org/10.1073/pnas.2024792118

United Nations D of E and SAPD (2019) World urbanization prospects the 2018 revision

Varentsov M, Vasenev V, Dvornikov Y et al (2023) Does size matter? Modelling the cooling effect of green infrastructures in a megacity during a heat wave. Sci Total Environ 902:165966. https://doi.org/10.1016/j.scitotenv.2023.165966

Voogt JA, Oke TR (2003) Thermal remote sensing of urban climates. Remote Sens Environ 86:370–384. https://doi.org/10.1016/S0034-4257(03)00079-8

Wang X, Li H, Sodoudi S (2022) The effectiveness of cool and green roofs in mitigating urban heat island and improving human thermal comfort. Build Environ 217:109082. https://doi.org/10.1016/j.buildenv.2022.109082

Xi C, Han L, Wang J et al (2023) How can greenery space mitigate urban heat island? An analysis of cooling effect, carbon sequestration, and nurturing cost at the street scale. J Clean Prod 419:138230. https://doi.org/10.1016/j.jclepro.2023.138230

Xiao XD, Dong L, Yan H et al (2018) The influence of the spatial characteristics of urban green space on the urban heat island effect in Suzhou Industrial Park. Sustain Cities Soc 40:428–439. https://doi.org/10.1016/j.scs.2018.04.002

Xu Z, Zhao S (2023) Scale dependence of urban green space cooling efficiency: a case study in Beijing metropolitan area. Sci Total Environ 898:165563. https://doi.org/10.1016/j.scitotenv.2023.165563

Yang Q, Huang X, Tang Q (2020) Irrigation cooling effect on land surface temperature across China based on satellite observations. Sci Total Environ 705:135984. https://doi.org/10.1016/j.scitotenv.2019.135984

Yao L, Sun S, Song C et al (2022) Recognizing surface urban heat 'island' effect and its urbanization association in terms of intensity, footprint, and capacity: a case study with multi-dimensional analysis in Northern China. J Clean Prod 372:133720. https://doi.org/10.1016/j.jclepro.2022.133720

Zhang Y, Wang X, Fan Y et al (2023) Urban heat dome flow deflected by the Coriolis force. Urban Clim 49:101449. https://doi.org/10.1016/j.uclim.2023.101449

Zheng T, Qu K, Darkwa J, Calautit JK (2022) Evaluating urban heat island mitigation strategies for a subtropical city centre (a case study in Osaka, Japan). Energy 250:123721. https://doi.org/10.1016/j.energy.2022.123721

Zhou W, Huang G, Cadenasso ML (2011) Does spatial configuration matter? Understanding the effects of land cover pattern on land surface temperature in urban landscapes. Landsc Urban Plan 102:54–63. https://doi.org/10.1016/j.landurbplan.2011.03.009

Part III
NBS—Traditional Methods and Community Driven Approaches

Chapter 10
Comprehending Drought Mitigation Probabilities Through Integrated Techniques of Using NbS and Traditional Disaster Risk Management

Integrating NbS with Traditional Disaster Risk Management Approaches

Divya S. Agarwal and Alka Bharat

Abstract The study aims to comprehend the drought conditions and provide integrated solutions that are nature based as well as traditional for the study case of Banaskantha district, Gujarat. Standardized Precipitation Index (SPI) was used to determine the duration of dry and wet conditions whereas Vegetation Condition Index (VCI) was used to determine the spatial extent of plant-based soil moisture for the study case. Integrating Nature-based Solutions (NbS) along with the traditional disaster risk management approaches and engineered solutions have been suggested through the development of scenarios. Sample-based scenario building combines four different surface roughness and soil types. A variation in surface roughness through the application of integrated solutions increases the soil moisture content whereas reduces the runoff depth. Agricultural cultivation, plantations in agricultural land, restoration of native plant species, restoration of flood plains, and revival of wetlands were assessed as surface roughness modulators. The runoff depth decreased in a range of 8–54.76% for different soil types and the soil moisture content increased in a range of 41.28–92.57%.

Keywords Comprehensive approach · Groundwater recharge potential zones · Integrated disaster risk mitigation · Nature-based solutions · Traditional disaster risk management

D. S. Agarwal (✉)
Assistant Professor (temporary), Department of Architecture and Planning, MANIT, Bhopal, M.P., India
e-mail: divyasampada@gmail.com

A. Bharat
Professor, Department of Architecture and Planning, MANIT, Bhopal, M.P., India
e-mail: alka_bharat@yahoo.com

M. Firoz C et al. (eds.), *Nature-Based Solutions for Urban and Peri-Urban Areas*, Disaster Risk Reduction, https://doi.org/10.1007/978-981-96-5933-3_10

10.1 Introduction

About two-thirds of the world's population will live in the cities by the year 2050 (Gupta and Bharat 2022a; Nações Unidas 2019), which would increase challenges in meeting the needs of the growing population (UN 2018). Urban sprawl has increased the percentage of impervious surfaces (Gupta and Bharat 2022a, b; Strohbach et al. 2019) thereby, reducing the water infiltration capacity (Chithra et al. 2015).

As per the Indian Meteorological Department (IMD), the instances of drought have increased by 57% since 1997 (Rai 2022). The endless variations ranging from extreme heat and drought to extreme rain and floods have affected 1.4 billion people (Masson-delmotte 2018). India has faced frequent drought events in the last few decades (Mishra et al. 2022). Drought leads to a rise in temperature which causes the soil temperature to rise thus, reducing the moisture content of soil (Al-Kaisi 2017; Qing et al. 2023; Zhang et al. 2020). It leads to reduced growth rates, increased stress on the vegetation, and alters the entire ecosystem (Seleiman et al. 2021).

Drought-affected areas pose a threat to food and water security (Husbands 2023; Suri and Sharma 2022). A thorough Bibliometric study was conducted using the Web of Science database, where only 7 articles were found when Nature-based solutions (NbS) AND disaster risk reduction were used as keywords. When keywords Nature-based solutions (NbS) AND drought were searched, only 1 article was found. This created a need for a study that could mitigate drought with solutions that are natural and traditional.

NbS approaches such as Ecosystem-based disaster risk reduction (Eco-DRR) and ecosystem-based adaptation (EbA) can be used as natural approaches for mitigating drought-like disasters (Matić and Karleuša 2022; Nehren et al. 2023; Renaud et al. 2016; UNDRR 2020). Over and above, traditional disaster mitigation approaches such as the construction of Tals, Virdas, rainwater harvesting, restoration of native plant species etc. (Datta 2015; Govil and Kanchan 2023; Machiwal et al. 2018; More 2021) along with the Nature-based Solutions would provide co-benefits such as promoting biodiversity, soil health, and sustainable land management, which are beyond disaster risk reduction (Agarwal et al. 2024; Robotham et al. 2022; Ommer et al. 2022; Simelton et al. 2021).

The study lays down the importance of water infiltration into the soil surface using surface roughness modulators based on Nature-based Solutions along with traditional disaster mitigation approaches to mitigate drought conditions. The study emphasizes the groundwater recharge zones, which allow excess water percolation through relatively porous, unconsolidated, or fractured materials (Fauzia et al. 2021; Hartmann 2022). These zones help maintain the flow of springs, wetlands, and rivers, which are lifelines to plant and animal species during drought conditions (The Nature Conservancy 2017).

10.2 Materials and Methods

Drought can be identified through various indicators such as deficiency in precipitation, variation in vegetation cover, the soil moisture content, and reduced infiltration capacity of the soil surface (Boogaard 2022; Ho-Hagemann et al. 2015; Kumar and Viswanadh 2017; Satheeshkumar et al. 2017; Wang et al. 2020).

10.2.1 Methodology

The study focuses on using the excess water to fulfil the deficit during drought conditions (Fig. 10.1).

Soil Conservation Service—Curve Number (SCS CN) method was used to compute runoff (Ara and Zakwan 2018; Shi and Wang 2020; Soulis 2021). Vegetation Condition Index (VCI) was used to spatially map drought zones (Baniya et al. 2019; Mupepi and Matsa 2023; Senamaw et al. 2021) whereas the Standardized precipitation index (SPI) computed the duration of wet and dry days for the study case (Cerpa Reyes et al. 2022; Saini et al. 2022; Shah et al. 2015). The soil texture and its properties were explored to understand the maximum absorption capacities

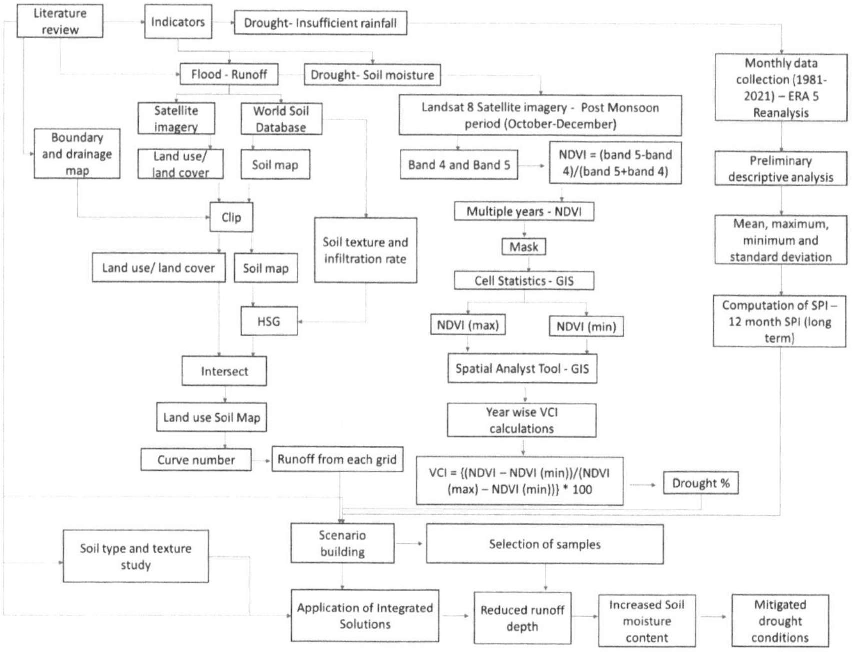

Fig. 10.1 Methodology of the study

and volumetric water content of each soil type (Guo et al. 2022; Saha et al. 2020). To mitigate drought, 'Nature-based Solutions' along with traditional disaster mitigation approaches were suggested.

10.2.2 Study Case and Data Computation

Gujarat has been affected by some of the nation's deadliest disasters (Parkash et al. 2022; Saha 1997; Singh et al. 2022). In the last seven years, there have been two floods and two drought episodes in the Banaskantha district in North Gujarat (GSDMA 2016; Prajapati and Bansal 2017). The years 2016 and 2018 (Mittal Patel 2019) were extremely prone to drought, and the years 2015 and 2017 had significant flooding because of excessive rainfall in the area and overflowing dams (GSDMA 2016). The district is vulnerable to water accumulation (Mistry and Suryanarayana 2019). According to the district's morphology, it lies in the semi-arid zone and is prone to drought (Bandyopadhyay et al. 2020; Kruti et al. 1997). The data for 2021 was chosen for the study to compute both dry and wet conditions.

Groundwater recharge potential zones were computed using weighted overlay analysis (Fig. 10.2g). DEM (Fig. 10.2c), slope (Fig. 10.2b), drainage density (Fig. 10.2a), soil (Fig. 10.2e), rainfall (Fig. 10.2f), and contour map were selected as input parameters to develop the recharge potential zones (Agarwal and Bharat 2023; Yeh et al. 2016) (Fig. 10.2E).

The digital elevation model (DEM) was developed in ArcGIS through extracted satellite imagery from USGS Aster data for 15th October 2021 (Fig. 10.2c) (Abrams et al. 2020). A supervised classification was conducted using the Landsat 8 Collection 1 (15th October 2021) satellite imagery downloaded from the USGS database—EarthExplorer (USGS 2021) to compute land use land cover map Fig. 10.2d (Alam et al. 2020). The soil map for the district was developed using Food and Agricultural Organization (FAO) World Soil data (Fig. 10.2e) (Nachtergaele et al. 2008). The World hydrological soil group was extracted from Global Hydrologic Soil Groups (HYSOGs250m) to get two major soil groups—Group B and Group C for the study case (DAAC 2018).

Monthly precipitation data for the past 40 years (1982–2021) was extracted from the ERA 5 Reanalysis database developed by the European Centre for Medium-Range Weather Forecasts (ECMWF 2018) for the district's geographical coordinates.

10.2.3 Methods Used

Different methods were incorporated to compute and map water deficit conditions for the study case.

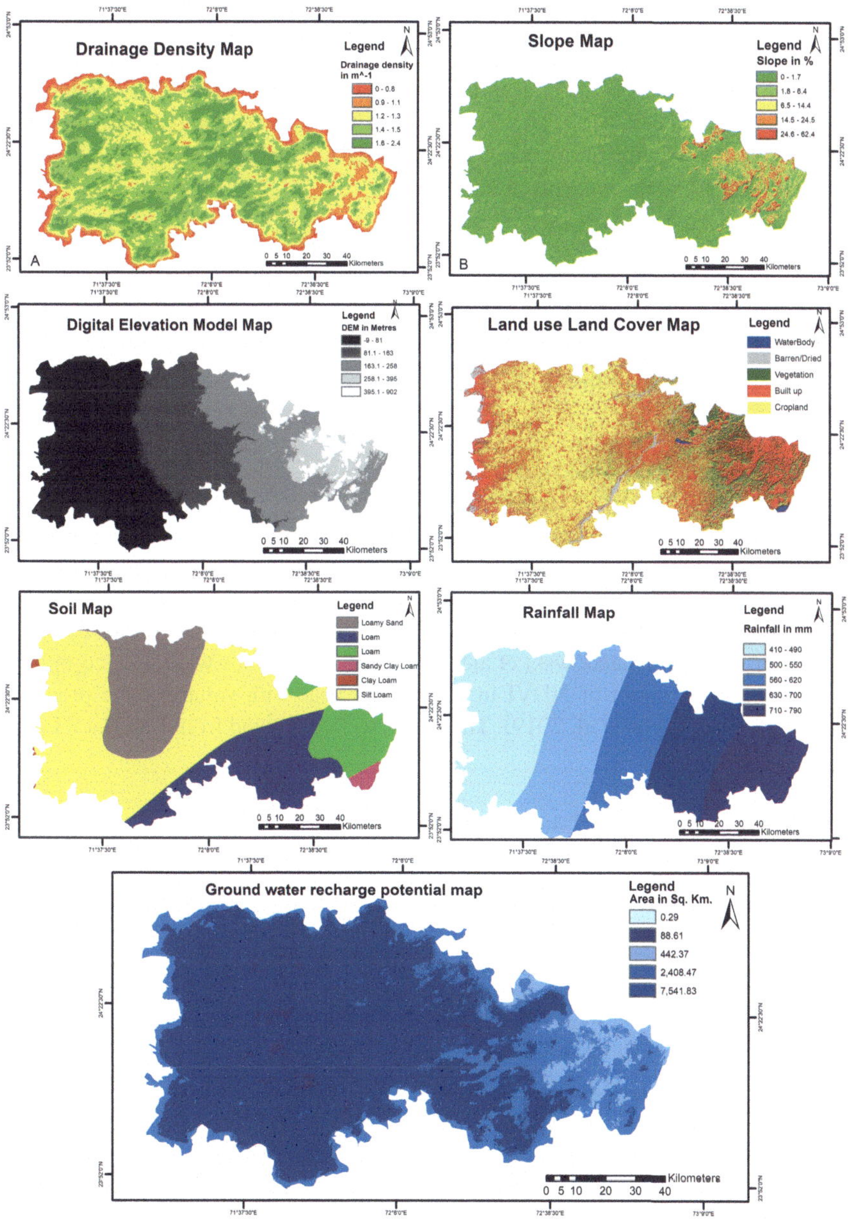

Fig. 10.2 Maps of Banaskantha district, Gujarat (Agarwal and Bharat 2023)

10.2.3.1 Soil Conservation Service–Curve Number (SCS–CN Method)

The method was used to compute runoff depth (Satheeshkumar et al. 2017; Shahadha et al. 2022; Shi and Wang 2020). It relies upon two parameters—precipitation (P) (mm) and potential maximum retention capacity (S) (mm) of the soil (Eq. 10.1) (Patel 2020; Suryakant and Pravendra 2022).

$$Q = (P - 0.2\,S)^2/(P + 0.8\,S)^2 \quad (10.1)$$

(Bartlett et al. 2016; Kumar et al. 2021a; Raju et al. 2018); where Q indicates runoff in mm, P indicates precipitation in mm and S indicates potential maximum retention (S = (1000/CN)–10).

10.2.3.2 Vegetation Condition Index (VCI)

Vegetation Condition Index (VCI) is useful for monitoring drought conditions and vegetation stress (Jiang et al. 2022; Liang et al. 2017; Mishra et al. 2022; Wolteji and Garbaba 2023). It is based on past Normalized Differential Vegetative Index (NDVI) values (Mishra et al. 2022; Un-spider org 2014). The NDVI was calculated by using Eq. 10.2 in ArcGIS (Blachowski et al. 2023). The VCI values were then computed by comparing the effect of NDVI in the current year with values of the previous years (Eq. 10.3) (Eyoh et al. 2019). The bands were extracted from Landsat 8 Level 1 satellite imagery.

$$\text{NDVI} = (\text{Band 5} - \text{Band 4})/(\text{Band 5} + \text{Band 4}) \quad (10.2)$$

$$\text{VCI} = 100 * (\text{NDVI} - \text{NDVI min})/(\text{NDVI max} - \text{NDVI min}) \quad (10.3)$$

The VCI is expressed in (%) (Baniya et al. 2019; Jalayer et al. 2023). Lower values indicate bad vegetation conditions indicating drought conditions.

10.2.3.3 Standardized Precipitation Index (SPI)

SPI characterizes the drought at various time scales (Eq. 10.4) (Almedeij 2014; Jiang et al. 2022; Moccia et al. 2022; Palchaudhuri and Biswas 2013; Zeleňáková et al. 2023). The SPI computation requires long-term precipitation data (Guenang and Francois 2014; Tsesmelis et al. 2023; Vélez-Nicolás et al. 2022).

$$\text{SPI} = (\text{xi} - \text{xm})/\text{Sx} \quad (10.4)$$

where xi is initial precipitation data, xm is the mean data, and Sx is the standard deviation.

The values range from extreme wet to extreme drought. The values below 0 depict drought (Moccia et al. 2022).

10.2.4 Soil Type and Texture

Soils can regulate the water dynamics (Adams et al. 2020; Cheng et al. 2021). They trap the runoff and then release the stored water geologically when required during dry periods in order to maintain the soil moisture content (Bhattacharjee and Behera 2017; Qin 2020; Rasheed et al. 2022; Saco et al. 2021). As the district falls under 5 soil types– loamy sand, silt loam, loam, sandy clay loam, and clay loam (Chaudhari et al. 2020), different soil textures at different locations in the district have different water capturing capacities (Dhanalakshmi et al. 2020). To capture the flowing or stagnated water into the soil profile, the area and depth of the soil (unsaturated soil) along with the maximum capacity to hold water was studied and later analysed for their maximum capacity.

10.3 Analysis

Excess water computed as a result of generation of runoff was calculated for different surface roughness (Land covers) (Feldmann et al. 2023) for the case of Banaskantha district. Drought was studied under two aspects—lack of precipitation and deficit soil moisture (Akhtar et al. 2021; Mullapudi et al. 2023). Place-based solutions were suggested to reduce the deficit due to drought.

10.3.1 Flood Analysis

66.02 mm was taken as the average annual rainfall (P) for the year 2021. Curve number (CN) values were generated using look-up table and CN map in ArcGIS. Computed S values were inputted in Eq. 10.1 to compute different runoff (Q) depth values in millimetres (mm.).

10.3.2 Drought Analysis

Vegetation Condition Index (VCI) computed the drought conditions (post-monsoon) spatially of the district for six years (2016–2021) whereas the Standardized Precipitation Index (SPI) computed the severity and duration of drought in the district for 40 years (1982–2021).

10.3.2.1 Vegetation Condition Index (VCI)

Equation 10.3 computed VCI values in a range of 0 to 100. It ranged from low drought (80–100%) to high drought (0–10%).

10.3.2.2 Standardized Precipitation Index (SPI)

Using Eq. 10.4, different values were computed for a 12-month period for 40 years. The computed SPI values ranged from extremely wet (> = 2.00) to extremely dry (< = –2.00).

10.3.3 Soil Analysis

The spatial extent of each soil was marked using polygon-based geometry in ArcGIS in square metres. Maximum capacity to hold water till saturation was calculated for each soil type using the volume of soil and its maximum volumetric water content.

10.3.4 Mitigating Solutions

Percolation of excess water into the soil surface helps in maintaining soil moisture levels, which in turn reduces the chances of drought by maintaining optimum water levels. There have been many engineered solutions that were used in the past to reduce drought situations such as check dams, retaining walls, reservoirs, rainwater harvesting, and construction of canals (Angelakıs et al. 2020; Baba and Ahmed 2018; Yimer et al. 2024), where a foreign structure was added to the ecosystem. With co-benefits for nature, Nature-based Solutions can help mitigate, adapt to, and be resilient to climate change. (IUCN 2020a; Seddon et al. 2020; UNFCCC 2020; Walters 2016).

Application of Nature-based Solutions (NbS) along with traditional disaster mitigation approaches restores the natural areas along with the ecosystem. It reduces the environmental challenges connected with rapid urban expansion and impervious surface development (Asare et al. 2023; Cohen-Shacham et al. 2016; Walters 2016). Nature-based Solutions put together natural features and processes to increase long-term human, ecological, and infrastructure resilience towards climate change and other different environmental impacts (Anderson and Gough 2022; Gonzalez-Ollauri et al. 2023). Re-vegetation of the hill slopes, increasing soil infiltration to increase water table, afforestation, development of retention ponds and wetlands, agroforestry, and revival of flood plains (Ghimire et al. 2014; IUCN 2020b; Kumar et al. 2021b; UNDRR 2021a, b) are some of the Nature-based Solutions that can be incorporated along with traditional approaches such as construction of Virdas, storage tanks, and rainwater harvesting (Ariyananda 2021; Machiwal et al. 2018) to increase the infiltration capacity of the soil surface.

10.4 Results and Discussions

Excess water as runoff (Q) ranged from 0.919 mm to 0.996 mm for different surface roughnesses. A total runoff volume of 9965.18×10^{12} mm^3 was generated for the total district area. A runoff map (Fig. 10.3a) was developed with a range of very high runoff to very low runoff for the year 2021. The Vegetation Condition Index (VCI) computed drought-prone areas in % ranging from no drought to severe drought for the year 2021 (Fig. 10.3b).

Standardized Precipitation Index (SPI) computed numerical values ranging from a maximum of 3.29 to a minimum of −1.00 for a 12-month scale for the years 1982–2021. The values below 0 indicated dry months whereas the values above 0 indicated wet months (Fig. 10.4).

The maximum capacity to hold water in the soil was found in loamy sand soil while the least was found in sandy clay loam soil (Table 10.1).

To mitigate and reduce the runoff generated and increase the infiltration, scenarios were built by selecting different samples with varied characteristics.

10.4.1 Development of Scenarios

Situation-based scenarios were built using basic physical characteristics of the study case where appropriate solutions were provided to achieve a balanced and water-secured system.

The developed CN value-based runoff map (Fig. 10.3a) and VCI-based drought map (Fig. 10.3b) were overlaid with land cover (surface roughness) (Fig. 10.2) and soil map (Fig. 10.2) to get zones of excess and deficit water. Several locations were marked where excess water could be infiltrated to increase the soil moisture content (Fig. 10.5).

Four sample land patches (Sample 1 to Sample 4) with an area of 1 square metre each (Fig. 10.5) were selected. All four samples were different in physical characteristics such as different surface roughness (CN value dependent), varied soil textures, constant precipitation (66.02 mm) for the year 2021 (Frame et al. 2022), and the presence of different locations of groundwater potential recharge zones.

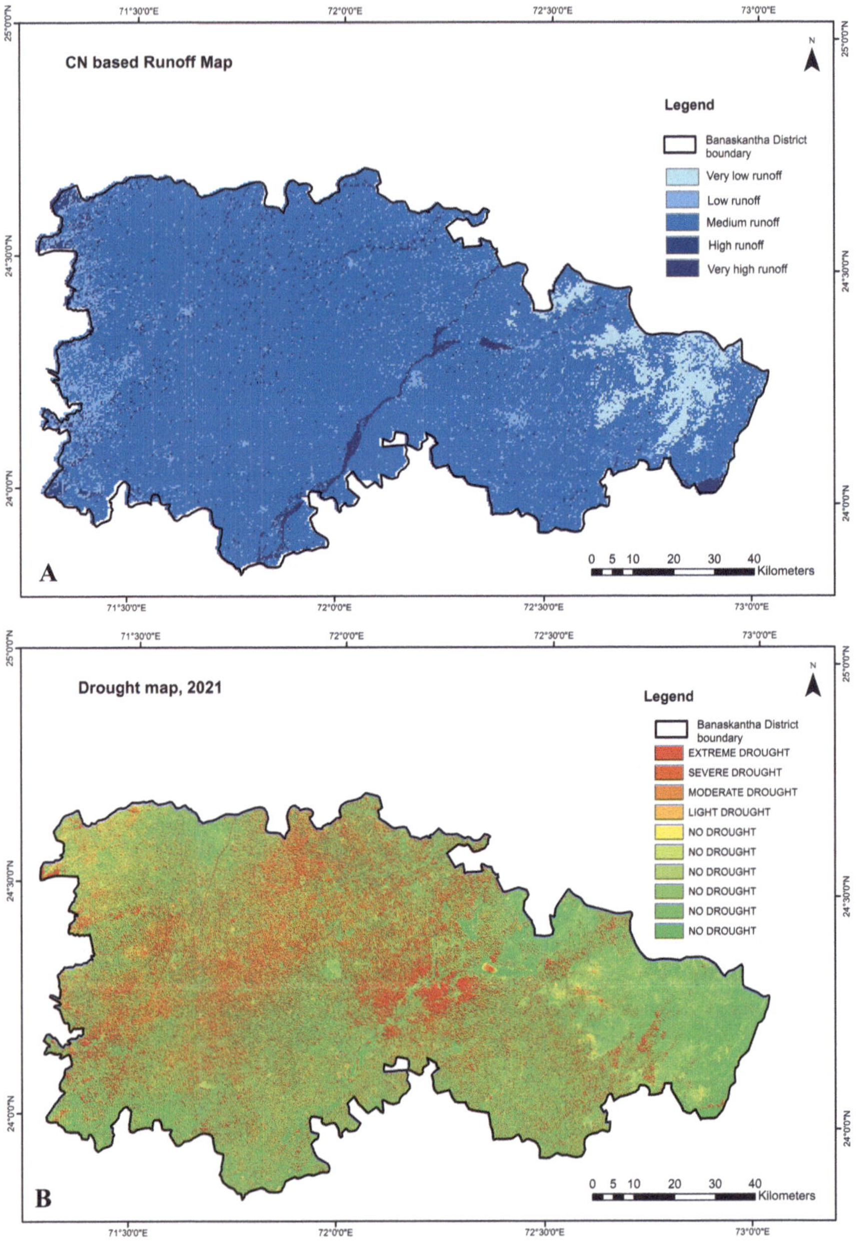

Fig. 10.3 Runoff map (**a**) and drought (dry) map (**b**), Banaskantha district, 2021

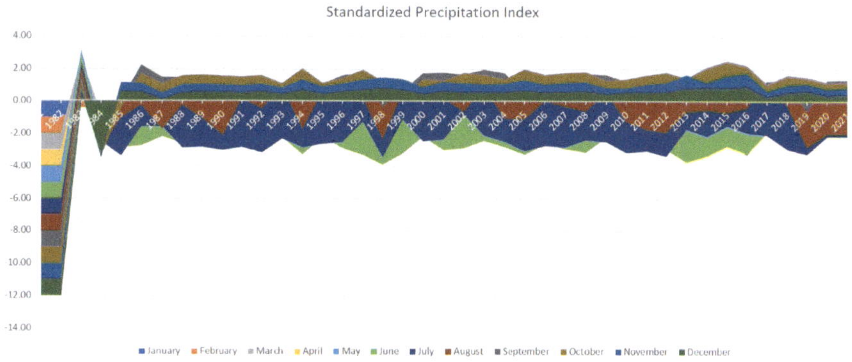

Fig. 10.4 Standardized Precipitation Index (SPI), 1982–2021

Table 10.1 Saturation capacities of each soil type

Soil type	Area of soil (sq. m.)	Depth of soil (m.)	Volumetric water content (%)	Max. capacity to hold water and turn saturated (cu. m.)
Loamy sand	141.13×10^9	1	35	20.59×10^5
Silt loam	51.17×10^9	1	45	9.59×10^5
Loam	29.18×10^9	1	35	4.25×10^5
Sandy clay loam	8.92×10^9	1	30	1.11×10^5
Clay loam	16.65×10^9	1	45	3.12×10^5

10.4.2 Assessment of Scenarios

The selected samples were analysed to compute different runoff depths and soil moisture contents (Table 10.2). Since the precipitation was constant, the runoff varied with the varied surface roughness. The surface roughness varied according to the land cover and soil texture.

The recharge zones play an important role in infiltrating water. The presence of recharge zones increases the potential of the surface to absorb the flowing water (Maqsoom et al. 2022).

To reduce the drought situation, runoff had to be decreased and soil moisture had to be increased to reduce drought-like situation to achieve a balanced hydrological system. Various solution combinations were explored for the given excess-deficit situations along with the soil-surface roughness combinations.

For Sample 1, agricultural cultivation modified the surface roughness, thus reducing the runoff depth and maximizing the capacity of the soil to maintain soil moisture (Table 10.3) (Sonneveld et al. 2022). Agricultural cultivation is a traditional approach where a combination of conservation tillage, crop rotations, and cover crops have been practiced (Farmaha et al. 2022).

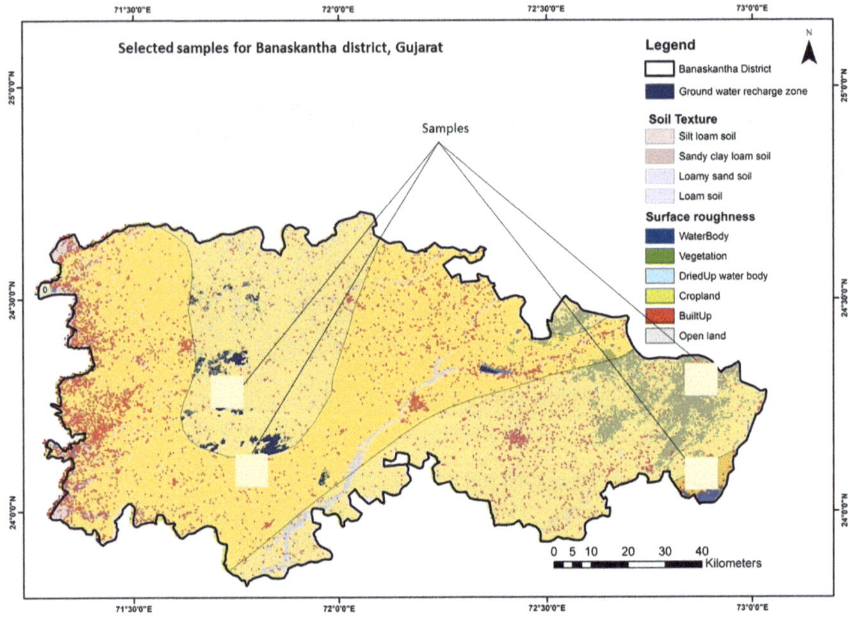

Fig. 10.5 Selected samples for scenario building

Table 10.2 Scenario 1 for sample 1–4

Scenario 1	Initial results			
	Precipitation (mm)	Curve number (CN)	Runoff depth (mm)	Soil moisture content (mm)
Sample 1 (Silt loam)	66.02	72	0.9	0.18
Sample 2 (Loamy sand)	66.02	74	0.89	0.044
Sample 3 (Loam)	66.02	66	0.858	0.085
Sample 4 (Sandy clay loam)	66.02	91	0.97	0.145

Table 10.3 Scenario 2 for sample 1–4

Scenario 2	Modified results			
	Precipitation (mm)	Curve number (CN)	Runoff depth (mm)	Soil moisture content (mm)
Sample 1 (Silt loam)	66.02	61	0.828	0.3726
Sample 2 (Loamy sand)	66.02	58	0.808	0.1212
Sample 3 (Loam)	66.02	30	0.524	0.1834
Sample 4 (Sandy clay loam)	66.02	32	0.553	0.2212

Table 10.4 % variation in runoff depth and soil moisture content

Samples	Variation in runoff depth (%)—Scenario 1 to Scenario 2	Variation in soil moisture content (%)—Scenario 1 to Scenario 2
Sample 1 (Silt loam)	8% (de)	69.7% (in)
Sample 2 (Loamy sand)	9.65% (de)	92.57% (in)
Sample 3 (Loam)	48.33% (de)	72.5% (in)
Sample 4 (Sandy clay loam)	54.76% (de)	41.28% (in)

**Note* de—decrease; in—increase

For Sample 2, plantations in agricultural land were suggested (Sonneveld et al. 2022; UNEP and FAO 2020). The effectiveness of the suggested NbS was assessed through the modification in surface roughness as it increased the soil organic content by focusing on the agricultural practice thereby increasing the capacity to hold soil moisture (Table 10.3).

For Sample 3, the restoration of native plant species was suggested as the existing surface roughness was majorly forested land. Ecological restoration has been a traditional way of re-vegetation which has been practiced for years. This practice helps in increasing the water-holding capacity of the soil.

Lastly, for Sample 4, restoration of flood plains was suggested as it lies in proximity to a water body. Restoration through the development of storage tanks (Virdas or kunds) and revival of wetlands helps in blocking the flow of water. This water increases the soil moisture, thus moisture content (Table 10.3).

Scenarios 1 and 2 were compared to get variations in the runoff depth and soil moisture contents. A variation in % was achieved where the runoff depth % was decreased and the soil moisture content % was increased for the four selected samples (Table 10.4).

10.5 Conclusion

The study investigated a drought situation in the Banaskantha district of Gujarat. Varying precipitation patterns affect the dry and wet days whereas modified surfaces lead to runoff and reduce infiltration of water. Excess water was reduced using surface roughness modulators through the incorporation of NbS along with traditional disaster mitigation approaches. These solutions helped in reducing runoff depth and increasing the soil moisture content by varying the surface roughness with constant soil types. Scenarios were developed on 4 samples where the runoff depth decreased in a range of 8–54.76% for different soil types and the soil moisture content increased in a range of 41.28–92.57% (Table 10.4).

Acknowledgements The authors want to thank "Maulana Azad National Institute of Technology, Bhopal" for providing us a platform for conducting this research. Both authors have contributed equally to the submitted chapter.

References

Abrams M, Crippen R, Fujisada H (2020) ASTER Global Digital Elevation Model (GDEM) and ASTER Global Water Body Dataset (ASTWBD). Remote Sens

Adams MB, Archer VA, Bailey S, McGuire K, Miniat CF, Neary DG et al (2020) Soils and Water BT—forest and rangeland soils of the United States under changing conditions: A comprehensive science synthesis. In: Pouyat RV, Page-Dumroese DS, Patel-Weynand T, Geiser LH (eds). Springer International Publishing, Cham, pp 33–49

Agarwal DS, Bharat A (2023) Nature-based solutions for flood–drought mitigation using a composite framework: A case-based approach. J Water Clim Chang, jwc2023369

Agarwal DS, Bharat A, Kjeldsen TR, Adeyeye K. Assessing impact of nature based solutions on peak flow using HEC-HMS. Water Resour Manag

Akhtar MP, Faroque FA, Roy LB, Rizwanullah M, Didwania M (2021) Computational analysis for rainfall characterization and drought vulnerability in Peninsular India. Gritli H, editor. Math Probl Eng Hindawi, 2021:5572650

Alam A, Bhat MS, Maheen M (2020) Using Landsat satellite data for assessing the land use and land cover change in Kashmir valley. GeoJournal 85(6):1529–1543

Al-Kaisi M (2017) How drought affects soil health. Integrated Crop Management. Iowa State Univ, pp 2017–2020

Almedeij J (2014) Drought analysis for Kuwait using standardized precipitation index. Sci World J 20(2014):9

Anderson V, Gough WA (2022) A typology of nature-based solutions for sustainable development: An analysis of form, function, nomenclature, and associated applications. Land

Angelakıs AN, Zaccaria D, Krasilnikoff J, Salgot M, Bazza M, Roccaro P et al (2020) Irrigation of world agricultural lands: Evolution through the Millennia. Water

Ara Z, Zakwan M (2018) Estimating runoff using SCS curve number method. Int J Emerg Technol Adv Eng 8(5):195–200

Ariyananda TN (2021) Rainwater Harvesting Systems as Disaster Risk Reduction (DRR) tool for drought and flood prone areas of Sri Lanka BT—sustainable climate action and water management. In: Mishra RK, Singh RB, Dubey A (eds). Springer Singapore, Singapore, pp 19–26

Asare P, Atun F, Pfeffer K (2023) Nature-Based Solutions (NBS) in spatial planning for urban flood mitigation: The perspective of flood management experts in Accra. Land Use Policy 133:106865

Baba A, Ahmed A (2018) Developments in water dams and water harvesting systems throughout history in different civilizations. Int J Hydrol 26:2

Bandyopadhyay N, Bhuiyan C, Saha AK (2020) Drought mitigation: Critical analysis and proposal for a new drought policy with special reference to Gujarat (India). Prog Disaster Sci, The Author(s) 5:100049

Baniya B, Tang Q, Xu X, Haile GG, Chhipi-Shrestha G (2019) Spatial and temporal variation of drought based on satellite derived vegetation condition index in Nepal from 1982–2015. Sensors (Switzerland) 19(2)

Bartlett MS, Parolari AJ, McDonnell JJ, Porporato A (2016) Beyond the SCS-CN method: A theoretical framework for spatially lumped rainfall-runoff response. Water Resour Res 52(6):4608–4627. John Wiley & Sons, Ltd

Bhattacharjee K, Behera B (2017) Forest cover change and flood hazards in India. Land Use Policy 67:436–448

Blachowski J, Dynowski A, Buczyńska A, Ellefmo SL, Walerysiak N (2023) Integrated spatiotemporal analysis of vegetation condition in a complex post-mining area: Lignite mine case study. Remote Sens

Boogaard FC (2022) Spatial and time variable long term infiltration rates of green infrastructure under extreme climate conditions, drought and highly intensive rainfall. Water

Cerpa Reyes LJ, Ávila Rangel H, Herazo LC (2022) Adjustment of the Standardized Precipitation Index (SPI) for the evaluation of drought in the Arroyo Pechelín Basin, Colombia, under zero monthly precipitation conditions. Atmosphere (Basel)

Chaudhari H, Jat J, Kumar S, Malav J, Pavaya R (2020) Physical and chemical properties of soils in Banaskantha District under groundnut cultivation 10(9):382–387

Cheng K, Xu X, Cui L, Li Y, Zheng J, Wu W et al (2021) The role of soils in regulation of freshwater and coastal water quality. Philos Trans R Soc B Biol Sci Royal Society 376(1834):20200176

Chithra SV, Harindranathan Nair M, Amarnath A, Anjana NS (2015) Impacts of impervious surfaces on the environment. Int J Eng Sci Invent 4(5):27–31

Cohen-Shacham E, Walters G, Janzen C, Maginnis S (2016) What are Nature-based Solutions? IUCN, Nature-based Solut. to address Glob. Soc. challenges

DAAC O (2018) Global Hydrologic Soil Groups (HYSOGs250m) for curve number-based runoff modeling [Internet]. Earth Data. https://daac.ornl.gov/SOILS/guides/Global_Hydrologic_Soil_Group.html

Datta P (2015) Traditional rainwater harvesting at crossroads Time to seek ethical opportunities for reform

Dhanalakshmi A, Vijayakumari KK, Marimuthu S, Surendran U (2020) Evaluation of different soil textures in combination with growing media on growth, yield, and water productivity of Blackgram. Commun Soil Sci Plant Anal, 51(21):2670–2682. Taylor & Francis

ECMWF (2018) ERA5 hourly data on single levels from 1959 to present [Internet]. Copernicus. https://cds.climate.copernicus.eu/cdsapp#!/dataset/reanalysis-era5-single-levels?tab=form

Eyoh A, Okeke F, Ekpa A (2019) Assessment of the effectiveness of the Vegetation Condition Index (VCI) as an indicator for monitoring drought condition across the Niger Delta region of Nigeria using AVHRR/MODIS NDVI. Eur J Earth Environ 6(1):12–18

Farmaha BS, Sekaran U, Franzluebbers AJ (2022) Cover cropping and conservation tillage improve soil health in the southeastern United States. Agron J 114(1):296–316. John Wiley & Sons, Ltd

Fauzia, Surinaidu L, Rahman A, Ahmed S (2021) Distributed groundwater recharge potentials assessment based on GIS model and its dynamics in the crystalline rocks of South India. Sci Rep 11(1):11772

Feldmann D, Laux P, Heckl A, Schindler M, Kunstmann H (2023) Near surface roughness estimation: A parameterization derived from artificial rainfall experiments and two-dimensional hydrodynamic modelling for multiple vegetation coverages. J Hydrol 617:128786

Frame JM, Kratzert F, Klotz D, Gauch M, Shalev G, Gilon O (2019) Deep learning rainfall—runoff predictions of extreme events. (2019):3377–3392

Ghimire C, Bruijnzeel L, Bonell M, Coles N, Lubczynski M (2014) The effects of sustained forest use on hillslope soil hydraulic conductivity in the Middle Mountains of Central Nepal. Ecohydrology

Gonzalez-Ollauri A, Mickovski SB, Anderson CC, Debele S, Emmanuel R, Kumar P et al (2023) A nature-based solution selection framework: Criteria and processes for addressing hydro-meteorological hazards at open-air laboratories across Europe. J Environ Manage 331:117183

Govil A, Kanchan MR (2023) Traditional & indigenous practices for climate resilience in India

GSDMA (2016) Gujarat State disaster management plan—Volume 1

Guenang G, Francois M (2014) Computation of the Standardized Precipitation Index (SPI) and its use to assess drought occurrences in Cameroon over recent decades. J Appl Meteorol Climatol 1(53):2310–2324

Guo Z, Li P, Yang X, Wang Z, Lu B, Chen W et al (2022) Soil texture is an important factor determining how microplastics affect soil hydraulic characteristics. Environ Int 165:107293

Gupta P, Bharat A (2022a) Developing sustainable development Index as a tool for appropriate urban land take. Environ Dev Sustain 24(11), 13378–13406. Springer Netherlands

Gupta P, Bharat A (2022) An integrated approach for capturing intermediate demographic changes. J Urban Plan Dev 148(2):1–20
Hartmann A (2022) The hydrology of groundwater systems—from recharge to discharge. In: Mehner T, Tockner KBT-E of IW (eds). Oxford, Elsevier, pp 324–330
Ho-Hagemann HTM, Hagemann S, Rockel B (2015) On the role of soil moisture in the generation of heavy rainfall during the Oder flood event in July 1997. Tellus, Ser A Dyn Meteorol Oceanogr 6(1)
Husbands S (2023) How water shortages impact food security [Internet]. Earth. Org. https://earth.org/how-water-shortage-impact-food-security/
IUCN (2020a) Nature-based solutions for water infrastructure at your service [Internet]. IUCN. https://www.iucn.org/news/water/202003/nature-based-solutions-water-infrastructure-your-service
IUCN (2020b) IUCN Global standards for nature based solutions
Jalayer S, Sharifi A, Abbasi-Moghadam D, Tariq A, Qin S (2023) Assessment of spatiotemporal characteristic of droughts using in situ and remote sensing-based drought indices. IEEE J Sel Top Appl Earth Obs Remote Sens 16:1483–1502. IEEE
Jiang W, Niu Z, Wang L, Yao R, Gui X, Xiang F et al (2022) Impacts of drought and climatic factors on vegetation dynamics in the Yellow River basin and Yangtze River basin, China. Remote Sens
Robotham J, Gasca-Tucker D, Old J, McKnight D, GOPRDSETJB (2022) Nature–based solutions enhance sediment and nutrient storage in an agricultural lowland catchment. Earth Surf Process Landf
Kruti D, Parmar M, Sharma AK, Rajawat AS, Pandya (1997) Land degradation assessment and mapping in Banaskantha district, Gujarat using remote sensing and GIS techniques
Kumar PA, Viswanadh GK (2017) Estimation of runoff by using SCS curve number method integrated with GIS. Int Adv Res J Sci Eng Technol 4(7):34–38
Kumar A, Kanga S, Taloor AK, Singh SK, Đurin B (2021a) Surface runoff estimation of Sind river basin using integrated SCS-CN and GIS techniques. HydroResearch. The Authors 4:61–74
Kumar P, Debele SE, Sahani J, Rawat N, Marti-Cardona B, Alfieri SM et al (2021b) Nature-based solutions efficiency evaluation against natural hazards: Modelling methods, advantages and limitations. Sci Total Environ 784:147058
Liang L, Sun Q, Luo X, Wang J, Zhang L, Deng M et al (2017) Long-term spatial and temporal variations of vegetative drought based on vegetation condition index in China. Ecosphere 8(8):e01919. John Wiley & Sons, Ltd
Machiwal D, Kumar S, Sharma G, Jat S, Dayal D (2018) Studying an indigenous rainwater harvesting system in Banni grassland of Kachchh, India. Indian J Tradit Knowl 1:17
Masson-Delmotte V, Zhai P, Pörtner HO, Roberts DC, Skea J, Shukla PR, Pirani A, Moufouma-Okia W, Péan C, Pidcock R, Connors S (2018) Global Warming of 1.5° C: Summary for Policy Makers. IPCC—The Intergovernmental Panel on Climate Change
Maqsoom A, Aslam B, Khalil U, Mehmood MA, Ashraf H, Siddique A (2020) An integrated approach based earthquake risk assessment of a seismically active and rapidly urbanizing area in Northern Pakistan. Geocarto Int 37(27):16043–16073. https://doi.org/10.1080/10106049.2022.2105404
Matić BB, Karleuša B (2022) Ecosystem-based disaster risk reduction framework as a tool for improved river basin natural water retention capacity and environmental hazard resilience. Environ Sci Proc
Mishra D, Goswami S, Matin S, Sarup J (2022) Analyzing the extent of drought in the Rajasthan state of India using vegetation condition index and standardized precipitation index. Model Earth Syst Environ 8(1):601–610. Springer International Publishing
Mistry P, Suryanarayana T (2019) Categorization of drought during twentieth century using precipitation in Banaskantha District, Gujarat, India. In: Proceedings of ICIIF 2018. Adv Intell Syst Comput, pp 267–274
Mittal Patel (2019) Water-management-to-avoid-disasters-caused-by-climate-change

Moccia B, Mineo C, Ridolfi E, Russo F, Napolitano F (2022) SPI-based drought classification in Italy: Influence of different probability distribution functions. Water
More H (2021) A review of Indian traditional method of rain water harvesting. Int J Innov Res Sci Eng Technol 9:2080–2086
Mullapudi A, Vibhute AD, Mali S, Patil CH (2023) A review of agricultural drought assessment with remote sensing data: methods, issues, challenges and opportunities. Appl Geomatics 15(1):1–13
Mupepi O, Matsa MM (2023) A combination of vegetation condition index, standardized precipitation index and human observation in monitoring spatio-temporal dynamics of drought. A case of Zvishavane District in Zimbabwe. Environ Dev 45:100802
Nachtergaele F, Velthuizen H Van, Verelst L (2008) Harmonized world soil database. Food Agric:43
Nehren U, Arce-Mojica T, Barrett AC, Cueto J, Doswald N, Janzen S et al (2023) Towards a typology of nature-based solutions for disaster risk reduction. Nature-Based Solut. 3:100057
Nações Unidas (2019) World population prospects 2019. Dep Econ Soc Aff World Popul Prospect
Ommer J, Bucchignani E, Leo LS, Kalas M, Vranić S, Debele S et al (2022) Quantifying co-benefits and disbenefits of nature-based solutions targeting disaster risk reduction. Int J Disaster Risk Reduct 75:102966
Palchaudhuri M, Biswas S (2013) Analysis of meteorological drought using standardized precipitation index—a case study of Puruliya District, West Bengal, India. Int J Environ Ecol Geol Geophys Eng 15(7):119–126
Parkash S, Kaur H, Bindal MK (2022) Floods challenges and its management case study of Gujarat Floods—2017
Patel (2020) SCS Runoff curve number method—Introduction—Professor Patel
Prajapati P, Bansal PDN (2017) Banaskantha flood 2017: Flood risk assessment. Int J Trend Sci Res Dev 2(1):893–895
Qin Y (2020) Urban flooding mitigation techniques: A systematic review and future studies. Water
Qing Y, Wang S, Yang Z-L, Gentine P (2023) Soil moisture−atmosphere feedbacks have triggered the shifts from drought to pluvial conditions since 1980. Commun Earth Environ 4(1):254
Rai D (2022) More and more droughts_ How both India and the world is getting affected and what it will lead to—India Today. India Today
Raju RS, Raju GS, MR (2018) Estimation of rainfall runoff using SCS-CN Method with RS and GIS techniques for Mandavi Basin in YSR Kadapa District of Andhra Pradesh, India. Hydrospatial Anal 2(1):1–15
Rasheed MW, Tang J, Sarwar A, Shah S, Saddique N, Khan MU et al (2022) Soil moisture measuring techniques and factors affecting the moisture dynamics: A comprehensive review. Sustainability
Renaud F, Sudmeier-Rieux K, Estrella M, Nehren U (2016) Ecosystem-based disaster risk reduction and adaptation in practice
Saco PM, McDonough KR, Rodriguez JF, Rivera-Zayas J, Sandi SG (2021) The role of soils in the regulation of hazards and extreme events. Philos Trans R Soc B Biol Sci 376 (1834)
Saha A, Rattan B, Sekharan S, Manna U (2020) Quantifying the interactive effect of water absorbing polymer (WAP)-soil texture on plant available water content and irrigation frequency. Geoderma 368:114310
Saha S (1997) Management of Drought in Gujarat : (Case Study of Banaskantha District)
Saini D, Singh O, Bhardwaj P. Standardized precipitation index based dry and wet conditions over a dryland ecosystem of northwestern India. Geol Ecol Landscapes 6(4):252–264. Taylor & Francis
Satheeshkumar S, Venkateswaran S, Kannan R (2017) Rainfall–runoff estimation using SCS–CN and GIS approach in the Pappiredipatti watershed of the Vaniyar sub basin, South India. Model Earth Syst Environ 3(1):1–8. Springer International Publishing
Seddon N, Chausson A, Berry P, Girardin CAJ, Smith A, Turner B (2020) Understanding the value and limits of nature-based solutions to climate change and other global challenges. Philos Trans R Soc B Biol Sci 375(1794):20190120. https://doi.org/10.1098/rstb.2019.0120
Seleiman MF, Al-Suhaibani N, Ali N, Akmal M, Alotaibi M, Refay Y et al (2021) Drought stress impacts on plants and different approaches to alleviate its adverse effects. Plants 10(2):1–25

Senamaw A, Addisu S, Suryabhagavan KV (2021) Mapping the spatial and temporal variation of agricultural and meteorological drought using geospatial techniques, Ethiopia. Environ Syst Res 10(1):15

Shah R, Bharadiya N, Manekar V (2015) Drought index computation using Standardized Precipitation Index (SPI) method for Surat District, Gujarat. Aquat Procedia 4:1243–1249

Shahadha A, Haitham M, Afan A, Hatem A, Khamis K, Sayl N. Runoff mapping using the SCS—CN method and artificial neural network algorithm, Ratga Basin, Iraq Arab J Geosci. Springer International Publishing, pp 1–10

Shi W, Wang N (2020) Improved SMA-based SCS-CN method incorporating storm duration for runoff prediction on the Loess Plateau. China Hydrol Res 51(3):443–455

Simelton E, Carew-Reid J, Coulier M, Damen B, Howell J, Pottinger-Glass C et al (2021) NBS framework for agricultural landscapes. Front Environ Sci 9(August):1–16

Singh G, Jakhar R, Sachar P (2022) Impact on people's lives and livelihoods in Gujarat due to Natural hazards. Int J Emerg Trends Eng Res 10(4):191–194

Sonneveld BGJS, Amani A, Merbis MD (2022) Nature-based solutions for agricultural water management. Palgrave Handb Clim Resilient Soc, 1–2

Soulis KX (2021) Soil Conservation Service Curve Number (SCS-CN) method: Current applications, remaining challenges, and future perspectives. Water

Strohbach MW, Döring AO, Möck M, Sedrez M, Mumm O, Schneider AK et al (2019) The "hidden urbanization": Trends of impervious surface in low-density housing developments and resulting impacts on the water balance. Front Environ Sci 7(MAR):1–10

Suri S, Sharma N (2022) India: Climate change poses biggest risk to country's food security [Internet]. Prev Web. https://www.preventionweb.net/news/climate-change-poses-biggest-risk-food-security-india

Suryakant T, Pravendra B (2022) Quantification of impact of spatio—temporal variability of land use/land cover on runoff generation using modified NRCS—CN method. Springer International Publishing, Arab J Geosci

The Nature Conservancy (2017) Wetlands and groundwater: It's all connected. Wisconsin Wetl Assoc

Tsesmelis DE, Leveidioti I, Karavitis CA, Kalogeropoulos K, Vasilakou CG, Tsatsaris A et al (2023) Spatiotemporal application of the Standardized Precipitation Index (SPI) in the Eastern Mediterranean. Climate

UN (2018) Around 2.5 billion more people will be living in cities by 2050, projects new UN report. United Nations

UNDRR (2020) Ecosystem—based disaster risk reduction: Implementing Nature-based solution for resilience

UNDRR (2021a) Reducing flood risk through the modelling of Nature-Based Solutions. Prevention Hub

UNDRR (2021b) Nature-based solutions for disaster risk reduction

UNEP, FAO (2020) The UN decade on ecosystem restoration 2021–2030. UNEP

UNFCCC (2020) What do adaptation to climate change and climate resilience mean? UNFCCC

Un-spider org (2014) Recommended practice: Drought monitoring using the Vegetation Condition Index (VCI). Un-spider org, p 1

USGS (2021) Earth Explorer [Internet]. USGS. https://earthexplorer.usgs.gov/

Vélez-Nicolás M, García-López S, Ruiz-Ortiz V, Zazo S, Molina JL (2022) Precipitation variability and drought assessment using the SPI: Application to long-term series in the strait of Gibraltar Area. Water

Walters GM (2016) Nature-based solutions to address global societal challenges

Wang P, Qiao W, Wang Y, Cao S, Zhang Y (2020) Urban drought vulnerability assessment—A framework to integrate socio-economic, physical, and policy index in a vulnerability contribution analysis. Sustain Cities Soc 54:102004. Elsevier

Wolteji BN, Garbaba FG (2023) A geo-spatial assessment of drought impacts on forest cover in yabello forest, in the semi-arid region of Ethiopia. SN Appl Sci 5(5):148

Yeh HF, Cheng YS, Lin HI, Lee CH (2016) Mapping groundwater recharge potential zone using a GIS approach in Hualian River, Taiwan. Sustain Environ 26(1):33–43. Res Elsevier Ltd

Yimer EA, De Trift L, Lobkowicz I, Villani L, Nossent J, van Griensven A (2024) The underexposed nature-based solutions: A critical state-of-art review on drought mitigation. J Environ Manage 352:119903

Zeleňáková M, Sol'áková T, Purcz P, Abd-Elhamid HF (2023) Analysis of hydrological drought in the Eastern Part of Slovakia using standardized precipitation index BT—water resources management and sustainability: Solutions for Arid Regions. In: Sherif M, Singh VP, Sefelnasr A, Abrar M (eds). Springer Nature Switzerland, Cham, pp 119–28

Zhang Z, Pan Z, Pan F, Zhang J, Han G, Huang N et al (2020) The change characteristics and interactions of soil moisture and temperature in the Farmland in Wuchuan County, Inner Mongolia, China. Atmosphere (Basel) 14(11):503

Chapter 11
Urban Flood Prevention Through Community-Centered Green Stormwater Infrastructure Planning

Mohammad Newaz Sharif, Claire Connolly Knox, Dominic Boyer, and Jessica Eisma

Abstract Floods in highly urbanized coastal areas pose significant risks to the well-being and resiliency of the inhabitants. The ability of a community to recover from disasters is directly influenced by social vulnerability where underserved populations are disproportionately affected by hazards. One way to reduce this impact is to incorporate hazard mitigation measures—including green stormwater infrastructure (GSI)—into land use plans, especially following a major disaster. Research funded by the National Oceanic and Atmospheric Administration provides a novel approach that incorporates stakeholder data, as well as historical and projected storm events, into the planning of GSI projects for two underserved neighborhoods of Kashmere Gardens and Trinity/Houston Gardens in Texas, United States. This chapter discussed the results of (1) a multiple plan analysis focused on resiliency planning, hazard mitigation, and lessons learned from Hurricane Harvey in 2017, and (2) an analysis of engagement with residents through semi-structured interviews and at various events in the community to highlight how land use decisions have impacted the disaster experiences of residents and, ultimately, their readiness to participate in the planning process. The chapter concludes by outlining the major findings and implications for flood management strategies in vulnerable coastal and non-coastal communities facing similar challenges.

M. N. Sharif (✉) · C. C. Knox
School of Public Administration, University of Central Florida, Orlando, USA
e-mail: mo604188@ucf.edu

C. C. Knox
e-mail: Claire.Knox@ucf.edu

D. Boyer
Center for Coastal Futures and Adaptive Resilience, Rice University, Houston, USA
e-mail: Dominic.C.Boyer@rice.edu

J. Eisma
Department of Civil Engineering, The University of Texas at Arlington, Arlington, USA
e-mail: jessica.eisma@uta.edu

M. Firoz C et al. (eds.), *Nature-Based Solutions for Urban and Peri-Urban Areas*, Disaster Risk Reduction, https://doi.org/10.1007/978-981-96-5933-3_11

Keywords Green Stormwater infrastructure · Flood prevention · Hazard mitigation planning · Community resilience · Disaster risk reduction

11.1 Introduction

Among all natural hazards that affect the United States (U.S.), flooding remains the most common and costly. Between 1980 and 2023, 44 flood events exceeded $1 billion in economic losses, costing a total of $201 billion (Adam 2024). Urban floods have become more common due to human-induced climate change and the rapid expansion of urban areas in the U.S. and abroad (Dharmarathne et al. 2024). Floods in highly urbanized coastal areas pose significant risks to the well-being and resiliency of the inhabitants. The community's ability to recover from disasters is directly influenced by social vulnerability as underserved populations are disproportionately affected by hazards (Tyler et al. 2023). Allowing haphazard development in flood-prone areas, especially floodplains, increases residents' vulnerability and decreases a community's resiliency (Malecha et al. 2021). One way to reduce this impact is to incorporate hazard mitigation measures—including green stormwater infrastructure (GSI)—into land use plans. However, what happens when a jurisdiction's lack of mandates for zoning and a comprehensive land use plan requires reliance on public–private partnerships for the design and implementation of GSI projects as disaster reduction strategies?

Research funded by the National Oceanic and Atmospheric Administration (NOAA) provides a novel approach that incorporates stakeholder data, as well as historical and projected storm events, into the planning of GSI projects for two underserved neighborhoods located in northeast Houston, Texas: Kashmere Gardens and Trinity/Houston Gardens (hereinafter the Gardens). These neighborhoods are increasingly prone to flooding due to climate change-related factors, including rising sea levels and more frequent and intense rainfall events. Moreover, Gardens' residents have been afflicted by the presence of vacant lots, decaying buildings, environmental injustices, state-identified cancer clusters, and residences that have been adversely affected by flooding due to inadequate drainage infrastructure. A significant dimension of GSI planning in Houston is the city's lack of zoning which reduces the ability of public authorities to plan for current and future land uses. Thus, a more complex model of private and public stakeholder alignment is needed.

GSI scholars have emphasized the necessity for interdisciplinary research (te Wierik et al. 2020), particularly from social science perspectives (Venkataramnan et al. 2020), to support equitable and inclusive GSI program implementation, especially in underserved communities (Thorne et al. 2018). Responding to this call, a multidisciplinary team conducted research focusing on community-centered initiatives for urban flood prevention in the Gardens. Collaborating with local government agencies, community leaders, and residents, the project aims to co-create GSI plans tailored to the neighborhoods' unique challenges and needs, a decision-making framework for assessing GSI projects, and guidelines for community engagement in GSI planning (NEHRC, personal communication, November 17, 2020).

This chapter discusses the results of a multiple plan analysis of city plans and reports focused on land use and resilience planning, hazard mitigation, and lessons learned from Hurricane Harvey in 2017 with a specific focus on flood risk, stormwater, and GSI-related projects. The analysis of semi-structured interviews highlights how land use decisions have impacted the disaster experiences of residents and, ultimately, their readiness to participate in the land use planning process. The chapter concludes by outlining the major findings and implications for flood management strategies in vulnerable communities facing similar challenges.

11.2 Literature Review

11.2.1 Hazard Mitigation Planning for Flood Risks

While U.S. federal programs exist to reduce flooding risks(e.g., Federal Emergency Management Agency's National Flood Insurance Program), this risk is primarily planned for and managed by emergency managers, urban planners, and floodplain managers at the local levels of government (e.g., city, county/parish). Communities that include hazard reduction related information in their land use plans experience less damage from a disaster and rebound more quickly; essentially the definition of resilience (Knox 2017; Wu et al. 2024). While managing this risk occurs at the local level of government, a comprehensive effort requires interdependent ecological, physical, political, and social interactions with multiple stakeholders (Wu et al. 2024).

Communities can adopt structural and nonstructural mitigation strategies to reduce flood risk. Structural examples primarily include gray infrastructure, namely seawalls, levees, and dams, while nonstructural examples include updating flood maps and plans (e.g., land use, hazard mitigation, local mitigation strategy), educating community members, and implementing GSI projects. Quantifying the impacts of flood mitigation efforts in relation to resiliency is essential. For example, the National Institute of Buildings (2020) reports that for every $1 spent on flood mitigation, a community saves $6 post-disaster. This chapter focuses on one strategy—creation and implementation of GSI projects in flood-prone areas.

11.2.2 Green Stormwater Infrastructure

As a nature-based solution, GSI mimics natural hydrological processes and improves flood resilience while providing social and environmental benefits (Green et al. 2021; Kim and Park 2016; Venkataramanan et al. 2020). As highlighted in Table 11.1, GSI examples include vegetated buffer strips, green roofs, permeable pavement, rain gardens, and restored wetlands, which intercept and store precipitation, reducing runoff and the risk of floods (Roberts et al. 2012; Schubert et al. 2017; Sharma and Malaviya 2021). Benefits include effective pollutant removal in urban areas (Biswal et al. 2022), like heavy metals (>60%), nutrients (50–80%), and hydrocarbons from stormwater (Hamouz and Muthanna 2019; Sharma and Malaviya 2021; Zhou et al. 2018;). They also increase vegetative cover and improve biodiversity, recharge groundwater, mitigate urban heat island effects, improve aesthetics, provide community and recreational spaces, and improve mixed-use land use patterns (Kim and Park 2016; Mei et al. 2018; Narzis and Eisma 2023; Woodruff et al. 2021). Specific to this study's context, scholars researching communities in coastal Texas concluded that the most effective measures for reducing flood losses, include protecting and restoring naturally occurring wetlands, and acquiring and conserving open space (Kim and Park 2016).

While countries around the world are prioritizing GSI and gradually shifting their focus from traditional gray infrastructure to green infrastructure, implementation in the U.S. has remained mixed (La Bissoniere 2022). Several factors need to be considered throughout the design and implementation of GSI projects, including technical acumen, site feasibility, maintenance needs, budget and financing limitations, conformity with regulations, public opinion and support, tracking efficiency over time, and durability against the adverse impacts of the changing climate (Johns 2019;

Table 11.1 Benefits of green stormwater infrastructure in a flood prone region

Type of green stormwater infrastructure	Common practices	Influence on disaster preparedness	Implications for building community resilience
Vegetation-based approaches	• Trees and vegetation (e.g., community gardens, forests) (Ko 2018) • Bioswales/rain gardens (Lee 2019) • Vegetated buffer strips (Roberts et al. 2012)	• Natural barriers against storm surges and coastal flooding • Enhances biodiversity and ecosystem services • Improved agricultural productivity • Enhances aesthetics	· Reduces the risk of property destruction through calamities · Reduces flooding and enhances valuable habitat · Supports economic development · Improves public health · Engages communities in raising awareness, emergency preparedness, and sustainable use of resources
Built interventions	• Green roofs and walls (Im 2019) • Green streets (Im 2019) • Permeable pavements (Santhanam and Majumdar 2020) • Restored wetlands (Santhanam and Majumdar 2020)	• Mitigates erosion • Enhances floodplain connectivity and storage reservoirs • Reduces heat island effect • Improves air and water quality • Reduces flooding • Reduces energy consumption	

Li et al. 2021). Incorporating environmentally friendly GSI structures with conventional gray infrastructure is imperative for effective flood risk reduction (Green et al. 2021). However, it faces technological, monetary, and political obstacles. Sociopolitical barriers (e.g., limited knowledge and understanding) hinder the adoption of GSI which emphasizes the need for ongoing monitoring, sharing of knowledge, and defining stakeholder responsibilities (Li et al. 2021; Venkataramanan et al. 2020).

11.2.3 Shifting Priorities from Gray to Green Stormwater Infrastructure

Recently, there has been a notable transformation in the process of administrative decision-making pertaining to flood management. Historically, the prevalence of "gray infrastructure" solutions has been prominent due to local land use laws and cost–benefit analysis practices that only considered direct monetary outcomes (Rosenbloom 2018). Those projects were specifically devised to effectively manage the flow of floodwaters through massive structures (e.g., dams, barrages, levees, concrete channels) (Jovanovic et al. 2019).

The shift towards prioritizing GSI for flood management is being driven by key factors that span multiple academic domains, including urban development, design, ecological research, and policymaking (Johns 2019). First, the limited effectiveness of traditional gray infrastructure has become increasingly apparent (Hendricks and Dowtin 2023). While these structures offer some protection, they can lead to unintended consequences (i.e., false sense of security, loss of critical ecosystem services), ultimately reducing community resilience (Rosenbloom 2018). Second, the rising costs associated with building and maintaining gray infrastructure, exacerbated by the escalating frequency and intensity of floods, have also prompted a reevaluation of stormwater management strategies. Finally, the adverse environmental impacts of gray infrastructure (i.e., habitat destruction, water quality degradation) have underscored the need for more sustainable alternatives (Onuma and Tsuge 2018).

Shifting priorities in states like Maryland, Arizona, Oregon (McPhillips and Matsler 2018) and Texas (Woodruff et al. 2021) demonstrate a growing recognition of the value of green infrastructure in promoting resilience and sustainability. Maryland's Department of Natural Resources (n.d.) has enacted the "Green Infrastructure Resilience Program" for stormwater management that emphasizes funding future construction projects adopting GSI principles in three phases (understand, plan, and implement) to reduce dependence on traditional drainage systems and promote environmentally friendly flood protection practices. Meanwhile, the Texas Commission on Environmental Quality (n.d.) encourages the use of best management practices that include GSI practices for managing stormwater although it does not have a specific GSI policy.

Despite challenges (e.g., upfront costs, land availability, need for education, and awareness), the momentum towards embracing GSI signifies a positive step towards building more resilient and sustainable communities in the face of climate change-induced flood risks (Taguchi et al. 2020). Research findings conclude that improving GSI in conjunction with traditional gray infrastructure may improve flood resistance and offer various advantages to communities (Green et al. 2021). Political buy-in is essential and demands backing from influential stakeholders, as well as helpful policies and legislation (Grabowski et al. 2023; Hopkins et al. 2018). Yet, the extensive implementation of GSI is impeded by gaps in comprehending the wider societal, environmental, and technological nexus. There remains a lack of research examining how demographic or socio-economic elements impact people's perceptions towards GSI. Therefore, it is important to incorporate social aspects into environmental and technical designs to produce efficient and enduring GSI projects (Venkataramanan et al. 2020) (see Table 11.1).

11.3 Community-Engaged Urban Design

Community-engaged urban design has a long history in the U.S., dating back to the 1980s (Greenbaum 1991). More recently, participatory design practices have been recognized as a promising element of urban climate resilience planning. Although viewed globally, especially outside of Europe, they remain at "an early developmental stage" (LopezDeAsiain and Díaz-García 2020). Today, most interventions are short-termed, top-down, and aimed at single, specific initiatives (often based on specific state and federal programs); they fail to build the long-term trust required for the social sustainability of holistic efforts at urban and environmental reform (Samaddar et al. 2022). These interventions run the risk of misinformation by soliciting opinions on complex problems without necessarily providing laypersons with the baseline education required to offer meaningful input (Anguelovski et al. 2019; Finewood et al. 2019). However, imagining the transformation of urban and environmental infrastructure systems requires expertise in which many engaged communities are willing to invest when given the chance (Campbell-Arvai and Lindquist 2021).

Stormwater infrastructure solutions take years to complete; therefore, the length and depth of community engagement matters. Short-term engagements are less likely to yield the necessary, sustained, collective effort required to achieve optimal results and often focus on current problems rather than on dynamic long-term problems (i.e., climate change) (Schepers et al. 2022). Strategies like participatory design visioning (Gaziulusoy and Ryan 2017) help engage communities not only about their present circumstances and risks but also the legacies that their actions will leave to future residents. Future anticipation is critical given the relatively short residency cycle of American residential neighborhoods (2.5 years for home renters, 8 years for homeowners on average). Moreover, participatory design planning has the potential to tie climate change adaptation to future local economic development opportunities.

11.4 Case Study: Mitigating Urban Flooding in the Gardens

The City of Houston is in Harris County, Texas, located in the southern U.S. along the Gulf of Mexico, and has an estimated population of 2.3 million in 2024. It is the largest city in Texas and the fourth largest in the U.S. (behind New York, Los Angeles, and Chicago) (World Population 2024). Houston is one of the fastest sinking cities in the world and is comprised of 22 bayous and waterways extending 2,500 miles. It remains the largest U.S. city without a zoning ordinance in a state that continues to avoid mandating comprehensive land use planning. Although building ordinances do exist in Houston, they tend to exist at the subdivision level in more affluent areas and focus on aesthetics rather than safety issues.

The lack of planning efforts has resulted in spontaneous, rapid, and haphazard urban development, which has increased the city's significant flooding risks. Most notable is the increase in impervious cover that accompanies urban growth, specifically rooftops and transportation systems (e.g., roads, parking lots, sidewalks) and subsequential loss of natural ecosystems (e.g., wetlands, prairies), which historically absorbed stormwater and recharged the aquifer (Lynn 2017). This urbanization has increased runoff and overwhelmed drainage systems, leading to flooding even beyond floodplains (City of Houston, 2020). This is mainly due to the city's flat topology and soil rich in clay, which slows the infiltration of precipitation during rain events (City of Houston, 2020) Intense rainfall events, as witnessed during Hurricane Harvey in 2017, exacerbate the situation, causing catastrophic flooding (Blake and Zelinsky 2018).

Specific to this study, the Gardens neighborhoods in northeastern Harris County occupy approximately one-third of the Hunting Bayou watershed with nearly all the properties located in a floodplain (see Fig. 11.1). Due to rapid urbanization, poor local planning decisions, and high-intensity rainfall events, these neighborhoods have faced an increase in repeated flooding as they lack effective gray and green stormwater infrastructure (Blackburn et al. 2021). Moreover, environmental injustices have plagued residents in these neighborhoods for over 100 years. A Texas Department of State Health Services investigation confirmed increased respiratory cancer risk among residents due to contamination from a nearby railyard, multiple landfills, concrete batch plants, and garbage incinerators. Combined, these injustices have resulted in devalued property values, vacant lots, dilapidated buildings, and flood-damaged homes (Lynn 2017). Implementing climate-informed GSI plans can enhance the resilience of the Gardens to flooding and improve air quality by mitigating air pollutants.

With climate change projections indicating increased rainfall variability and sea level rise by 2100, experts expect the vulnerability of the Gardens to flooding to worsen. The average social vulnerability index for these neighborhoods is already high at 0.90 out of 1.00. In 2017, Hurricane Harvey highlighted the extreme vulnerability of residents in these neighborhoods, prompting some support from public and private entities. However, interest in addressing the issues facing the gardens has

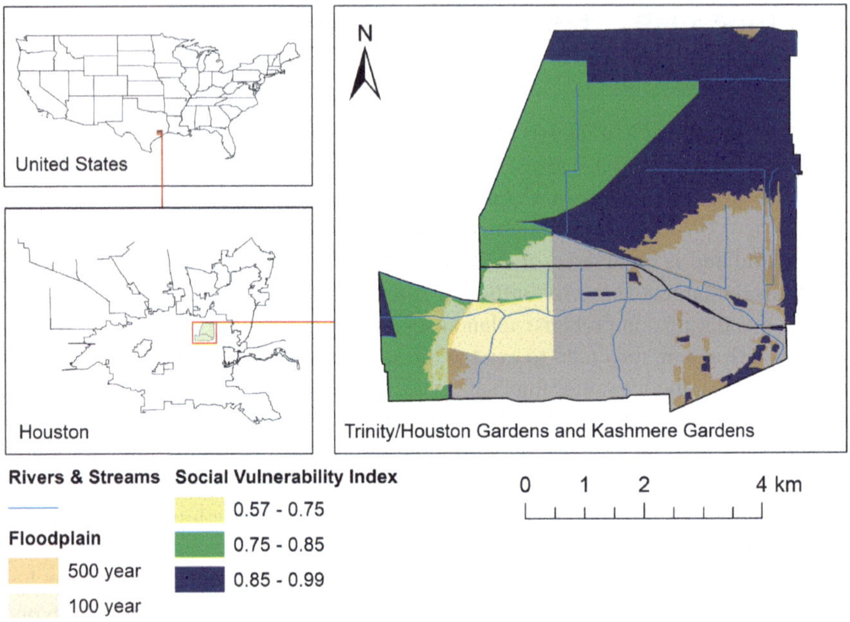

Fig. 11.1 Study area map showing the location of the Gardens neighborhoods in Houston, Texas

waned. Residents prioritize addressing flooding and environmental injustices, exacerbated by inadequate stormwater infrastructure and the disproportionate presence of open stormwater drains. These poorly maintained drains are less effective than traditional buried systems and contribute to flooding even during moderate rainfall events. The Northeast Houston Redevelopment Council (NEHRC) is advocating for GSI to address both flooding and environmental injustices.

11.5 Methods

11.5.1 Plan Analysis Methods

To analyze the extent to which Harris County and the City of Houston are planning for GSI projects, especially considering lessons learned from the extensive flooding following Hurricane Harvey, the research team collected and analyzed five city plans and reports focused on land use and resiliency planning, hazard mitigation, and lessons learned from the hurricane (see Table 11.2). The team opened and axial coded the data using codes based on the literature and related codes that arose from the plans, which allowed for an inductive and deductive coding process using Dedoose, a qualitative software package (Corbin and Strauss 2015).

Table 11.2 Analyzed plans and reports

Plan/report	Date	Pages	Summary	Example language
Plan Houston: Opportunity, diversity, community, home final report	2015	28	Houston's first general plan that includes the vision, 32 goals organized under 9 categories, and 12 core strategies for the jurisdiction; Texas has no requirement/mandate for cities or counties to have a comprehensive land use plan	Core strategy: Sustain quality infrastructure Actions: • "Encourage development of green infrastructure and low impact development • Align infrastructure investment with neighborhood needs • Use infrastructure improvements strategically to enhance economic development • Minimize risk of damages resulting from catastrophic events" (p. 8)
City of Houston Hazard mitigation plan	2018	245	Houston's hazard mitigation plan created by the city's office of emergency management and approved by the Federal Emergency Management Agency in accordance with the Disaster Mitigation Act of 2000	Guidelines: • "Control the alteration of natural floodplains, stream channels, and natural protective barriers, which are involved in the accommodation of floodwater • Prevent or regulate the construction of flood barriers which will unnaturally divert floodwaters or which may increase flood hazards to other lands" (p. 52)
Hurricane Harvey after-action report and improvement plan, Harris county office of homeland security and emergency management	2018	64	Harris County's after-action report from Hurricane Harvey that captures best practices, strengths, areas of improvement, and policy and institutional recommendations	N/A

(continued)

Table 11.2 (continued)

Plan/report	Date	Pages	Summary	Example language
Resilient Houston	2020	188	City-wide plan that includes a framework of 18 resilience goals under five objectives; created as part of the 100 Resilient Cities funded by The Rockefeller Foundation	Goal 11: "We will modernize Houston's infrastructure to address the challenges of the future" (p. 118) • "Integrate green stormwater infrastructure (GSI) into Houston's built environment • Advance comprehensive construction and maintenance of GSI • Build capacity for the design, construction, and maintenance of GSI • Apply a resilience quotient to GSI projects" (p. 119)
Kashmere gardens complete communities action plan	2020	78	A complete community action plan created through a stakeholder participatory design, which includes the vision, 32 goals, 80 projects, policies, and programs	• "Develop incentives and policies to promote green stormwater infrastructure" (p. 36)

11.5.2 Interview Methods

To gain insights into attitudes and decision-making processes concerning GSI, the research team conducted 14 semi-structured interviews with community leaders, community members, and technical experts whose activities directly impact GSI planning in the Gardens. These interviews averaged 45 min and were conducted by telephone and Zoom. Additionally, the research team engaged in nearly 30 informal conversations with community members at (1) public events focused on GSI in the Gardens and (2) site-specific GSI prototypes, including the creation of Northeast Houston's first rain garden. Interviews focused on eliciting community attitudes, experiences, and aspirations concerning GSI specifically but inevitably touched on broader issues of interest (e.g., community relationships with local government and industry, public health, environmental racism, conventional stormwater infrastructure, climate change).

11.6 Results/Findings/Discussion

11.6.1 Plan Analysis Results

The oldest plan, Plan Houston, included a vision statement, 32 goals and 12 core strategies. As the city's first general plan in a state, it does not mandate local jurisdictions to have a comprehensive land use plan. Of the codes used in this analysis, critical infrastructure (drainage) and ecological features (tree canopy, green infrastructure) were the only ones included in this plan. Specifically, "Sustain Quality Infrastructure", a core strategy, identified four action items regarding improving drainage, green infrastructure, etc. (see Table 11.1). This plan lacks many best practices. For example, there is no reference to flooding, risk/hazard information, climate change/global warming, relocation, vulnerable populations, or the Gardens. Meanwhile, Houston's Hazard Mitigation Plan, updated in 2018, thoroughly identifies and details high, medium, and low hazard risks that could potentially impact the city. Flooding is the top high hazard in the plan's risk assessment. Yet, many of the specific recommendations included in this hazard mitigation plan are not reflected in Plan Houston.

Following Hurricane Harvey, Houston's Office of Emergency Management completed an after action report. These reports are not required following a disaster but are considered a best practice to notate what went well and did not go well before, during, and after a disaster. While this report provides a concise account of the event, successes, and challenges, most of the report is internal facing with recommendations detailing identified gaps in the response efforts by city departments. For example, while the report included flooding 12 times and documented its impact, recommendations focused on internal struggles including city departments needing a continuity of operations plan, employees stranded in flood waters, and employees lacking training for executing water evacuations. While this hurricane was a major flooding event, the report does not mention drainage, stormwater, or the Gardens.

Two more recent plans, Resilient Houston and Kashmere Gardens Complete Communities Action Plan, were completed post-Hurricane Harvey in 2020. The first, created as part of a nationally funded initiative is the longest plan with 18 goals across 5 objectives. Goal 11 focuses on modernizing the city's infrastructure to face future challenges. GSI is incorporated into this goal with a completion time of 2030:

> Create a multifaceted green stormwater infrastructure (GSI) program across private and public projects in order to mimic natural water flow conditions, minimize localized urban flooding, improve stormwater quality, and maximize co-benefits such as urban heat island effect reduction, neighborhood beautification, and improved air quality (p. 119).

The second plan was an intensive community-focused planning initiative—Complete Communities—led by Houston's Mayor; Kashmere Gardens was one of ten selected communities. The planning process included participation from 54 neighborhood leaders, 244 residents, and representatives from 33 city departments. Community members rated flood resilience as a high priority, which resulted in two primary goals: community preparedness and building flood resilience. The plan concluded with multiple goals and associated objectives addressing flood risks, climate change, and equity. As highlighted in Table 11.2, an example is Goal 11, which included the integration of GSI projects in the community. The community wanted city-led projects focused on "ensuring high quality drainage systems, including additional storm drains and well-maintained ditches and tributaries, and advocating for policies that promote green stormwater infrastructure" (p. 33).

11.6.2 Interview Results

All community leaders and members in this study agreed that climate change was real and impacting the Gardens in increasingly dramatic and damaging ways via more intense storms, rainfall, and flooding. All further agreed that more action was needed regarding stormwater management. They were dissatisfied with city and county-led efforts to create conventional stormwater infrastructure. Indeed, the only major conventional stormwater infrastructure project intended to help the Gardens, Project Hunting, was completed only in 2023 (despite having been first proposed in 1948). Several interlocutors expressed the opinion that Northeast Houston had always been a "low priority" area of the city in terms of infrastructural investment for economic and racial reasons.

There was also strong interest expressed in GSI as a complementary model of stormwater management. One community leader stated "*we're always telling people about GSI now. It's definitely part of the solution here.*" There was an appreciation for how some GSI techniques (e.g., rain gardens) could be accomplished with modest resources, beyond time and labor. This allowed the community to engage in GSI practices even without new external funding sources. However, some community members felt that time and labor were in short supply; they felt it would require more concerted efforts by local and federal government entities to realize the promise of GSI in the Gardens.

To add specificity to these general interview results, two brief studies, drawn from the semi-structured interviews, offer a sense of the range of community opinions concerning GSI.

Retired police officer VS (aged 72) said that flooding and "*the many environmental problems*" Gardens' residents faced like concrete batch plants poisoning the air concerned her the most. "*They're not even supposed to be within 500 feet of a residence, so how do they keep getting permits*?" Her home had flooded two to three feet during Tropical Storm Allison in 2001 and Hurricane Harvey in 2017; she blamed the inundation on releases from an upstream detention pond, "*we were flooded not by nature but by man.*" VS felt the drainage problems were related to Houston's lack of zoning ordinances which meant that no one was really paying attention to how new developments would impact what she described as an already "*fractured system.*" VS was enthusiastic about GSI because she thought it would bring trees back and give the community a way to control the problem of open drainage ditches, which she called "*orphan ditches.*" "*We're inundated with orphan ditches around here. They call them orphans because the city doesn't think they are responsible to maintain them. And the county isn't sure they should maintain them. So, they're in a donut and no one's doing anything.*" The most recent update from the community's leadership is that the city has finally reassumed maintenance responsibilities after more than 20 years of neglect.

A community organizer, BP (aged 55) expressed more skepticism about what GSI could do for the Gardens. His reasoning was not that people would have any specific objection to GSI, just that there "*is a high level of mistrust here because people haven't gotten the services, the level of respect they think they deserve. It would be easy for these [GSI] initiatives to gain acceptance in an affluent area like The Heights. These are individuals that have paid significant amounts for their homes. They've been well cared for it and they'll return the favor. Hell, they'll go run a 5 k for you to fundraise for it.*" BP said the Gardens by contrast had experience with "*people selling snake oil*" promising wonderful solutions that never materialized. They felt like they were being conned or made into guinea pigs to test ideas that would never be used to improve their lives. BP noted however that he thought people would accept small, scalable solutions like GSI if there was a practical demonstration that they worked. He said it was clear to everyone that the status quo of flood control was a broken model. "*We need a reset, need to get everyone into a room because it's an all hands-on deck situation. Here's the fundamental truth that we need to get our head around. Houston was built out of a swamp. We've put concrete everywhere. We've built roads and bridges and skyscrapers. You can drive over the swamp now. But it's still a swamp with swamp weather. We got to pay the freight for that decision.*"

To offer the kind of practical demonstration that BP and other community members called for, the research team, with the assistance of local landscape architect Keiji Asakura, helped organize the creation of the first community rain garden in Northeast Houston in 2022 and 2023 (see Image 1). The success of this rain garden prototype has led to a further collaboration and proposal to create a rain garden at the local Kashmere High School to help teach students and their families GSI techniques along with hydrology, botany, and other relevant scientific skills. Although this project has been delayed by the state takeover of the Houston Independent School District, the research team remains optimistic that progress will be made in 2024 toward this second project.

11.7 Conclusion

Urban flooding poses significant challenges for communities, particularly those with high social vulnerability indices. Despite the effectiveness of GSI in reducing flood risk, its benefits have eluded under-resourced neighborhoods (Baker et al. 2019). In response to this disparity, t a multidisciplinary team of scholars in two underserved neighborhoods in Houston aimed to design cost-effective, community-supported

Image 1 Trinity gardens community rain garden event (October 10, 2022)

GSI plans tailored to mitigate flooding exacerbated by climate change. By incorporating community input and leveraging local knowledge, the study aimed to foster acceptance and ownership of GSI initiatives. By integrating scientific expertise with community-driven solutions, this study sought to empower vulnerable communities to withstand the growing threats of urban flooding and climate change, setting a precedent for inclusive and sustainable urban development practices, and ensuring the resilience and well-being of communities.

When dealing with state and local authorities, community leaders in the Gardens often express their frustrations. The residents are organized, engaged, and trying to hold government officials accountable. They are clear in their requests for specific types of projects for increasing their resilience to flooding risks—gray and green stormwater infrastructure projects—as highlighted in the 2020 Kashmere Gardens Complete Community Action Plan. Yet, government officials are either dismissive or implementing other types of projects the community does not want.

Complicating the issue is that these residents have a history of documented injustices that eroded their trust in the government, but they are organized and actively involved in community development efforts (Lynn 2017). Specifically, they value owning land and passing it down to the next generation. Gentrification, environmental justice, and buy-out programs in Houston have stressed this culturally important tradition (Chilukuri 2024). Therefore, the nexus between research and implementation of projects is crucial for addressing issues, such as flood mitigation in underserved communities like the Gardens, where the lack of zoning emphasizes community-driven initiatives.

The analysis of the five major development plans reveals that Houston's approach to post-Hurricane Harvey planning exhibits notable strengths in its utilization of data-driven methodology, forward-thinking perspective, and capacity to draw lessons from previous disasters. Yet, these are multiple planning efforts; without a state mandate or a legally binding comprehensive land use plan, these efforts remain disjointed. Moreover, limitations are impeding the efficacy of these projects, including a deficiency in implementation, insufficient political support, limited involvement of the community, and unreliable financing channels. In cases like northeastern Houston where conventional, extensive infrastructure projects have encountered delays spanning decades, GSI initiatives present a more flexible and potentially community-oriented approach. Community members also think that these initiatives align with their goal for increased ownership in stormwater management, which is a shift from a bureaucratic system that residents feel has disregarded their perspectives.

To effectively address these problems, it is suggested that the implementation of public–private partnerships, the provision of incentives for green development, accelerating approval processes for GSI projects, and the establishment of specialized finance channels be considered as supplementary measures to bolster disaster resilience endeavors. Moreover, attention to the social component of the socio-ecological-technical system is crucial for ensuring equitable, inclusive, and sustainable GSI programs (Venkataramanan et al. 2020). To achieve wider acceptance of GSI solutions, it is imperative to maintain a persistent emphasis on community education and the exhibition of successful outcomes, while concurrently endeavoring to restore confidence in public authorities.

Acknowledgements This work was supported in part by the National Oceanic and Atmospheric Administration (grant no. NA21OAR4310270).

References

Adam B (2024) U.S. Billion-dollar weather and climate disasters, 1980—present (NCEI Accession 0209268). [19.19]. NOAA National Centers for Environmental Information. Dataset. https://doi.org/10.25921/stkw-7w73. Accessed 17 July 2024

Anguelovski I, Connolly JJ, Garcia-Lamarca M, Cole H, Pearsall H (2019) New scholarly pathways on green gentrification: What does the urban 'green turn' mean and where is it going? Prog Hum Geogr 43(6):1064–1086

Blake ES, Zelinsky DA (2018) National Hurricane center tropical cyclone report: Hurricane Harvey. National Hurricane Center, National Oceanographic and Atmospheric Association

Biswal BK, Bolan N, Zhu YG, Balasubramanian R (2022) Nature-based Systems (NbS) for mitigation of stormwater and air pollution in urban areas: A review. Resour, Conserv Recycl 186:106578

Blackburn J, Chenevert C, Hodder J, Swackhamer S, Fucile-Sanchez E and Luey G (2021) A proposal for equitable flooding solutions in Northeast Houston. Bayou City Initiative and Northeast Super Neighborhoods United

Campbell-Arvai V, Lindquist M (2021) From the ground up: Using structured community engagement to identify objectives for urban green infrastructure planning. Urban For & Urban Green 59:1–13. https://doi.org/10.1016/j.ufug.2021.127013

Chilukuri S (2024) Industry poisoned a vibrant Black neighborhood in Houston. Is a buyout the solution? "Do you have to lose your history, your culture, or your identity in that process?" Grist. Retrieved from https://grist.org/equity/industry-poisoned-black-neighborhood-houston-is-buyout-solution/

City of Houston (2020) Resilient Houston: an equitable, community-driven strategy for strengthening Houston's future. Office of the Mayor, City of Houston

Corbin J and Strauss A (2015) Basics of qualitative research: Techniques and procedures for developing grounded theory, 4th ed. Sage Publications

Dharmarathne G, Waduge AO, Bogahawaththa M, Rathnayake U, Meddage DPP (2024) Adapting cities to the surge: A comprehensive review of climate-induced urban flooding. Results Eng, 102123

Finewood MH, Matsler AM, Zivkovich J (2019) Green infrastructure and the hidden politics of urban stormwater governance in a Postindustrial city. Ann Am Assoc Geogr 109(3):909–925

Gaziulusoy I, Ryan C (2017) Shifting conversations for sustainability transitions using participatory design visioning. Des J 20(sup1):1916–1926. https://doi.org/10.1080/14606925.2017.1352709

Grabowski ZJ, McPhearson T, Pickett ST (2023) Transforming US urban green infrastructure planning to address equity. Landsc Urban Plan 229:104591

Green D, O'Donnell E, Johnson M, Slater L, Thorne C, Zheng S, Stirling R, Chan FKS, Li L, Boothroyd RJ (2021) Green infrastructure: The future of urban flood risk management? Wiley Interdiscip Rev Water 8(6):e1560

Greenbaum J (1991) A design of one's own: Towards participatory design in the US. DAIMI Report Series 20 (375). https://doi.org/10.7146/dpb.v20i375.6607

Hamouz V, Muthanna TM (2019) Hydrological modelling of green and grey roofs in cold climate with the SWMM model. J Environ Manage 249:109350

Hendricks MD, Dowtin AL (2023) Come hybrid or high water: Making the case for a Green-Gray approach toward resilient urban stormwater management. JAWRA J Am Water Resour Assoc 59(5):885–893

Hopkins KG, Grimm NB, York AM (2018) Influence of governance structure on green stormwater infrastructure investment. Environ Sci Policy 84:124–133

Im J (2019) Green streets to serve urban sustainability: Benefits and typology. Sustainability 11(22):6483

Johns CM (2019) Understanding barriers to green infrastructure policy and stormwater management in the City of Toronto: a shift from grey to green or policy layering and conversion? J Environ Planning Manage 62(8):1377–1401

Jovanovic T, Hale RL, Gironás J, Mejia A (2019) Hydrological functioning of an evolving urban stormwater network. Water Resour Res 55(8):6517–6533

Kim HW, Park Y (2016) Urban green infrastructure and local flooding: The impact of landscape patterns on peak runoff in four Texas MSAs. Appl Geogr 77:72–81

Knox CC (2017) A football field lost every 45 minutes: evaluating local capacity to implement Louisiana's coastal master plan. Coast Manag J 45(3):233–252

Ko Y (2018) Trees and vegetation for residential energy conservation: A critical review for evidence-based urban greening in North America. Urban For & Urban Green 34:318–335

La Bissoniere BP (2022) Evaluating strategies, opportunities, and tensions for implementing green stormwater infrastructure at the site scale. Doctoral dissertation

Lee J (2019) Green infrastructure as a solution to hydrological problems: bioswales and created wetlands. UF J Undergrad Res, 21(1)

Li C, Liu M, Hu Y, Zhou R, Wu W, Huang N (2021) Evaluating the runoff storage supply-demand structure of green infrastructure for urban flood management. J Clean Prod 280:124420

LopezDeAsiain M, Díaz-Garcia V (2020) The importance of the participatory dimension in urban resilience improvement processes. Sustainability 12(18):7305

Lynn KA (2017) Who defines' whole': an urban political ecology of flood control and community relocation in Houston. Texas. J Polit Ecol 24(1):951–967

Malecha ML, Woodruff SC, Berke PR (2021) Planning to exacerbate flooding: Evaluating a Houston, Texas, network of plans in place during Hurricane Harvey using a plan integration for resilience scorecard. Nat Hazard Rev 22(4):04021030

Maryland Department of Natural Resources (n.d.) Green infrastructure resilience. https://dnr.maryland.gov/ccs/Pages/GIR.aspx [Accessed 21 July 2024]

McPhillips LE, Matsler AM (2018) Temporal evolution of green stormwater infrastructure strategies in three US cities. Front Built Environ 4:26

Mei C, Liu J, Wang H, Yang Z, Ding X, Shao W (2018) Integrated assessments of green infrastructure for flood mitigation to support robust decision-making for sponge city construction in an urbanized watershed. Sci Total Environ 639:1394–1407

Narzis A, Eisma J (2023) Developing green stormwater infrastructure strategies to reduce flood runoff and improve water quality for the vulnerable coastal communities of Houston. In: World environmental and water resources congress 2023, pp 675–687

National Institute of Buildings (2020) Natural hazard mitigation saves: 2020 Report. Retrieved from https://www.nibs.org/files/pdfs/ms_v4_overview.pdf

Onuma A, Tsuge T (2018) Comparing green infrastructure as ecosystem-based disaster risk reduction with gray infrastructure in terms of costs and benefits under uncertainty: A theoretical approach. Int J Disaster Risk Reduct 32:22–28

Roberts WM, Stutter MI, Haygarth PM (2012) Phosphorus retention and remobilization in vegetated buffer strips: a review. J Environ Qual 41(2):389–399

Rosenbloom J (2018) Fifty shades of gray infrastructure: Land use and the failure to create resilient cities. Wash. l. Rev. 93:317

Samaddar S, Si H, Jiang X, Choi J, Tatano H (2022) How participatory is participatory flood risk mapping? Voices from the flood prone Dharavi Slum in Mumbai. Int J Disaster Risk Sci 13:230–248. https://doi.org/10.1007/s13753-022-00406-5

Santhanam H, Majumdar R (2020) Permeable pavements as sustainable nature-based solutions for the management of Urban Lake Ecosystems. Nat-Based Solut Resilient Ecosyst Soc, 329–345

Schepers S, Schoffelen J, Zaman B, Dreessen K (2022) Going beyond short-term, 'reduced' PD: Towards an encompassing typology for children's participation in infrastructuring processes. Int J Child-Comput Interact 33. https://doi.org/10.1016/j.ijcci.2022.100484

Schubert JE, Burns MJ, Fletcher TD, Sanders BF (2017) A framework for the case-specific assessment of Green Infrastructure in mitigating urban flood hazards. Adv Water Resour 108:55–68

Sharma R, Malaviya P (2021) Management of stormwater pollution using green infrastructure: The role of rain gardens. Wires Water 8(2):e1507

Taguchi VJ, Weiss PT, Gulliver JS, Klein MR, Hozalski RM, Baker LA, Nieber JL (2020) It is not easy being green: Recognizing unintended consequences of green stormwater infrastructure. Water 12(2):522

te Wierik SA, Gupta J, Cammeraat EL, Artzy-Randrup YA (2020) The need for green and atmospheric water governance. Wiley Interdiscip Rev Water 7(2):e1406

Texas Commission on Environmental Quality (n.d.) Best management practices: A tool for protection [Online]. https://www.tceq.texas.gov/drinkingwater/SWAP/bmp.html [Accessed 21 July 2024]

Thorne CR, Lawson EC, Ozawa C, Hamlin SL, Smith LA (2018) Overcoming uncertainty and barriers to adoption of Blue-Green Infrastructure for urban flood risk management. J Flood Risk Manag 11:S960–S972

Tyler J, Entress RM, Sun P, Noonan D, Sadiq AA (2023) Is flood mitigation funding distributed equitably? Evidence from coastal states in the Southeastern United States. J Flood Risk Manag 16(2):e12886

Venkataramanan V, Lopez D, McCuskey DJ, Kiefus D, McDonald RI, Miller WM, Packman AI, Young SL (2020) Knowledge, attitudes, intentions, and behavior related to green infrastructure for flood management: A systematic literature review. Sci Total Environ 720:137606

Woodruff S, Tran T, Lee J, Wilkins C, Newman G, Ndubisi F, Van Zandt S (2021) Green infrastructure in comprehensive plans in coastal Texas. J Environ Planning Manage 64(9):1578–1598

World Population Review (2024) Houston, Texas Population 2024. Retrieved from https://worldpopulationreview.com/us-cities/houston-tx-population

Wu Y, Lyles W, Overstreet K, Sutley E (2024) Planning for adaptation? Examining the planning integration for hazard risk reduction. Sustainability 16(10):3999

Zhou L, Shen G, Woodfin T, Chen T and Song K (2018) Ecological and economic impacts of green roofs and permeable pavements at the city level: the case of Corvallis, Oregon. J Environ Plan Manag 61(3), 430–450

Chapter 12
Shifting Sands and Rising Seas: An Exploratory Study of Community-Centric Nature-Based Solutions for Coastal Hazards and Coastal Erosion in Alappad, Kerala, India

David Brown, Souresh Cornet, Sheikh Omar Fye, and Sudha Arlikatti

Abstract Coastal areas are on the frontline of climate change, facing alarming sea level rise, with communities in developing countries being particularly vulnerable. In India, erosion mitigation measures primarily involve structural projects implemented by the government. A case study was conducted in Alappad, a community severely affected by the 2004 Indian Ocean Tsunami, where coastal erosion is now a considerable challenge. The multifaceted impacts of coastal erosion were investigated through a qualitative social study involving interviews with various stakeholders using participatory action research. The study aimed to understand the structural and non-structural mitigation measures implemented since 2004, assess their effectiveness, and explore the potential for hybrid solutions that incorporate more nature-based approaches. Scientific literature indicates that an integrated approach, combining structural and nature-based strategies, effectively mitigates coastal erosion. The findings highlight the severe impact of erosion on local livelihoods and the community's survival. The community regarded existing structural and non-structural mitigation solutions positively. As a result, a design integrating the strengths of both structural and nature-based elements is proposed. It consists of three layers: a TetraPOT-Mangrove defence on the absolute coastline, followed by a coconet-vetiver system for sand dune rehabilitation, and a bio-shield made of native tree species. This community-centric mitigation strategy is likely to safeguard Alappad's coastline and provide capacity-building opportunities. A stepwise implementation plan is recommended to ensure its sustainability.

D. Brown
School of Engineering & Aviation, Central Queensland University, Rockhampton, Australia

S. Cornet (✉) · S. O. Fye · S. Arlikatti
Amrita School for Sustainable Futures, Amrita Vishwa Vidyapeetham, Amritapuri, India
e-mail: soureshcornet@am.amrita.edu

M. Firoz C et al. (eds.), *Nature-Based Solutions for Urban and Peri-Urban Areas*, Disaster Risk Reduction, https://doi.org/10.1007/978-981-96-5933-3_12

Keywords Coastal erosion · Nature based solutions · Climate change · Community engagement · Sustainable development · Participatory research

12.1 Introduction

With a changing climate, there has been an increase in extreme weather events the world over, but developing countries have claimed the majority of casualties (WMO Atlas of Mortality 2021). In coastal areas, adverse impacts such as rising sea levels and increasing rainfall have been felt clearly for decades, constantly threatening the livelihood of coastal areas and communities (Lehman et al. 2015; UNCC 2021). In India, more than 200 million people live within 100 km of the coasts (Sharma 2023) and are climate-vulnerable.

The Indian Ocean tsunami in 2004 had a devastating impact, claiming the lives of approximately 7,000 people along the coasts of Kerala (Ramesh 2023). As a result, many structural and non-structural mitigation projects have been introduced by government agencies, non-governmental organisations (NGOs), and citizens along Indian coastlines, including the State of Kerala. Such mitigation measures, like shore protection and coastal engineering systems, produced diverse results (Sujatha and Benny 2010).

Studies on post-tsunami coastal hazard mitigation measures along the Kerala coast concluded that no single measure or uniform strategy effectively mitigated coastal hazards (Sujatha and Benny 2010). Many mitigation and rehabilitation programs focused primarily on economic recovery, largely ignoring social and environmental dimensions. These programs were launched through a bureaucratic, top-down system without adequately considering the needs of the communities (Joseph 2015), highlighting the need to explore more holistic, hybrid approaches.

12.1.1 Aim and Objectives

This study aimed to address this gap by understanding the experience and the perceived effectiveness of mitigation measures in Indian coastal communities where coastal hazards and erosion have dramatically impacted local lives.

This chapter reports on the outcomes of a qualitative case study conducted in the Alappad community in Kerala, India, which was affected by the 2004 Tsunami and has since experienced ongoing coastal erosion. Twenty years after the tsunami, it was timely to review the structural and non-structural mitigation measures implemented in the community since 2004, as well as the lived experiences of the people of Alappad. This review aimed to assess the effectiveness of these strategies in reducing coastal erosion and improving local lives, before developing recommendations for a better future for Alappad.

12.1.2 Scope

While an effort was made to gather representative primary data from coastal communities, the scope of the qualitative research completed in this study was limited. The engineering design discussed in this chapter is strictly conceptual at this point. It was primarily developed through a limited sample of respondents and desk study. Further research and experimentation are required before progressing to a finalised design. Recommendations are based on a best-case scenario, requiring government support for implementation, in accordance with the policy context in India.

This chapter includes an overview of existing scientific literature on structural and nature-based solutions for mitigating coastal erosion, a summary of the research methodology used, and a discussion of the study's findings. Proposed hybrid models for coastal mitigation that incorporate community feedback and scientific literature are then presented.

12.2 Conceptual Framework

12.2.1 Understanding the Coastal Hazard Context in Alappad

Located in Kerala, India, Alappad is a coastal community of more than 20,000 people, living on a narrow strip of land between the Arabian Sea and the backwaters (see Fig. 12.1). Most families rely on fishing for their livelihood and are predominantly Hindu. On December 26, 2004, a tsunami triggered by an earthquake in Indonesia devastated large areas of the village and shoreline. Alappad was Kerala's worst-affected village, with over 100 people killed. Additional bridges were built following the catastrophe. The Alappad coastline is strewn with black sand that is currently being mined due to its rich mineral content (see Fig. 12.2). Sand extraction became a major industry in Alappad in the 1960s, and has raised serious concerns in recent years (Deccan Chronicle 2018). Mining combined with the local terrain, makes the area especially sensitive to erosion caused by coastal hazards. Furthermore, previous coastal catastrophes have severely altered the aquatic ecosystem, destroying fish habitats and pushing local fishermen to relocate further to sea to make a living (Fye et al. 2024). This, along with rising seas and changing climate, has people questioning the security of their future.

In Kerala, various social groups, NGOs, and government agencies undertook the rehabilitation and reconstruction effort post-tsunami. Significant investments were made into structural mitigation systems, with the construction of seawalls and groins to protect the coastline (Kurian et al 2006; (Sujatha and Benny 2010). In contrast, implemented nature-based mitigation solutions were not documented much. In the immediate few years after the disaster, it was reported that government authorities closely monitored the rehabilitation process, but did not involve the communities in the initiatives (Joseph, 2015). As a result, some of the mitigation measures

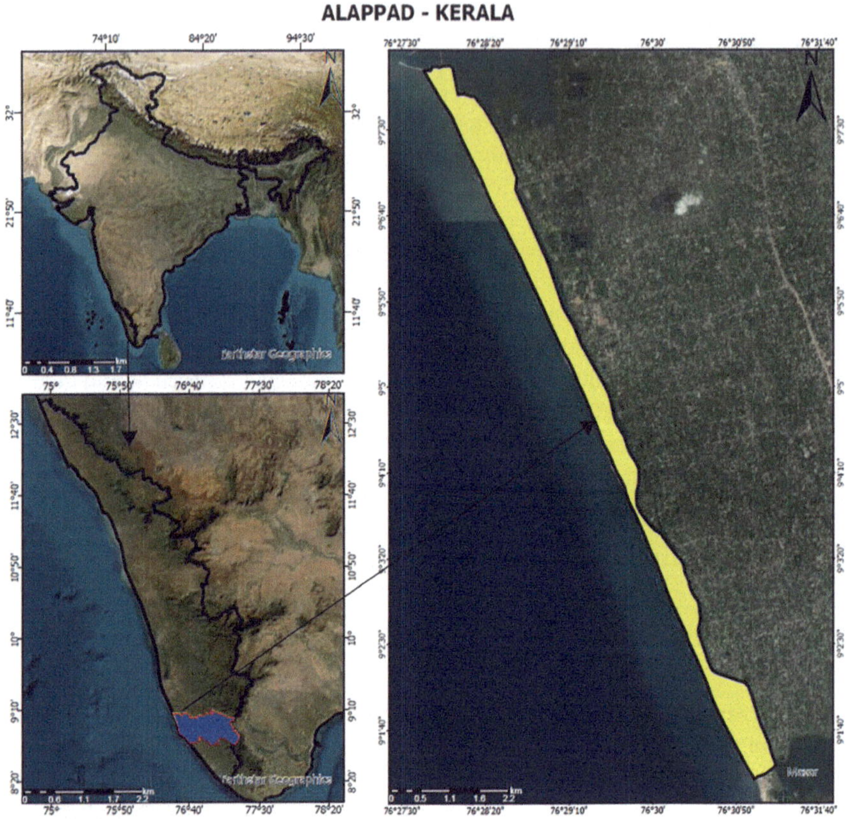

Fig. 12.1 Alappad study area map

had adverse effects. For instance, in the Alappad district, seawalls combined with backwaters and canals increased inundations along the coast (Sujatha and Benny 2010).

12.2.2 Coastal Erosion Management Strategies

Mitigation measures differ according to the climate, political scenario, and availability of material. This section reviews some of the most common strategies for addressing coastal hazards in the literature, distinguishing between structural and nature-based mitigation measures.

Fig. 12.2 Sand mining in Alappad

12.2.2.1 Structural Mitigation Measures

Hard Seawall Structures Seawalls are the most common structural coastal erosion mitigation measure. In Alappad a blocking wall and seawall groins have been implemented along the coastline to act as wave-slowing elements. These contribute to coastal defence and restoration of eroded beach materials (Oetjen et al. 2022). Permeability in structural designs is another important factor in coastal erosion control. Staggered non-continuous layouts in mitigation structures have been shown to reduce by 40% hydrodynamic forces during wave impact (Oetjen et al. 2022).

Tetrapods are a type of coastal engineering equipment intended to prevent wind- and wave-induced erosion. It has gained popularity due to its defensive properties. Tetrapods have exceptional stability in terms of hydraulic qualities, material strength, and manufacturing techniques. Tetrapods are made of concrete and have a tetrahedral structure that allows seawater to circulate between them, dispersing waves' impact. Tetrapods effectively limit displacement by interlocking their legs to protect a structure's surface. Tetrapods do not require steel reinforcement and are easily controllable while the concrete is still drying, which is a significant advantage for marine construction (Suwannarat et al. 2020).

12.2.2.2 Nature-Based Solutions

Mangroves Mangroves, native to tropical and subtropical coastal intertidal zones, thrive in high temperatures and humidity. Their dense root systems and vegetation act as filters, capturing pollutants and preserving water quality. Mangrove ecosystems are areas of high biodiversity, providing a home for many aquatic species. Studies reveal that they safeguard populations and infrastructure during cyclones because of their wave attenuation properties (PreventionWeb 2022). For instance, in 2004 the tsunami did cause fewer casualties to locations that had mangrove vegetation (Tanaka et al. 2007).

Vetiver Grass This perennial grass, native to tropical Asia, has a high self-rising ability: Its roots keep growing as soil accumulates around its stems, therefore being an ideal choice for shoreline rehabilitation (National Academies of Sciences 1993). A single root of vetiver roots has an estimated tensile capability of 27 MPa, while soil reinforced with vetiver root is 5.1 times stronger (Badhon et al. 2021). Incorporating vetiver roots into cohesionless soils significantly altered their strength parameters, causing them to behave like cohesive soils and greatly enhancing accretion (Hoque et al. 2021). This illustrates vetiver's significant capabilities for preventing coastal erosion. Vetiver systems have been applied for coastal erosion control in India, the Philippines, and Brazil. In Brazil, a hybrid solution using vetiver grass and a low seawall was implemented to reestablish dunes and prevent further erosion. Despite a severe storm damaging the vetiver grass, no soil loss occurred, and the vetiver system naturally regrew (Truong 2017).

Coastal Trees Shelterbelts, or bio-shields, have been implemented in various regions of the world as a coastline risk reduction technique (NIDM 2008). Casuarina Trees, Coconut Palms, Cashew Trees, and Giant Calotrope are among the most commonly observed species employed as bio-shields in South India, and they all have their own advantages and downsides when utilised in coastal mitigation (Saudamini et al. 2014; De Brito et al. 2018; Bourdeix et al. 2021; Heuze et al. 2016), which have been assessed in Appendix A.

12.2.3 Rationale of the Study

The study's rationale lies in the urgency to address the ongoing threats faced by coastal communities in Kerala. In the case of Alappad, which was heavily affected by the 2004 tsunami and is now exposed to severe coastal erosion, the very survival of the community is at stake.

Implemented mitigation measures had contrasting results. Studies have found that during natural disasters, communicating with community members and deciphering their actual needs is critical to ensuring appropriate response and guide relief operations (Guntha et al. 2020; Amritanand et al. 2020). Unfortunately, after the 2004 Tsunami, the communities were not consulted. Lack of community inclusion was a common gap in the rehabilitation and reconstruction efforts (Joseph 2015).

In light of this, almost 20 years after the tsunami, and given the limits of structural mitigation measures that were implemented, it is worth assessing the communities' perspective on the mitigation measures that were implemented over the years, their effectiveness, and to explore alternative and more holistic approaches to ensure better sustainability and resilience.

12.3 Methodology

This study was essentially qualitative and exploratory in nature. The chart shown in Fig. 12.3 provides a breakdown of the 4-stage methodology employed, illustrating how each step in the process builds on the last.

12.3.1 Qualitative Study of the Alappad Community

In line with the community focus of the study, a qualitative inductive approach was preferred. Given the exploratory nature of this research in Alappad, a case study was an appropriate method to study this challenge in its context (Cornet et al. 2024). Participatory action research was used to gather information and explore the community's experience with the challenge of coastal erosion (Cornet and Barpanda 2021). As part of this, two main data collection tools were used.

The main information collection tool was a community survey of target groups representative of the Alappad community. These specific target populations were chosen by looking at the most prevalent local livelihood sources, those more knowledgeable about the historical context of coastline deterioration in Alappad, those most affected by this erosion, those who must think about departing Alappad in search of a more sustainable living, and those tasked with governing the community:

- Fishermen—the predominant industry in Alappad.

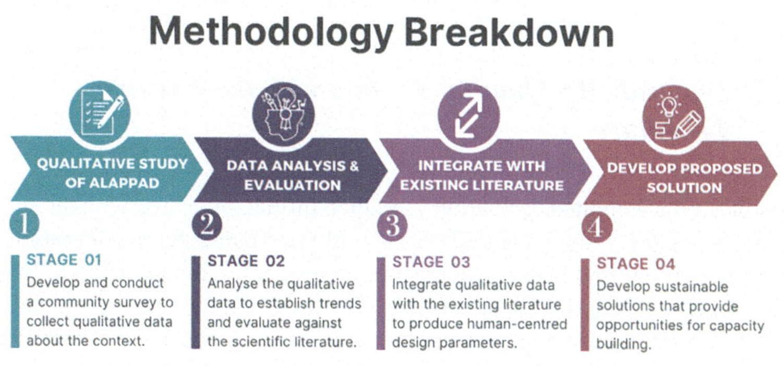

Fig. 12.3 Methodology breakdown chart

- Elderly—provide a glimpse of the past situation.
- Youth: to present an overview of Alappad's foreseeable future.
- Self-Help Groups (SHGs)—one of the greatest capacity-building opportunities in Alappad.
- Local Government Officials—to speak about local government support.

The questionnaire aimed to capture the personal experiences of community members regarding coastal hazards, assess their opinions on the effectiveness of current local mitigation measures, and understand their perceptions of Alappad's future prospects in terms of physical, social, and economic aspects. It consisted of three sections: Experience with coastal hazards, Mitigation measures, and Future prospects.

Where possible, the community survey was executed through interviews with focus groups of people from the selected target groups. Each focus group provided basic demographic information about the individuals sharing their experiences in response to the survey so that the data could be analysed categorically. The questionnaire was designed to require a minimal time commitment. It mainly contained semi-structured and open questions. This would make it accessible to the locals, achievable, and provide sufficient qualitative data to demonstrate trends.

In addition to this, extensive participant observation was conducted all along the coastline. It aimed at identifying on-site all the structural and nature-based solutions implemented in the communities where the survey was carried out. Observations were documented through field notes and photographs.

12.3.2 Data Analysis and Evaluation

Qualitative data was juxtaposed using Excel to identify patterns and trends, which were compiled in a graph to display their frequency. Based on the frequency of trend occurrence, a weighted evaluation was conducted to assess the priorities as perceived by the community.

12.3.3 Integrate the Qualitative Data with the Existing Literature

The results of the weighted evaluation were then further compared with the scientific literature. Where the human-centred priorities aligned with scientific literature, these priorities were confirmed. Where there was variance, the human-centred priorities were adjusted to reflect the impact projected by the integrated qualitative and scientific data. This integrative approach provided an evidence-based list of priorities that could be used to develop a human-centred design. Similarly, evidence of structural and nature-based mitigation methods collected through participant observations was assessed with related literature.

12.3.4 *Development of Proposed Solution*

Finally, evidence-based integrated priorities were used to propose a hybrid approach to coastal erosion mitigation and rehabilitation. Given the impacts of coastal erosion on livelihood in Alappad, the proposal seeks to include innovative approaches to enhance the local community's capacity. This includes creating new industries, rehabilitating the environment, and fostering economic development for a sustainable future in Alappad.

12.4 Results and Discussion

12.4.1 *Qualitative Study Results*

12.4.1.1 Community Qualitative Survey Results and Observed Trends

The qualitative study, surveying approximately 118 people in 5 focus groups and additional key informants through individual interviews, revealed the primary concerns held in common by the community members (Table 12.1).

Figure 12.4 graphically shows the percentage of respondents from the surveyed population that shared these concerns. Figure 12.5 shows the main trending responses. We can see that the first 3 major concerns were shared by all respondents: (1) the unsustainable black sand mining practices, (2) the lack of support from the government, and (3) the decline in livelihood. Next, (4) the constant fear of the threat of natural hazards and the impact of climate change were shared by 80% of respondents, and finally, 40% shared concerns regarding (5) the difficulty of reconstruction due to unsustainable mining practices leaving their land zoned unfavourably according to the Coastal Regulation Zones (CRZ) enforced by the Indian Government.

Additionally, the qualitative study indicated that the seawall groins constructed since the 2004 tsunami do provide some protection. Still, there is a general feeling

Table 12.1 Summary of the community members' respondents

	Focus group 1	Focus group 2	Focus group 3	Focus group 4	Focus group 5	Key informants
Fishermen	5	1	3			
Elderly	5	1	3	8	24	
Youth					5	
Ladies (SHG)				13	50	
Govt. Officials						2

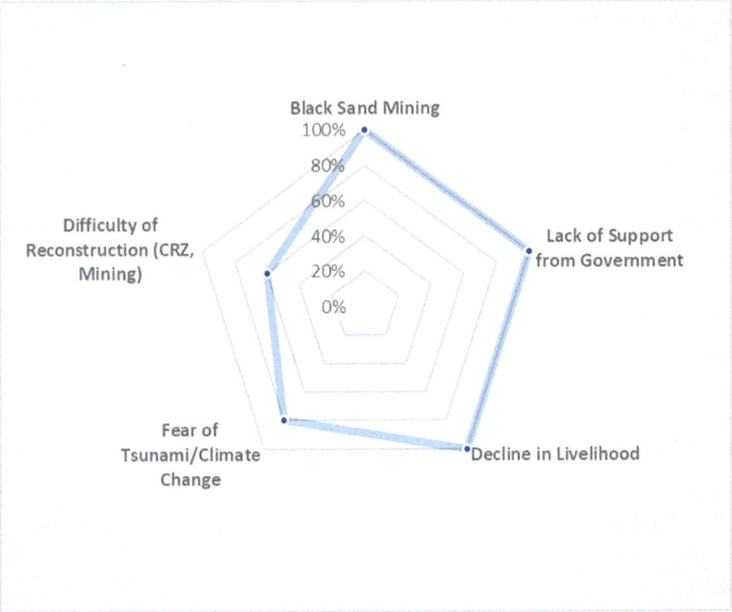

Fig. 12.4 Community survey results—primary concerns

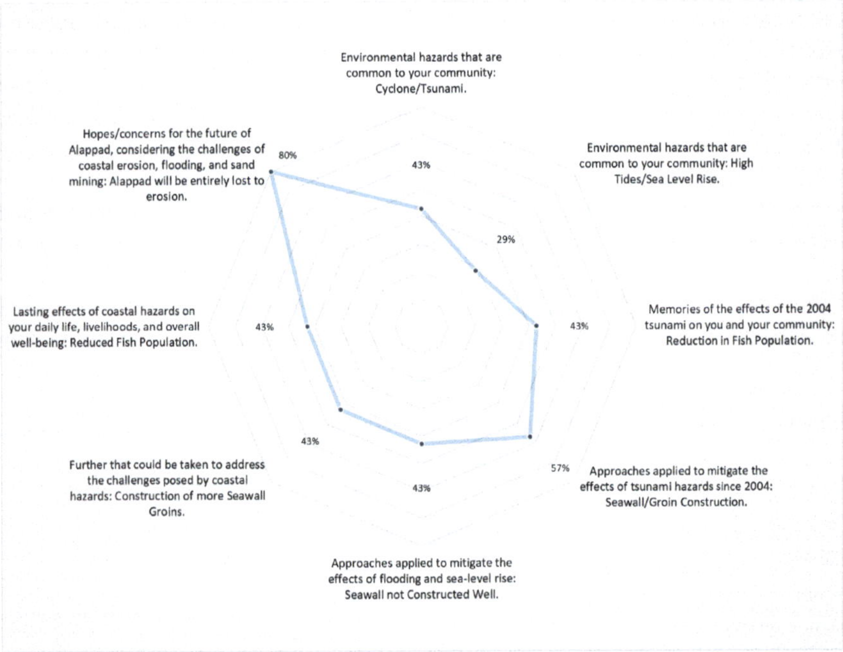

Fig. 12.5 Community survey results—trending responses

that they would be more effective if constructed longer. There was positive feedback about the areas where tetrapods have been employed in seawall construction, aligning with Suwannarat et al. (2020). Several of the fishermen indicated that the seawall construction technique was substandard. This was confirmed by on-site engineering analysis.

Another meaningful learning was that the fish population has reduced substantially since the erosion started. Where traditional manual fishing techniques once sustained the village's livelihood, a decline in fish stocks has forced fishermen to resort to mechanised fishing techniques several kilometres offshore to support their families. This significantly heightens the risks associated with their work, leading to considerable anxiety among their families, and is increasingly unprofitable due to the rising costs of non-renewable resources.

One group of elderly gentlemen also recalled a time when there were tall black sand dunes approximately 2–3 m high, with thick native forests along the coastline in Alappad that prevented erosion. Although there was consensus among the target groups that the future of Alappad appears bleak, with many foreseeing a potential exodus from the village, reflecting on the coastal environment's past could offer a glimpse of hope for the village's future.

12.4.1.2 Evaluation of Trends

The observed trends (Fig. 12.5) were compared with the literature, and several of them were confirmed. For instance, people recognised that building coastline seawall groins at periodic distances would play a role to absorb the waves and reinstate sand on the shores (Earth Science Australia 2024). Similarly, the need to improve the Seawall construction techniques to be more effective aligns with the literature (Australian Standards 2002). The overall positive feeling of the population towards revegetation is confirmed as a solution to address coastline erosion in the literature (NIDM 2008).

Secondary trends are also important to consider in the process of developing an appropriate solution design. In particular, livelihood is a primary concern due to a considerable drop in the fish population. Although it gives some locals jobs, sand mining is seen negatively since its detrimental environmental repercussions exceed the positive community benefits. Overall, there is a prevalent view that youth will most probably need to seek employment elsewhere due to the serious dangers facing the local fishing industry (Fig. 12.5).

Based on a weighted evaluation of qualitative trends, focusing on human-centred needs and community sustainability, the preferred mitigation methods were ranked as follows: (1) Improved Seawall Construction, (2) Enhanced Sand Mining Regulations, (3) Additional Seawall Groins, and (4) Revegetation. These findings align with previous studies in the area (Fye et al. 2024), offering guidance for developing sustainable mitigation strategies (Table 12.2).

Table 12.2 Weighted assessment of remediation techniques based on identified qualitative trends (Fye et al. 2024)

Mitigation methods	Physical security impact	Community sustainability impact	Environmental impact	Capacity building opportunity	Total score
Additional seawall groins	5	3	4	4	16
Improved seawall construction	5	4	5 (depending on the solution)	4	18
Revegetation	3	3	5	3	14
Improved sand mining regulations	5	5	5	2	17

12.4.2 *Observation and Evaluation of Structural and Nature-Based Solutions Implemented in the Study Area*

In addition to the community survey, on-site observation of the Alappad coastline enhanced understanding of the coastal erosion mitigation measures. While structural mitigation is crucial, Alappad's rural environment also features rich nature-based non-structural components. In this section, observations are presented and discussed in light of the literature.

12.4.2.1 Structural Components

Seawall Groins The local population appears to be quite unified in its support of seawall groins. These serve as breaker walls that extend perpendicular to the beach, limiting wave energy and reinstating sand along the shore (see Fig. 12.6). Nevertheless, heightened wave energy and a shifting climate worsen the scouring effects from overtopping and the erosion caused by seawater and sand penetrating voids, leading to further destabilisation of these structures. Thus, careful consideration in the construction of seawalls is crucial.

Overtopping causes the majority of damage observed to Alappad's seawall structures. To better prepare for future weather events, seawall construction techniques need to be enhanced (Oetjen et al. 2022). An integrated approach was observed in parts of the coastline, combining breakwater structures and seawalls to reduce wave energy, increasing sediment deposition, and preventing flooding and erosion.

Fig. 12.6 Seawall groins in Alappad

The literature supports the effectiveness of groins in mitigating wave energy, advocating for their construction at regular intervals along coastlines. However, determining the optimal dimensions and spacing of these structures to prevent hazardous currents in high-traffic areas is beyond the scope of this project.

Tetrapod Construction Because of the difficulties to supply rocks, cement tetrapods were set up at the northern Alappad break wall (see Fig. 12.7). The locals claim that these have successfully decreased wave energy and returned sand to the shore. The use of locally manufactured tetrapods also benefited the local economy.

The literature emphasises the effectiveness of tetrapods as permeable seawall structures. Studies have shown that the permeability of tetrapod structures helps to reduce wave energy and dissipate forces exerted on adjacent infrastructure (Suwannarat et al. 2020). This practical effect is particularly evident when optimising the balance between damage and stability by layering tetrapods in pairs (Suwannarat et al. 2020). Given their economic benefits and enhanced structural capacity, integrating tetrapods into future seawall construction and reconstruction is crucial.

12.4.2.2 Non-Structural Components

Engagement with the community revealed that after the 2004 Tsunami, several nature-based mitigation strategies were implemented, which gave overall positive results. Along some parts of the Alappad coastline, Casuarina and Coconut trees were grown as a secondary barrier. These trees act as a permeable buffer, slowing wave energy during tsunamis, stabilising the soil to mitigate erosion, and providing a windbreak during storms and cyclones. This highlights the potential inclusion of

Fig. 12.7 Tetrapod seawall construction in Alappad

non-structural measures in an integrated approach to combat coastal erosion. The following non-structural elements offer alternative strategies to enhance mitigation efforts in Alappad, beginning from the immediate coastal zone and extending inland.

Mangrove Development Historically, mangrove forests covered extensive areas along the Kerala coast. While they have largely disappeared from the coastline, their presence in the Alappad backwaters serves as evidence that they can still thrive.

The tremendous wave energy, especially during the monsoon season, poses a problem for establishing mangroves along the shore. This has an impact on plant establishment, making it impossible for them to take root and grow into healthy trees that can effectively mitigate erosion. The following strategy could be created and implemented to increase their likelihood of establishment:

1. Selecting planting dates such that the establishment occurs before the monsoon.
2. For best results when planting on the shore, raising seedlings in a mangrove nursery.
3. Creating a structural barrier component that serves as a mangrove planter and acts as a buffer to aid in the establishment and growth of the plants.

The community's requirements can potentially be satisfied by the creation of mangrove plantations across Alappad. It could provide a nature-based structural defence against coastal erosion, while at the same time improving the water quality and regenerating the fish population.

Sand Dune Development Community elders described dunes as a historical feature of Alappad's natural environment and as one of nature's defences against coastal erosion. However, recent beach washing and ongoing deep mining processes have destabilised the coastal sand. The lack of vegetation, exacerbated by coastal risks and mining activities, prevents sand consolidation and makes it susceptible to displacement. Addressing these issues is crucial for promoting dune redevelopment and providing safety and tranquillity to the local population.

12.4.3 An Integrated Approach to Coastal Erosion Mitigation

This study aimed to explore effective community-centric solutions to mitigate coastal hazards in Alappad. Based on extensive community interactions and field observation of structural and nature-based measures, an approach that integrates both solutions appears relevant. This could offer enough structural support. It could also incorporate elements derived from nature that can protect the natural environment from coastal hazards, restore habitat, while offering opportunities for capacity building. This observation aligns with recent findings on disaster management in the Indian context, stating that a diverse country like India necessitates integrated solutions for managing multiple hazards, and that empowering communities to build disaster resilience is essential in achieving this goal (Ramesh et al. 2021).

A broad concept design with three distinct levels of defence is depicted in Fig. 12.8. The TetraPOT-Mangrove defence, which is both structural and non-structural, is the initial level of security along the whole coastline. The following level of protection is the coconut-vetiver system, which strengthens the terrain, increases shear stress, and protects the shoreline from overtopping. Finally, strategically placed coastal trees stabilise the terrain against severe winds, as well as absorb wave power in the case of a major weather event such as a storm or tsunami.

12.4.3.1 First Layer: TetraPOT-Mangrove Defence

Two of the most efficient coastline protections in Alappad are the Tetrapod, and the Mangroves. A hybrid mitigation solution integrating these two components is proposed as a first layer of defence on the absolute coastline.

A hybrid design of a Tetrapod that can simultaneously function as a mangrove planter has recently emerged as a promising solution (Salam et al. 2023). As the mangrove takes root, the concrete structure mitigates structural damage. TetraPOTs supply cost is probably not going to be economically possible for the community, however, a similar design could be developed to suit the local context.

12.4.3.2 Second Layer: A Coconet-Vetiver System

Dune landscapes could be planted with vetiver grass throughout the coconet system and covered in a layer of coconut fibre geo-netting (coconet) behind a tetrapod seawall planted with mangroves. This non-structural nature-based solution can help mitigate

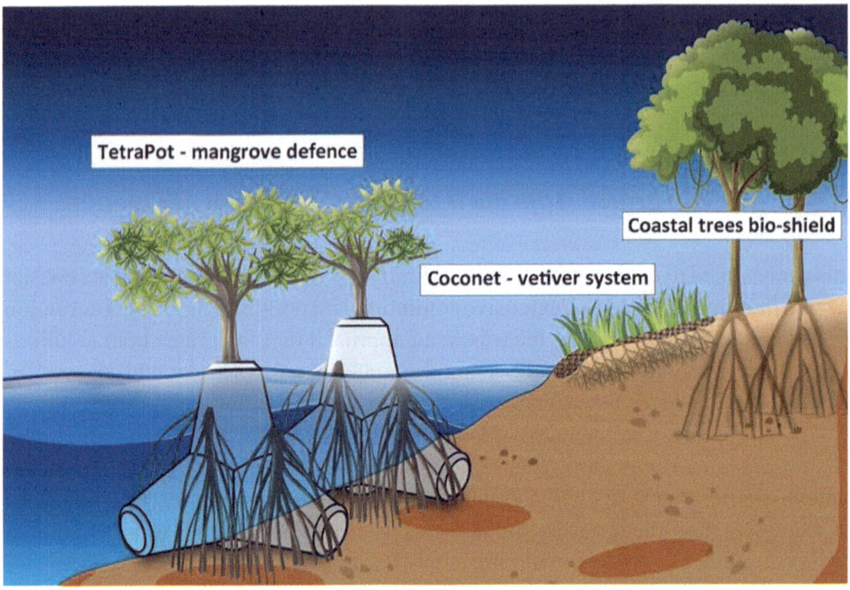

Fig. 12.8 Conceptual design of a hybrid solution based on the Alappad case scenario

the coastal erosion challenges and increase dune redevelopment on the Alappad coastline.

Studies have proven the structural effectiveness of vetiver grass as a sand-binding, naturally occurring technique for preventing beach erosion (Badhon et al. 2021). The cohesionless sandy soil's shear strength will be substantially enhanced by the vetiver grass, increasing its resistance to wave power. Vetiver grass is a natural dune builder since it also has a great capacity for self-rising (Encyclopaedia Britannica 2024). Truong (2017)'s rehabilitation efforts offer a theoretical example of how the Alappad dunes can gradually regenerate. According to earlier research, vetiver grass and coconets installed in dune preservation programmes significantly reduced erosion (Truong 2017). Given the local availability of coconut on the Kerala coast, fibres from coconut husk could be used to manufacture coconet as a capacity-building opportunity for locals relying solely on their fishing earnings to make ends meet.

12.4.3.3 Third Layer: Coastal Trees Bio-Shield

A significant volume of vegetation is necessary for a sustained coastal rehabilitation programme. When revegetating the Alappad coast, protecting against erosion, reestablishing natural habitats, and enhancing capacity are the main factors to take into account.

The community feels optimistic about revegetation efforts. However, given the low soil fertility and stability post-mining, only the plantation of coconut and casuarina trees has been thriving so far as a secondary defence behind the seawall. Additionally, several species of tree native to South India could be considered for the revegetation process as mitigation for coastal hazards and erosion: (1) Giant Calatrope, (2) Calophyllum Inophyllum, (3) Pongame Oiltree, (4) Pandanus, (5) Cashew Trees.

12.4.3.4 Social Dimension and Community Capacity Building

A central aspect of this approach is the societal dimension underpinning community-level initiatives. For long-term sustainability, community involvement and direct benefits are essential. Therefore, the societal dimension is integrated at all levels. Building community capacity and stabilising livelihoods can be achieved by focusing on two primary goals.

Firstly, diversifying the community's skills and introducing new livelihood opportunities, such as manufacturing Tetrapods and TetraPots is important. The community already has experience with this from the northernmost breaker. Installing TetraPots, coconets, and plant vegetation will create significant employment opportunities. Additionally, maintenance work can provide part-time jobs for several months after each section is completed.

Secondly, restoring the natural coastal ecosystem will help re-establish local fish populations, supporting a sustainable fishing industry in the future. Establishing mangroves and coastal flora will contribute to this effort. Coastal rehabilitation nurseries, focusing on species needed for revegetation, can also create opportunities for building community capacity by establishing local industries or employing Self-Help Groups (SHGs).

Each goal is interdependent. Understanding how the first aim affects the second is necessary for the community to fully support the first goal, which can only be accomplished by putting the first goal into action.

The extensive scope of this rehabilitation program makes it challenging to finance and implement all aspects simultaneously. Therefore, a stepwise approach is proposed to enhance the likelihood of acceptance and successful implementation:

1. Install seawall groynes on a regular basis, enhancing capacity by manufacturing and installing tetrapods.
2. Set TetraPOT equivalents into practice, plant mangroves across the coastline, and establish the coconet-vetiver system. This includes developing capacity through the cultivation of vetiver and mangroves, as well as designing, manufacturing, installing, and maintaining TetraPOTs.
3. Start a revegetation programme that adheres to predetermined planting patterns. This programme should include capacity training via cultivation, planting, and continuing upkeep.

12.4.4 Recommendations

This project offers insights into the social and environmental context of coastal erosion in the Alappad community and presents a high-level conceptual plan for mitigating coastal hazards. Based on research and community engagement, a stepwise rehabilitation program is proposed. The conceptual design outlined in this chapter demonstrates how this could be implemented. However, this preliminary design requires further development and implementation through a participatory process. This approach will ensure the solutions meet the community's needs and provide reassurance to residents facing the threats of tsunamis and climate change. Several recommendations must be addressed before moving forward with implementation:

- This strategy needs to be validated by additional research. In order to assess the forces acting on the structure and model the response to those forces, studies should gather data from real-world situations.
- The State Government should be consulted on mining policy and funding for capacity-building initiatives to support this program. Effective post-mining land rehabilitation is crucial for achieving meaningful change in Alappad and preventing similar issues from emerging elsewhere.

12.5 Conclusion

Twenty years after the Indian Ocean tsunami, this study aimed at assessing the effectiveness of coastal erosion mitigation strategies from the perspective of the communities, so that more community-centric solutions could be developed.

The qualitative study findings resonate with the distressing narrative of the community–coastal erosion has cast a long shadow over the residents of Alappad. Among the negative trends, the declining fish supply, exacerbated by coastal changes, has undermined the community's primary industry, forcing fishermen to cast their nets further out to sea, increasing stress for their families. Additionally, the constant threat of tsunamis and climate change has affected mental well-being, amplifying insecurity among villagers as they face an uncertain future.

Despite this, the qualitative study unearthed positive feelings about specific mitigation strategies observed along the coastline. Seawall groins, tetrapod seawall construction, revegetation efforts, and integrated solutions combining structural and nature-based elements have proved effective and garnered positive sentiment, which aligned with the scientific literature.

Consequently, a holistic mitigation solution was proposed, integrating the strengths of both structural and nature-based elements. The best sustainable method for gradually returning the coastline to its previous vegetated state appears to be a phased coastal rehabilitation programme.

This holistic, community-centric mitigation strategy emphasises the pivotal role of community engagement in developing solutions that integrate local perspectives.

It highlights the importance of capacity-building at every stage to create a resilient and prosperous future for Alappad. This approach aims to combat coastal erosion effectively while enhancing the well-being of its residents.

Acknowledgements The authors express their immense gratitude to Sri. Mata Amritanandamayi Devi, Chancellor of Amrita Vishwa Vidyapeetham, who has inspired them to perform selfless service to society.

This study was conducted as part of the Amrita School for Sustainable Futures and the UNESCO Chair on Experiential Learning for Sustainable Innovation & Development, at Amrita Vishwa Vidyapeetham in collaboration with the Civil & Humanitarian Engineering student program at Central Queensland University, Australia. The authors sincerely thank the Amrita Live-in-Labs® Academic program for providing all the support. The authors also thank all the community stakeholders for participating in the study and guiding them throughout the process. Finally, the authors thank the government officials of the Alappad district of Kerala.

Appendix A Evaluation of Coastal Tree Species Used as Shelterbelt

Tree names	Advantages	Disadvantages	Sources
Casuarina (She-oak) trees	·Regulates atmospheric carbon and hydrological flow of water, and retains soil within ecosystems · Is salt-tolerant and improves nutrient cycling and nitrogen fixation · Provides excellent shelterbelts to horticultural crops and good soil-borne disease suppression · Provides both a habitat and food source for birds · Can generate business in the timber, fuelwood, charcoal production, and paper pulp industries	·No hard evidence of erosion protection. Some data shows it may have exacerbated coastal erosion in India · Considerably higher death rates in Indian villages protected by Casuarina compared to mangrove trees · Better as an inland species—rather than a littoral species · Can threaten other coastal species	Saudamini, D. and Harpinder, S. (2014) 'Role of exotic vegetation in Coastal protection: An investigation into the ecosystem services of Casuarina', Economic and Political Weekly, 49(1), pp. 42–50

(continued)

(continued)

Tree names	Advantages	Disadvantages	Sources
Coconut palms	· Fibrous or adventitious root system · Source of food and industry · The trunk of the coconut palm has good tensile strength properties · Providing an emergency exit—through climbing	· Minimal deep root penetration reduces stability · Less stability in dwarf varieties due to root ball deficiency. Can be mitigated by deep planting · Root ball undermining uprooting and felling risk. Companion planting can mitigate this	Bourdeix, R., Sourisseau, J. M., & Lin, J. (Eds.) (2021) Coconut Risk Management and Mitigation Manual for the Pacific Region. Land Resources Division, SPC Available at: https://agritrop.cirad.fr/600537/1/ID600537.pdf (Accessed: 13 March 2024) Rajendra, I. A., & Sumariati, D. A. R. (2018). 'The role of coconut plants in relation to disaster management in the tropical coastal regions'. In MATEC Web of Conferences (Vol. 229, p. 01,012). EDP Sciences. (Accessed: 13 March 2024)
Cashew trees	· Initially brought to India as erosion mitigation because of its extensive root system · Provides moisture conservation · Can thrive in sandy soils	· Not as wind tolerant as the other species reviewed	De Brito, E. S., De Oliveira Silva, E., & Rodrigues, S. (2018) 'Caju—Anacardium occidentale'. In: Exotic fruits. Academic Press, pp. 85–89. (Accessed: 13 March 2024) Saroj, P. L., Kumar, N. K., & Janakiram, T. (2014). 'Converting wastelands into goldmine by cashew cultivation'. Indian Horticulture, 59(6) (Accessed: 13 March 2024)
Giant Calotrope (The Crown Flower)	· The extensive root system and tap root prevent erosion and provide coastal hazard resilience · Is tolerant of soil salinity and salt spray and provides sediment runoff prevention · An ornamental plant that provides medicinal, and industry benefits (fibres for ropes, cords, fishing nets and textiles)	· Contains toxic compounds in various parts of the plant · Considered a noxious weed in some countries—mitigation of this effect can be achieved by competition from other plants, especially grasses	Heuzé, V., Tran, G., Baumont, R., Bastianelli, D. (2016). 'Calotropis (Calotropis procera)'. Feedipedia, a programme by INRAE, CIRAD, AFZ and FAO. Available at: https://www.feedipedia.org/node/588 (Accessed: 13 March 2024)

(continued)

(continued)

Tree names	Advantages	Disadvantages	Sources
Pandanus	· Structural root system binds sand, protects against wind erosion, and provides dune stabilisation · Protects less tolerant dune plants from the damaging effects of salt-laden winds · Withstands moderate to severe tropical cyclones · Can be used as a living fence as a windbreak and prevents soil erosion, reducing maintenance costs	· Does not like long-term water-logged soil · Has limited ability to coppice · Growth is slowed and flowering ceases when the plant becomes shaded	Hyder, A.B.K. (2015) 'Conservation of screw pine 'pandanus' in Kerala, need for reestablishment of community – resource link', Meridian, Vol. 4(2), pp. 106–111 (Accessed: 13 March 2024) Pasiecznik, N. (2015). Pandanus Tectorius (screw pine), CABI Compendium. Available at: https://www.cabidigitallibrary.org/doi/https://doi.org/10.1079/cabicompendium.38447
Calophyllum Inophyllum (Tamanu)	· Part of the native ecosystem, it can protect against coastal erosion because it tolerates shallow and saline soils, salt spray and occasional inundation · An attractive tree with natural medicinal benefits · Strong ability to coppice and rapidly regenerate · Dense foliage makes this tree a good windbreak · The kernel can be used to produce biodiesel	· The tree is reportedly brittle · The tree has a shallow root system · Requires heavy watering and regular weeding in the first two years · Requires almost complete sun	Agoramoorthy, G., Sarnaik, J., Mungikar, R., Punde, S., & Hsu, M. J. (2012). 'Non-edible oil seed producing Calophyllum inophyllum ideal for India's future biofuel development'. GCB Bioenergy, 4(6), 708–712. (Accessed: 13 March 2024) Arumugam, A., & Ponnusami, V. J. R. E. (2019). 'Biodiesel production from Calophyllum inophyllum oil a potential non-edible feedstock: An overview'. Renewable Energy, 131, 459–471

(continued)

(continued)

Tree names	Advantages	Disadvantages	Sources
Pongame Oiltree	· Extensive root system with a tap root · The root spread is greater than most other species (9 m in 18 years) and profusely produces suckers · Tolerates moderate levels of salinity, binds dunes, and is often grown in conjunction with mangroves · Grass grows normally beneath this tree, making it ideal for companion planting · Produces industry opportunities (green manure production for other food crops, internal timber, paper production from pulp, biodiesel manufacture from the kernel, the press cake can be used as a pesticide against nematodes)	· If not managed well, it has the potential to become a weed · Growth of young trees is slow (only 1.3 m in 13 months)	Plants for a Future (2024). 'Pongamia pinnata - (L.) Pierre'. Retrieved from https://pfaf.org/User/Plant.aspx?LatinName=Pongamia+pinnata#:~:text=The%20plant%20has%20a%20long,root%20suckers%20profusely%5B303%20%5D Wiart, C. (2021). 'Antiparasitic Asian medicinal plants in the Clade Fabids'. In C. Wiart (Ed.), Medicinal plants in Asia and Pacific for Parasitic infections (pp. 121–232). Academic Press. ISBN 9,780,128,168,110. https://doi.org/https://doi.org/10.1016/B978-0-12-816,811-0.00007-X

References

Amritanand S, Anand S, Amritesh AR (2020) Dynamic and time critical emergency management for level three disaster: a case study analysis of Kerala floods 2018. In: Proceedings of the 21st international conference on distributed computing and networking, pp 1–6

Australian Standards (2002) Earth retaining structures. https://www.saiglobal.com/PDFTemp/Previews/OSH/AS/AS4000/4600/4678-2002(+A2).pdf (Accessed: 14 Mar 2024)

Badhon FF, Islam MS, Islam MA, Arif MZ (2021) A simple approach for estimating contribution of Vetiver roots in shear strength of a soil–root system. Innov Infrastruct Solut, 6(2). https://doi.org/10.1007/s41062-021-00469-1

Bourdeix R, Sourisseau JM, Lin J (Eds) (2021) Coconut risk management and mitigation manual for the Pacific region. Land Resources Division, SPC. https://agritrop.cirad.fr/600537/1/ID600537.pdf (Accessed 13 Mar 2024)

Cornet S, Barpanda S (2021) Exploring social innovation through co-creation in rural India using action research. Soc Enterp J 17(2):240–259

Cornet S, Barpanda S, Guidi MAD, Viswanathan PK (2024) Sustainability education and community development in higher education using participatory and case based approaches in India. Int J Sustain High Educ 25(9):78–93

De Brito ES, De Oliveira Silva E, Rodrigues S (2018) ‘Caju—Anacardium occidentale’. In: Exotic fruits. Academic Press, pp 85–89. (Accessed 13 Mar 2024)

Deccan Chronicle (2018) Alappad relay hunger stir against mining continues. https://www.deccanchronicle.com/nation/current-affairs/131218/alappad-relay-hunger-stir-against-mining-continues.html

Earth Science Australia (2024) Coastal environment. http://earthsci.org/education/teacher/basicgeol/coastal/coastal.html (Accessed 14 Mar 2024)

Encyclopedia Britannica, Inc (2024) ‘Vetiver’. Encyclopedia Britannica. Retrieved from https://www.britannica.com/plant/vetiver

Fye SO, Ekuban CAA, Brown D, Arlikatti S (2024) Exploring the effects of coastal erosion on livelihoods: The case of a fishing community in Kerala, India. In: Proceedings of geographic perspectives on climate change mitigation in urban and rural environments conference (In Press)

Guntha R, Rao SN, Ramesh MV (2020) Deployment experiences with Amrita Kripa: a user-friendly feature rich crowdsourced humanitarian application. Procedia Comput Sci 171:302–311

Heuzé V, Tran G, Baumont R, Bastianelli D (2016) ’Calotropis (Calotropis procera)’. Feedipedia, a programme by INRAE, CIRAD, AFZ and FAO. Retrieved from https://www.feedipedia.org/node/588

Hoque IL, Islam MS, Hoque E (2021) Effect of vetiver root on triaxial shear strength of a cohesionless soil. Geomech Geoengin 17(3):857–871. https://doi.org/10.1080/17486025.2021.1903089

Jose B, Harikumar K, Krishnan PN et al (2016) In vitro mass multiplication of screw pines (Pandanus spp.)-an important coastal bio-resource. J Coast Conserv 20:443–453

Joseph N (2015) Missing link between state and community: Post-tsunami reconstruction and rehabilitation of Alappad Panchayat, Kerala, India. In: Ha H, Fernando R, Mahmood A (eds) Strategic disaster risk management in Asia. Springer, New Delhi. https://doi.org/10.1007/978-81-322-2373-3_10

Kurian NP, Pillai AP, Rajith K, Krishnan BTM, Kalaiarasan P (2006) Inundation characteristics and geomorphological impacts of December 2004 tsunami on Kerala coast. Curr Sci 90(2):240–249

Lehmann J, Coumou D, Frieler K (2015) Increased record-breaking precipitation events under global warming. Clim Change 132:501–515

National Academies of Sciences, Engineering, and Medicine (1993) Vetiver Grass: A thin green line against erosion. In vetiver grass: A thin green line against erosion. The National Academies Press, Washington, DC. https://doi.org/10.17226/2077

NIDM (2008) National disaster management guidelines—management of cyclones. https://nidm.gov.in/PDF/pubs/NDMA/4.pdf (Accessed: 13 Mar 2024)

Oetjen J, Sundar V, Venkatachalam S, Reicherter K, Engel M, Schüttrumpf H, Sannasiraj SA (2022) A comprehensive review on structural tsunami countermeasures. Nat Hazards 113(3):1419–1449. https://doi.org/10.1007/s11069-022-05367-y

PreventionWeb (2022) Mangrove trees can act like a Sentinel against tsunamis. Retrieved from https://www.preventionweb.net/news/mangrove-trees-can-act-sentinel-against-tsunamis#:~:text=Mangrove%20trees%20can%20act%20like%20a%20sentinel%20against%20hurricane%2Dgenerated,of%20an%20incoming%20wave’s%20energy

Ramesh MV, Sudarshan VC, Harilal GT, Singh B, Sudheer A, Ekkirala HC (2021) Kerala floods 2018: causative factors that transformed single event to multi-hazard disaster. In: Civil engineering for disaster risk reduction. Springer Nature Singapore, Singapore, pp 61–82

Ramesh VM (2023) Embrace of compassion: 2004 Indian Ocean tsunami and a swift transformation & community resilience. Amrita Vishwa Vidyapeetham

Salam A, Prasad S, Noor NV, Nasma PP, Kumar A (2023) Study on effectiveness of TetraPOT as coastal protection structure. Retrieved from. https://doi.org/10.2139/ssrn.4559335

Saudamini D, Harpinder S (2014) Role of exotic vegetation in coastal protection: An investigation into the ecosystem services of Casuarina. Econ Pol Wkly 49(1):42–50

Sharma M, Khan S (2023) Coastal resilience and urbanization challenges in India. In: A. Singh (eds) International Handbook of Disaster Research. Springer Nature Singapore. https://doi.org/10.1007/978-981-16-8800-3_27-1

Sujatha CH, Benny N (2010) Impact of December 2004 tsunami on Indian coasts and mitigation measures. In: Jha MK (eds) Natural and anthropogenic disasters. Springer, Dordrecht. https://doi.org/10.1007/978-90-481-2498-5_4

Suwannarat W, Kositgittiwong D, Ekkawatpanit C, Sukjan R (2020) A study of the energy reduction of flow through concrete and natural rubber tetrapods. In: Paper presented at the 25th national convention on civil engineering

Tanaka N, Sasaki Y, Mowjood MIM et al (2007) Coastal vegetation structures and their functions in tsunami protection: experience of the recent Indian Ocean tsunami. Landsc Ecol Eng 3:33–45. https://doi.org/10.1007/s11355-006-0013-9 (Accessed: 15 Mar 2024)

Truong P (2017) Vetiver system for beach erosion rehabilitation. The vetiver network International. https://www.vetiver.org/TVN_VS_GAL_PUB/VS_Coastal_Erosion_control_o.pdf

UNCC (2021) State of Climate in 2021: Extreme events and major impacts. https://unfccc.int/news/state-of-climate-in-2021-extreme-events-and-major-impacts

WMO (2021) Atlas of mortality 2021. https://library.wmo.int/viewer/57564/download?file=1267_Atlas_of_Mortality_en.pdf&type=pdf&navigator=1

Chapter 13
Exploring the Impact of Place Attachment on Residents' Preferences for Implementing NbS in Urban Areas: A Case of Using Stormwater Recycled Through MAR in Taiyuan, China

Zhifang Wu, Ganesh Keremane, Shufen Guo, and Hanliang Fu

Abstract Background: Nature-based solutions (NbS) are seen as one of the approaches to address climate-related challenges as they protect, sustainably manage, and restore natural ecosystems and are uniquely suited for urban adaptation. While NbS can be applied in various shapes and sizes, this Chapter focuses on implementing NbS to address water security challenges in urban environments, acknowledging that place attachment and community engagement from designing to implementation are critical to the success of NbS. **Objectives:** The primary objective of this Chapter is to investigate the implementation of NbS, particularly stormwater recycling through Managed Aquifer Recharge (MAR), to augment water supply for non-potable uses in urban areas. **Methods**: This Chapter investigates one proposed stormwater recycling scheme in Taiyuan City, which shows extreme environmental degradation and faces severe water scarcity problems. The data were collected via a survey of urban residents using 'WeChat Moments'. Structural Equation Modelling was performed to test the hypotheses. **Results:** It was found that respondents reacted positively to using treated stormwater for residential purposes to alleviate the pressure on the freshwater supply. Place attachment (measured by length of residence) was found to significantly impact residents' preferences for treatment options, perceived health risks, perceived emotions, and social norms that relate to using treated stormwater for various residential purposes. **Conclusions**: The findings have important implications for decision-making about the input and scale of infrastructure investment

Z. Wu (✉) · G. Keremane
UniSA Business, University of South Australia, Adelaide, SA 5001, Australia
e-mail: zhifang.wu@unisa.edu.au

S. Guo
School of Public Administration, Shanxi University of Finance and Economics, Taiyuan 030006, China

H. Fu
School of Management, Xi'an University of Architecture and Technology, Xi'an 710055, China

M. Firoz C et al. (eds.), *Nature-Based Solutions for Urban and Peri-Urban Areas*, Disaster Risk Reduction, https://doi.org/10.1007/978-981-96-5933-3_13

for implementing NbS in urban areas. The study highlights the importance of place attachment and community engagement in the application of NbS.

Keywords Nature-based solutions (NbS) · Recycled Stormwater · Managed Aquifer Recharge (MAR) · Non-potable uses · Place attachment · Pro-environment behaviour

13.1 Introduction

Rapid urbanization, population growth, resource degradation, and climate change-induced natural disasters are major environmental challenges of the twenty-first century. The United Nations (2018) reports that over 50% of the world's population will face severe water shortages by 2050. Addressing these challenges requires nature-based solutions (NbS) to protect, manage, and restore ecosystems while providing multiple co-benefits. NbS leverages natural processes to develop sustainable solutions. One promising NbS for urban areas is stormwater capture, storage, and harvesting through Managed Aquifer Recharge (MAR), which replenishes groundwater supplies and provides sustainable water sources for urban needs (Dillon et al. 2019). For instance, in Australia, cities like Perth and Adelaide use MAR schemes with stormwater and treated wastewater to enhance urban water security (Vanderzalm et al. 2020).

The success of NbS depends on stakeholder inclusion, with community engagement and place attachment playing crucial roles from design to implementation. Place attachment refers to the emotional or psychological bond between residents and their environments, and this bond can enhance the effectiveness of NbS by fostering a sense of ownership and responsibility among the community members. Similarly, community engagement is essential for the successful implementation of NbS. Engaging the community from design to implementation phase helps in gaining their trust and cooperation, which is critical for the long-term success of NbS projects.

This Chapter focuses on implementing NbS to address urban water security challenges, including increasing water quantity, enhancing quality, and managing excess water through stormwater collection and MAR. It investigates a proposed stormwater reuse scheme in Taiyuan City, which faces severe water scarcity and environmental degradation. Based on a survey of potential users, it explores the impact of place attachment on residents' attitudes and intentions toward using treated stormwater for residential purposes.

13.2 Understanding NbS

In literature, the term NbS has been defined and used in different ways. However, this chapter adopts the definition provided by the IUCN (2016) where nature-based solutions are defined as *actions to protect, sustainably manage, and restore natural or modified ecosystems, that address societal challenges effectively and adaptively, simultaneously providing human well-being and biodiversity benefits*. The concept of NbS is embedded in a wider picture of other concepts and approaches to conservation, sustainable use, and management of natural systems such as ecosystem services and green infrastructure (Kabisch et al. 2017). All these concepts and approaches have been introduced to protect nature and harness its potential to create a more resilient and sustainable world, particularly in cities where rapid urbanisation is placing mounting pressure on the environment.

13.2.1 Nature-Based Solutions for Water Security

Ensuring water security is critical to achieving sustainable development which requires "the provision of an acceptable quantity and quality of water for health, livelihoods, ecosystems and production, coupled with an acceptable level of water-related risks to people, environments and economies" (Grey and Sadoff 2007). NbS offers effective and sustainable ways to enhance water security as these solutions leverage natural processes and ecosystems to address water-related challenges. NbS for water security can be applied in various shapes and sizes and encompass various approaches like source water protection, watershed management, wetlands restoration and protection, water harvesting, afforestation, sustainable drainage systems, mangrove protection, and so on. Irrespective of the approaches, the overarching goal is to ensure water is available in the right quantity, quality, time, and place, which essentially is the definition of water security. One such NbS for ensuring water security is groundwater replenishment using managed aquifer recharge (MAR)—recharging harvested water to aquifers purposefully and storing water below ground for subsequent recovery or environmental benefit. The sources of water that can be used to feed MAR projects include surface water, rainwater, wastewater, or stormwater. Irrespective of the type, all types of MAR projects have seven elements, as illustrated in Fig. 13.1 (Dillon et al. 2009).

13.2.2 Challenges to Implementing NbS

Undoubtedly, NbS plays a crucial role in enhancing urban resilience, improving quality of life, and addressing environmental challenges. Nevertheless, implementing NbS in urban areas faces several challenges. For example, Kabisch et al. (2022,

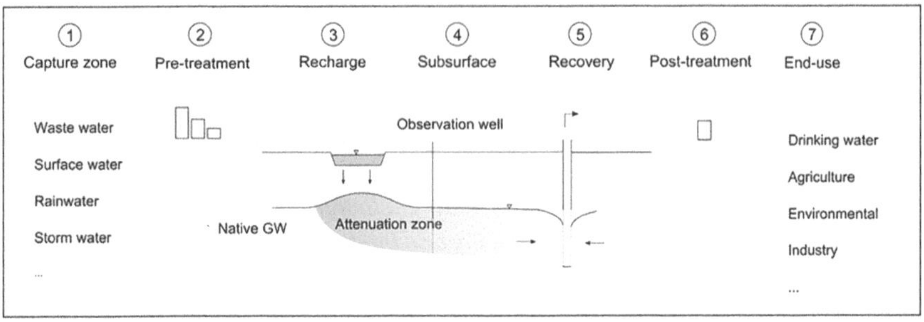

Fig. 13.1 Elements of MAR projects *Source* Modified from Dillon et al. (2009); https://demeau-fp7.eu/toolbox/what-managed-aquifer-recharge.html

p.1388) identify specific challenges to implementing NbS in urban areas including spatial conflicts with the urban system; specific urban biodiversity, fragmentation, and altered environments; value plurality, multi-actor interdependencies, and environmental injustices; path-dependencies with cultural and planning legacies; and a potential misconception of cities being artificial landscapes disconnected from nature. Similarly, Castelo et al. (2023) identified spatial, governance, assessment, finance, and socio-cultural factors as the five main barriers to implementing NbS in urban areas. Nelson et al. (2020) grouped the challenges under five coherent themes—participation and equity, governance, valuation, infrastructure integration, and scale and feedback. In summary, the key challenges include knowledge gaps, spatial constraints, biodiversity and fragmentation, social and economic factors, financial constraints, policy and regulation, public perception, and multi-actor coordination.

Despite these challenges, NbS offers immense potential for enhancing urban resilience, improving quality of life, and fostering sustainable development. In the context of this study, 'public perception/acceptance' of using recycled water for non-potable uses is the major barrier to implementing stormwater recycling via MAR (Dolnicar and Schafer 2009; Dolnicar et al. 2011; Wu et al. 2013; Dillon et al. 2019; Mankad et al. 2019). These perceptions can broadly be categorised into two broad themes that are significant for understanding the contextual and psychological factors of using recycled water (see Table 13.1).

Accordingly, this study sought to explore urban residents' attitudes toward acceptance and intention to use recycled stormwater, and the following research questions were formed:

- *Does place attachment (measured by length of residence) impact residents' preferences for the treatment options of stormwater used for residential purposes?*
- *Does place attachment (measured by length of residence) impact residents' attitudes toward acceptance and intention to use treated stormwater for residential purposes?*

First, we assume that the longer the resident resides in the local area, the more knowledgeable they are of the local natural environment (may relate to the water

Table 13.1 Contextual and psychological factors influencing using recycled water

Contextual factors	Psychological factors
• Water quality issues (relating to the source of recycled water) • Proposed uses of the water (relating to the proximity to human contact)	• Perceived value of the water • Trust in authorities • Altruism • Emotions • Perceived health risks • Social norms • Acceptance

Source Literature review

exploitation or contamination caused by coal mining in the study city) and that, they would be more favourable to the most complex or most advanced treatment option for the harvested stormwater when considering using it for residential purposes. Therefore, residents' preferences for the treatment options of recycled stormwater are examined.

Second, we assume that the longer the resident resides in the local area, the more likely they would be willing to use treated stormwater because it helps relieve the stress of freshwater supply that they have been facing for decades. People who own homes in their local area are likely to do good to the area with the perception of living there for the long term, which leads to place attachment (Bowe et al. 2020; Maricchiolo et al. 2021). It could therefore be hypothesised that the longer their residency is, the more positive attitude residents hold toward using treated stormwater for non-potable purposes. Consequently, the impact of length of residence on residents' attitudes toward acceptance and intention to use treated stormwater for residential purposes is examined.

13.3 Hypothesis Development

Place attachment is an emotional or psychological bond between residents and their environments (Brown and Raymond 2007). It is a complex construct measured by various dimensions, such as place dependence, place identity, social bonding, community attachment, rootedness, and bondedness (Sampson 1988; Hay 1998; Kyle et al. 2005). Due to the lack of universally agreed dimensions, researchers have created measures based on their context and Lewicka (2011) argues that this inconsistency hinders research generalization, especially in different cultural settings. Trentelman (2009) suggests simplifying place attachment to physical attachment, excluding social or economic aspects. Following this, our study measured place attachment by the length of residence.

Residency can be seen as an investment in the community, leading to stronger engagement in local activities (Toruńczyk-Ruiz and Martinović 2020; Maricchiolo et al. 2021). Such residents are more likely to engage in environmentally responsible behaviour, benefiting both the environment and their well-being (Zhang et al. 2018). Therefore, place attachment is crucial for improving residents' quality of life and promoting pro-environmental behaviours (Junot et al. 2018).

13.3.1 Impact of Length of Residence on Preferences for Stormwater Treatment Options

Past studies in this area have examined the community's attitudes towards acceptance and intention to use recycled water, but most of them omitted to include community attitudes to or preferences for 'treatment options' which is an important element of recycled water projects. The omission may directly affect the decision-making on recycled water's quality attributes including the odour, colour, and salt content of the water, subsequently impacting users' acceptance and intention to use these alternative sources of water (Wu et al. 2012). To address this gap, this study presented three stormwater treatment options (subject to the complex level of treatments) to the research respondents and explored their preferences for those options relating to using treated stormwater for various residential purposes (Keremane et al. 2011).

13.3.2 Impact of Length of Residence on Perceived Health Risks

Place attachment is a personal connection to a place (Lewicka 2011) and it relates to one's perception of the physical environment (Tartaglia 2012) and may affect their life quality (Bowe et al. 2020). Long-term residency is associated with a tendency to protect that place (Guo et al. 2018; Fornara et al. 2019). It is understood that place attachment encourages participation in local activities (Junot et al. 2018) and also tends to increase the feeling of safety in the local areas (Guo et al. 2018). This feeling of 'our community' could trigger responsibility for protecting or improving the local residential environment (Enqvist et al. 2018; Zhang et al. 2018). We could expect that the residents who live in the local area for a longer term will have stronger involvement and a more positive attitude toward new initiatives in the local area.

13.3.3 Impact of Length of Residence on Perceived Negative Emotions

Past studies on this topic have found that emotions of using recycled water for residential purposes are mostly negative (Marks et al. 2003; Guo et al. 2022). On the other hand, place attachment is a positive element as it can build people's sense of the place they are attached to, and thus, it is beneficial to residents' emotional well-being and happiness (Coelho et al. 2017; Junot et al. 2018). Therefore, we assumed that the longer the resident resides in the local area, the more affective bonds they'll have with the local area, and therefore fewer negative emotions regarding using stormwater recycled via the local aquifer.

13.3.4 Impact of Length of Residence on Social Norms

Individuals are often influenced by people around them in making decisions. It reflects how people around them will perceive the individual if they participate in certain behaviours. We expect to find stronger place attachment in more established neighbourhoods (Enqvist et al. 2018; Maricchiolo et al. 2021). While place attachment relates to the feeling that a person belongs to a place, it can also be perceived that a place belongs to the residents who live there (Lewicka 2011; Krys et al. 2019). Therefore, it is assumed that the longer the resident resides in the local area, the more likely they are influenced by their reference group in performing a certain behaviour.

13.3.5 Impact of Length of Residence on Trust in Government

In times of unpredictable or uncertain events, it is expected that governments will ensure the safety of their citizens (Toya and Skidmore 2014) which will build trust in the citizens regarding the government's capacity and expertise to deal with uncertainties (Kumagai and Iorio 2020). Furthermore, people's engagement experience with the government can also influence the level of trust (OECD 2013). We assume that the residents who reside in the local area for a longer period engage more with their government, and accordingly are more likely to trust the government to govern and guarantee the safety of the schemes that the government implements, like the stormwater recycling schemes in this study.

Based on the literature review, we propose the following hypotheses (see Table 13.2). In this study, length of residence performs as an independent variable which first affects psychological factors (serve as intermediate variables) and then in turn predicts residents' acceptance and intention to use treated stormwater (the dependent variable).

Table 13.2 Hypotheses tested in the study

H1: The longer the resident resides in the local area, the more likely they prefer the highest level of treatment for recycled stormwater
H2: Length of residence impacts residents' perceived health risks relating to using stormwater treated via MAR negatively
H3: Perceived health risks impact residents' acceptance and intention to use stormwater treated via MAR negatively
H4: Length of residence impacts residents' perceived negative emotions relating to using stormwater treated via MAR negatively
H5: Perceived negative emotions impact residents' acceptance and intention to use stormwater treated via MAR negatively
H6: Length of residence impacts social norms relating to using stormwater treated via MAR positively
H7: Social norms impact residents' acceptance and intention to use stormwater treated via MAR positively
H8: Length of residence impacts residents' trust in the government relating to the quality of stormwater treated via MAR positively
H9: Trust in government impacts residents' acceptance and intention to use stormwater treated via MAR positively

13.4 Materials and Methods

The study site Taiyuan, the capital city of Shanxi Province, China, in the past decades has shown extreme environmental degradation and faced severe water scarcity problems (Gaubatz and Hanink 2020). To deal with the problems, China's 2030 Agenda for Sustainable Development selected Taiyuan as one of the first three national sustainable development agenda innovation demonstration areas (Xinhuanet 2018). Following this agenda, the Taiyuan Municipal Government enacted the Development Plan (2017–2030) to solve the problem of water shortage by introducing alternative water sources (e.g., recycled stormwater) for domestic use.

The data were collected via a survey of urban residents in Taiyuan. Participants were firstly introduced to the aim and nature of the study, and the important concepts of stormwater recycling and treatment. They were then presented with three stormwater treatment options for residential uses (see Fig. 13.2), following the study by Keremane et al. (2011).

The proposed end uses of treated stormwater were based on existing Australian research (Wu et al. 2013; Fielding et al. 2015; Mankad et al. 2019). These uses were adapted to suit the survey population in China—city apartment dwellers with indoor plants and no lawns. Drinking was excluded as previous studies indicated public reluctance toward this use (Fielding et al. 2015; Mankad et al. 2019). Therefore, the proposed non-potable uses in this study included: personal washing, clothes washing, washing pets, watering indoor plants, car washing, toilet flushing, outdoor irrigation, and road spraying.

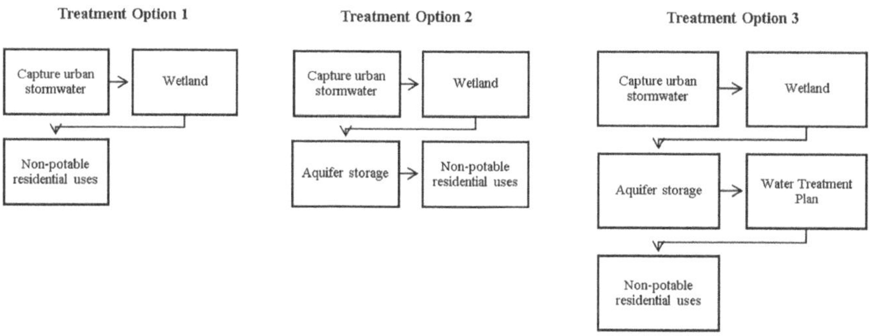

Fig. 13.2 Three treatment options for residential uses (Adapted from Keremane et al 2011)

Data were collected online via 'WeChat Moments' (Grazia and Marco 2023), yielding 669 valid responses. Demographics matched Taiyuan's population distribution, except for age and gender. Survey items were back-translated for accuracy, and missing values (<1%) were treated using regression imputation in AMOS 28.0. Kruskal–Wallis tests compared preferences against length of residence. Structural Equation Modelling (SEM) tested the theoretical framework, and Confirmatory Factor Analysis (CFA) assessed reliability and validity. The structural model fits well (see Table 13.3) with the data (Peugh and Feldon 2020), allowing for the interpretation of results.

13.5 Results and Discussion

13.5.1 Respondents' Preferences for Stormwater Treatment Options

Results demonstrated that respondents preferred option 3 (requiring the highest level of treatment) the most if the water was used for uses having close human contact (see Table 13.4).

Irrespective of the treatment options, it was observed that respondents' preferences for the proposed use(s) declined as the degree of human contact increased suggesting people are not yet comfortable with using treated stormwater for uses with close human contact such as personal washing, washing clothes, or cleaning pets. At the same time, it was observed that using treated stormwater for uses having limited human contact was well accepted. Respondents would be very tolerant of the treatment level if the water was used for uses that have limited human contact. The mean value of preferences for treatment option 1 (requiring the lowest level of treatment) was the highest and that for treatment option 3 (requiring the highest level of treatment) was the lowest for the uses including toilet flushing, outdoor green irrigation, and road spraying. This finding has important implications for stormwater

Table 13.3 Fit indices of structural equation model

Fit indicator	Fit standards	Personal washing	Clothes washing	Washing pets	Watering plants-indoor	Car washing	Toilet flushing	Outdoor irrigation	Road spraying
χ^2/df	$1 < \chi^2/\text{df} < 3$	1.43	1.71	1.95	1.63	1.70	1.87	1.82	1.29
RMSEA	< 0.08	0.025	0.033	0.038	0.031	0.032	0.036	0.035	0.021
GFI	> 0.90	0.986	0.984	0.983	0.987	0.987	0.987	0.989	0.992
TLI	> 0.90	0.994	0.991	0.986	0.990	0.990	0.988	0.988	0.996
CFI	> 0.90	0.997	0.995	0.993	0.995	0.995	0.995	0.995	0.998

Table 13.4 Respondents' rankings of stormwater treatment options

Proposed residential uses	Stormwater treatment options					
	Option 1		Option 2		Option 3	
	Urban stormwater/ wetland/end-use		Urban stormwater/ wetland/aquifer/ end-use		Urban stormwater/ wetland/aquifer/ water treatment plant/end-use	
	Mean	*SD*	*Mean*	*SD*	*Mean*	*SD*
Personal washing	2.75	1.940	3.56	2.106	4.67	2.122
Clothes washing	3.59	2.121	4.14	2.152	5.03	2.031
Washing pets	4.29	2.135	4.91	1.997	5.45	1.840
Watering plants-indoor	5.82	1.635	5.93	1.522	5.84	1.680
Car washing	5.61	1.735	5.85	1.605	5.78	1.815
Toilet flushing	5.99	1.668	5.94	1.673	5.81	1.865
Outdoor irrigation	6.13	1.452	6.07	1.513	5.90	1.742
Road spraying	6.01	1.550	5.99	1.593	5.84	1.840

Note These were 7 Likert-scale questions with 1 = least preferable and 7 = most preferable

recycling schemes on the investment input and scale of infrastructure constructions. At the early stage of implementing such schemes, it would be more feasible to start with uses that have limited human contact, thereby eliminating the more complex treatment options such as 'using water treatment plant'.

13.5.2 Impact of Length of Residence on Respondents' Preferences for Treatment Options

The results of Kruskal–Wallis tests indicate that the length of residence affects respondents' preferences for stormwater treatment options concerning the use of treated stormwater for uses with limited human contact, as reported in Table 13.5. Only significant values are reported given the length constraint of the chapter.

Post-hoc tests using Bonferroni adjusted alpha level of 0.01 (0.05/5) were then used to compare all pairs of groups. The results of Mann–Whitney U tests reveal no significant differences in respondents' preferences for treatment option 1 or treatment option 3 for all proposed uses. P-values were > 0.01 implying the length of residence had no significant impact on residents' preferences for treatment option 1 or treatment option 3 with the lowest and highest treatment levels respectively. Similarly, residents' acceptance level for treatment options 1 or 3 for all uses was the same regardless of how long they have lived in the local area.

However, significant differences were found in respondents' preferences for treatment option 2 in four proposed uses including watering indoor plants, toilet flushing,

Table 13.5 Differences in respondents' preferences for stormwater treatment options for residential uses by comparing length of residence—results of Kruskal–Wallis H tests

Treatment options	Proposed uses	*Kruskal–Wallis H*	*df*	*P*
Option 1	Outdoor irrigation	13.00	4 0.011	
Option 2	Watering plants-indoor	16.16	4	0.003
	Toilet flushing	19.04	4	0.000
	Outdoor irrigation	17.51	4s	0.002
	Road spraying	16.26	4	0.003
Option 3	Car washing	9.61	4	0.048
	Outdoor irrigation	11.11	4	0.025
	Road spraying	9.58	4	0.048

Note The five groups compared include: residency is < 5 years ($n = 122$), residency is 5–10 years ($n = 118$), residency is 11–15 years ($n = 81$), residency is 16–20 years ($n = 60$), and residency is > 20 years ($n = 288$). Only significant values ($p < 0.05$) are reported

outdoor irrigation, and road spraying (Table 13.5). The effect size (r) would be considered small using Cohen's (1988) criteria of 0.1 = small effect, 0.3 = medium effect, and 0.5 = large effect.

Further, for treatment option 2, the median and mean values of these four proposed uses were higher for respondents who have lived in the local area for more than 15 years compared to the respondents who have lived in the area for less than 10 years (Table 13.6). The results suggest that, for uses having limited human contact, the respondents with higher level of place attachment (lived longer in the local area) are more tolerant toward treatment option 2 compared with the respondents with lower level of place attachment. Therefore, Hypothesis 1—the longer the resident resides in the local area, the more likely they prefer the highest level of treatment for recycled stormwater- was rejected.

13.5.3 Results of SEM Tests on Hypotheses

The results of the eight SEM tests on hypotheses 2–9 are reported in Table 13.7.

Hypothesis 2—length of residence impacts residents' perceived health risks relating to using stormwater treated via MAR negatively—was partially accepted as length of residence was found to significantly impact the level of respondents' perceived health risks for uses that have close human contact. The respondents who lived in the local area longer perceived a lower level of health risk compared with those who lived for a shorter period. The finding supports the argument that longer-term residency leads to an increased feeling of safety in the local areas (Guo et al. 2018). We therefore can expect a positive attitude from residents toward implementing stormwater recycling initiatives and protecting or improving their local environment

Table 13.6 Differences in respondents' preferences for stormwater treatment option 2 for residential uses by comparing length of residence—results of Mann–Whitney U tests

	< 5 years	5–10 years	11–15 years	16–20 years	> 20 years				
Proposed uses	*Median (mean)*					*Mann–Whitney U*	*z*	*p*	*r*
Indoor plant watering		6(5.60)			7(6.08)	13,662	−3.47	< 0.001	0.13
Toilet flushing	7(5.99)			7(6.53)		2859	−2.78	0.005	0.11
		6(5.49)		7(6.53)		2424	−3.85	< 0.001	0.15
		6(5.49)			7(6.03)	14,069	−3.10	0.002	0.12
Outdoor green irrigation		6(5.75)		7(6.55)		2532	−3.47	< 0.001	0.13
		6(5.75)			7(6.16)	13,970	−3.24	0.001	0.13
Road spraying		6.5(5.59)		7(6.43)		2642	−3.08	0.002	0.12
		6.5(5.59)			7(6.10)	14,025	−3.18	0.002	0.13

Note Only significant values ($p < 0.01$) are reported

in areas where they have lived for a longer term. For uses that have limited human contact, residents would support using treated stormwater regardless of how long they have lived in the local area.

As perceived health risks did not have a significant impact on respondents' acceptance and intention to use treated stormwater via MAR for all proposed uses, Hypothesis 3 was rejected. The finding supports the findings of Po et al. (2005) and Guo et al. (2022) but contradicts that of Dolnicar and Grun (2009) or Wu et al. (2013), thereby adding to the inconsistency in the literature and suggesting perceived health risks is not yet an established factor that affects public acceptance and intention to use alternative sources of water. It also suggests that perceived health risks are a weak intermediate variable in moderating the relationship between the length of residence and acceptance and intention to use treated stormwater.

Hypothesis 4-length of residence impacts residents' perceived negative emotions relating to using stormwater treated via MAR negatively—was partially accepted because the length of residence was found to have a significant impact on respondents' emotions if treated stormwater was used for uses with limited human contact. The respondents who lived in the local area for more than 15 years perceived a lower level of negative emotions compared with those who lived for a shorter period. The finding supports the argument that place attachment is a positive element that increases people's affective sense of the place they are attached to, thus promoting their emotional well-being and happiness (Coelho et al. 2017; Junot et al. 2018). For

Table 13.7 Standardized path coefficients for structural equation models

	Proposed uses							
	Personal washing	Clothes washing	Washing pets	Watering plants-indoor	Car washing	Toilet flushing	Outdoor irrigation	Road spraying
H2 Length of residence → Perceived health risks	−0.089*	−0.080*	−0.094*	−0.075	−0.060	−0.054	−0.069	−0.061
H3 Perceived health risks → Acceptance and intention to use	−0.055	−0.047	−0.063	−0.153	−0.166	0.026	−0.033	−0.034
H4 Length of residence → Perceived negative emotion	−0.019	0.003	−0.042*	−0.075*	−0.064	−0.092*	−0.101*	−0.084*
H5 Perceived negative emotion → Acceptance and intention to use	−0.039	−0.058	−0.012	−0.015	−0.013	−0.154	−0.011	−0.006
H6 Length of residence → Social norms	−0.017	−0.046	0.053	0.065	0.056	0.079*	0.083*	0.086*
H7 Social norms → Acceptance and intention to use	0.778***	0.788***	0.804***	0.733***	0.705***	0.777***	0.840***	0.857***
H8 Length of residence → Trust in government	−0.024	−0.014	−0.025	0.030	0.016	0.017	−0.001	0.006
H9 Trust in government → Acceptance and intention to use	0.069**	0.090***	0.041*	0.063	0.063*	0.063*	0.028	−0.009

Note *** significant at $p < 0.001$; ** significant at $p < 0.01$, * significant at $p < 0.05$

the two uses having very close human contact (personal washing& clothes washing), respondents felt emotionally uncomfortable regardless of their length of residence.

Perceived negative emotions did not illustrate significant impacts on respondents' acceptance and intention to use treated stormwater via MAR for all proposed residential uses. Hence, Hypothesis 5 was rejected. This finding supports the empirical studies in Australia which found that negative emotions did not affect people's intention to use recycled stormwater (Marks et al. 2003; Wu et al. 2013) and suggests that negative emotions may not be a meaningful influencing factor in studying public acceptance and intention to use alternative waters. It also suggests that perceived negative emotions are a weak intermediate variable in moderating the relationship between the length of residence and acceptance and intention to use treated stormwater.

Length of residence had a significant impact on social norms relating to using treated stormwater for three proposed uses that have limited human contact (toilet flushing, outdoor green irrigation, and road spraying). Therefore, Hypothesis 6-length of residence impacts social norms relating to using stormwater treated via MAR positively- was partially accepted. The assumption that the longer the resident resides in the local area, the more likely they are influenced by people around them regarding using treated stormwater for residential purposes was partially verified. A possible explanation could be that these end-uses typically have more exposure to others in the communities. Hence, people tend to engage in environmentally friendly behaviours as they perceive it is socially expected or valued. The reason that length of residence did not demonstrate a significant relationship with social norms for uses 1–5 might be due to the limited exposure of the activities to others around them.

Hypothesis 7—social norms impact residents' acceptance and intention to use stormwater treated via MAR positively—was accepted in the tests for all proposed uses. Social norms illustrated very significant impacts on respondents' acceptance and intention to use treated stormwater via MAR for all proposed residential uses. The finding supports previous empirical studies (Geng et al. 2017; Guo et al. 2022) indicating that the impact of social norms is especially helpful in explaining consumer attitudes and behaviours in collectivist societies. It also suggests that social norms are a strong intermediate variable in moderating the relationship between the length of residence and acceptance and intention to use treated stormwater.

Length of residence did not demonstrate significant impacts on respondents' trust in the government to ensure the quality of treated stormwater for residential uses. So, Hypothesis 8 was rejected in the tests for all proposed uses. The finding could be explained by Wu's (2021) survey results that about 91% of Chinese respondents trust their government to do what is right for its citizens, implying that residents would trust their government to ensure the safety of treated stormwater regardless of how long they have lived in the local area and how engaged they are with the government. Another possible explanation relates to respondents' cultural characteristics. People in collective societies are more likely to demonstrate trust in the government or proceed with cooperative behaviour upon government calls (Guo et al. 2022).

Hypothesis 9—trust in government impacts residents' acceptance and intention to use stormwater treated via MAR positively—was partially accepted. Residents' trust in their government has been found to significantly impact their acceptance and intention to use treated stormwater for five proposed uses that have close human contact. The finding is consistent with that of previous studies (Hurlimann et al. 2008) and proves that trust in the government is an important influencing factor that affects public acceptance and intention to use alternative waters (Wu et al. 2013; Mankad et al. 2019). The insignificant influence of trust in government on residents' acceptance and intention to use treated stormwater for uses that have limited human contact was assumed to be associated with respondents' sound understanding of and great support for using alternative waters to cope with the water stress problems they face. Regardless of whether or how much they trust the government to monitor or ensure the safety of the water quality, they would be willing to use the water for those purposes. The findings also suggest that trust in government has some role in moderating the relationship between the length of residence and acceptance and intention to use treated stormwater.

13.6 Conclusion and Implications

This study investigated the impact of place attachment (measured by the length of residence) on residents' preferences for implementing NbS, i.e., stormwater recycling via MAR for non-potable uses in Taiyuan, China. The study found that the urban residents in Taiyuan responded positively to using recycled stormwater for residential purposes to alleviate the pressure of fresh water supply. Using treated stormwater for uses that have limited human contact was greatly acceptable. For uses that have close human contact, the highest level of treatment would be expected. Among the three treatment options, treatment option 1 (requiring the lowest level of treatment) was ranked as the most preferred treatment option for uses that have limited human contact, and treatment option 3 (requiring the highest level of treatment) was ranked as the least preferred for those uses. For each of the 3 options, respondents' preferences for the proposed uses declined as the degree of human contact increased. Length of residence was found to significantly impact residents' preferences for treatment options, perceived health risks, perceived emotions, and social norms that relate to using treated stormwater for various residential purposes.

13.6.1 Policy Implications

As part of a transition to sustainable societies, as evidenced in China by the recently established sustainable development agenda innovation demonstration areas, this study has sought to test empirically the responses of urban residents to implementing NbS for urban water security by using non-conventional waters (treated stormwater) for residential purposes.

The findings of this study have important implications on the investment input and scale of stormwater recycling schemes and NbS for decision-making and help to achieve cost-effectiveness. It is often found that policymakers prefer engineering solutions over NbS because with NbS it is difficult to prove the benefits compared to the alternatives (Castelo et al. 2023). It is clear from the results of this study that, at the early stage of stormwater recycling schemes, it would be more feasible to start with uses that have limited human contact, and for such purposes, the lowest level of treatment (i.e., via wetlands) would work well and more complex treatments would not be necessary. This is important because one of the biggest challenges of implementing NbS is finance (Castelo et al. 2023). We also recommend starting stormwater recycling schemes with more established communities since the residents who have lived in the local area for a longer time of period present more positive attitudes toward such schemes. In addition, the findings about the strong influence of social norms and trust in government on the community's positive attitudes toward stormwater reuse set signposts for the ease of implementation of NbS in general and particularly stormwater recycling schemes in Chinese collectivistic societies.

13.7 Research Limitations and Future Research Opportunities

Although the findings of this study make important contributions, there are a few limitations. The study is based on a sample of a city in northern China, where water scarcity has been a critical issue in past decades. While the robustness of the findings of this study would warrant a larger and heterogeneous sample across Northern China, Taiyuan is not representative of all of China's cities. Rather, we use the example of Taiyuan to provide concrete illustrations of broader concepts. Furthermore, data collection via WeChat Moments has the limitation of potential exclusion of elderly people from participating in such studies. Another research limitation was that the financial dimension (e.g., price factor) of treated stormwater was not included in this study. Using treated stormwater for residential purposes is still at the very initial stage of exploration in China; therefore, the pricing information for treated stormwater is not yet available. It is strongly recommended to include price factor in future investigations when pricing information becomes available. Finally, due to time and resource constraints, we were unable to collect and analyse cross-city data or data over time in this study, which can be future research opportunities.

Acknowledgements This research was funded by Scientific Research Special Fund Project (ZX2019307)—"New Ideas for Water Environment Governance in Taiyuan Sustainable Development Agenda Innovation Demonstration Zone", Resource-based Economy Transformation Collaborative Innovation Centre, Shanxi University of Finance and Economics.

References

Bowe M, Gray D, Stevenson C, McNamara N et al (2020) A social cure in the community: a mixed-method exploration of the role of social identity in the experiences and well-being of community volunteers. Eur J Soc Psychol 50:1523–1539. https://doi.org/10.1002/ejsp.2706

Brown G, Raymond C (2007) The relationship between place attachment and landscape values: Towards mapping place attachment. Appl Geogr 27:89–111. https://doi.org/10.1016/j.apgeog.2006.11.002

Castelo S, Amado M, Ferreira F (2023) Challenges and opportunities in the use of nature-based solutions for urban adaptation. Sustainability 15(9):7243. https://doi.org/10.3390/su15097243

Coelho F, Pereira M, Cruz L et al (2017) Affect and the adoption of pro-environmental behaviour: A structural model. J Environ Psychol 54:127–138. https://doi.org/10.1016/j.jenvp.2017.10.008

Dillon P, Stuyfzand P, Grischek T et al (2019) Sixty years of global progress in managed aquifer recharge. Hydrogeol J 27(1):1–30. https://doi.org/10.1007/s10040-018-1841-z

Dillon P, Pavelic P, Page D, Beringen H, Ward J (2009) Managed aquifer recharge. Waterlines Report Series No. 13, National Water Commission, Canberra

Dolnicar S, Hurlimann A, Grun B (2011) What affects public acceptance of recycled and desalinated water? Water Res 45(2):933–943. https://doi.org/10.1016/j.watres.2010.09.030

Dolnicar S, Grün B (2009) Environmentally friendly behaviour: Can heterogeneity among individuals and contexts/environments be harvested for improved sustainable management? Environ. Behav 41:693–714. https://doi.org/10.1177/0013916508319448

Enqvist J, West S, Masterson V et al (2018) Stewardship as a boundary object for sustainability research: Linking care, knowledge and agency. Landsc Urban Plan 179:17–37. https://doi.org/10.1016/j.landurbplan.2018.07.005

Fielding K, Gardne J, Leviston Z et al (2015) Comparing public perceptions of alternative water sources for potable use: The case of rainwater, stormwater, desalinated water, and recycled water. Water Resour Manag 29:4501–4518. https://doi.org/10.1007/s11269-015-1072-1

Fornara F, Lai A, Bonaiuto M et al (2019) Residential place attachment as an adaptive strategy for coping with the reduction of spatial abilities in old age. Front Psychol 10:856. https://doi.org/10.3389/fpsyg.2019.00856

Gaubatz P, Hanink D (2020) Learning from Taiyuan: Chinese cities as urban sustainability laboratories. Geogr. Sustainability 1(2):118–126. https://doi.org/10.1016/j.geosus.2020.06.004

Geng D, Liu J, Zhu Q (2017) Motivating sustainable consumption among Chinese adolescents: An empirical examination. J Clean Prod 141:315–322. https://doi.org/10.1016/j.jclepro.2016.09.113

Grazia M, Di M (2023) Exploring the use of WeChat for qualitative social research: The case of Italian digital diaspora in Shanghai. Front. Social 8:1144507. https://doi.org/10.3389/fsoc.2023.1144507

Grey D, Sadoff CW (2007) Sink or swim? Water security for growth and development. Water Policy 9(6):545–571. https://doi.org/10.2166/wp.2007.021

Guo S, Wu Z, Wen L (2022) Urban residents' acceptance intention to use recycled stormwater—an examination of values, altruism, social and cultural norms, and perceived health risks. Int J Environ Res Public Health 19:2825. https://doi.org/10.3390/ijerph19052825

Guo M, O'Connor D, Shebl F et al (2018) The effects of length of residence and exposure to violence on perceptions of neighborhood safety in an urban sample. J Urban Health 95(2):245–254. https://doi.org/10.1007/s11524-018-0229-7

Hay R (1998) Sense of place in a developmental context. J Environ Psychol 18:5–29. https://doi.org/10.1006/jevp.1997.0060

Hurlimann A, Hemphill E, McKay J et al (2008) Establishing components of community satisfaction with recycled water use through a structural equation model. J Environ Manag 88:1221–1232. https://doi.org/10.1016/j.jenvman.2007.06.002

International Union for Conservation of Nature (IUCN), Members' Assembly (2016) Resolution 6.069: Defining Nature-based Solutions, WCC2016-Res-069. IUCN, Gland, Switzerland

Junot A, Paquet Y, Fenouillet F (2018) Place attachment influence on human well-being and general pro-environmental behaviors. J Theo Soc Psychol 2:49–57. https://doi.org/10.1002/jts5.18

Kabisch N, Frantzeskaki N, Hansen R (2022) Principles for urban nature-based solutions. Ambio 51:1388–1401. https://doi.org/10.1007/s13280-021-01685-w

Kabisch N., Korn H., Stadler J., Bonn A (2017) Nature-based solutions to climate change adaptation in urban areas—linkages between science, policy and practice. Springer Cham. https://doi.org/10.1007/978-3-319-56091-5

Keremane G, McKay J, Wu Z (2011) No stormwater in my tea cup: An internet survey of residents in three major Australian cities. J Aust Water Assoc 38(2):118–124. https://papers.ssrn.com/sol3/papers.cfm?abstract_id=2474501

Kumagai S, Iorio F (2020) Building trust in government through citizen engagement. World Bank Group. https://openknowledge.worldbank.org/handle/10986/11866

Kyle G, Graefe A, Manning R (2005) Testing the dimensionality of place attachment in recreational settings. Environ Dev 37:153–177. https://doi.org/10.1177/0013916504269654

Lewicka M (2011) Place attachment: How far have we come in the last 40 years? J Environ Psychol 31:207–230. https://doi.org/10.1016/j.jenvp.2010.10.001

Mankad A, Walton A, Gardner J (2019) Psychological predictors of public acceptance for urban stormwater reuse. J Hydrol 572:414–421. https://doi.org/10.1016/j.jhydrol.2019.03.018

Maricchiolo F, Mosca O, Paolini D et al (2021) The mediating role of place attachment dimensions in the relationship between local social identity and well-being. Front Psychol 12:645648. https://doi.org/10.3389/fpsyg.2021.645648

Marks J, Cromar N, Fallowfield H et al (2003) Community experience and perceptions of water reuse. Water Supply 3:9–16. https://doi.org/10.2166/ws.2003.0002

Nelson DR, Bledsoe BP, Ferreira S, Nibbelink NP (2020) Challenges to realizing the potential of nature-based solutions. Curr Opin Environ Sustain 45:49–55. https://doi.org/10.1016/j.cosust.2020.09.001

OECD (2013) Government at a Glance 2013. OECD Publishing, Paris. https://doi.org/10.1787/gov_glance-2013-en

Peugh J, Feldon D (2020) How well does your structural equation model fit your data?: Is Marcoulides and Yuan's equivalence test the answer? CBE Life Sci Educ 19(3):es5. https://doi.org/10.1187/cbe.20-01-0016

Po M, Nancarrow B, Leviston Z et al (2005) Predicting community behaviour in relation to wastewater reuse: What drives decisions to accept or reject? CSIRO, Perth. https://doi.org/10.4225/08/5866a174964d6

Sampson R (1988) Local friendship ties and community attachment in mass society: A multilevel systemic model. Am Sociol Rev 53:766–779. https://doi.org/10.2307/2095822

Tartaglia S (2012) Different predictors of quality of life in urban environments. Soc Indic Res 113(3):1045–1053. https://doi.org/10.1007/s11205-012-0126-5

Toruńczyk-Ruiz S, Martinović B (2020) The bright and dark sides of the length of residence in the neighbourhood: Consequences for local participation and openness to newcomers. J Environ Psychol 67:101383. https://doi.org/10.1016/j.jenvp.2019.101383

Toya H, Skidmore M (2014) Do natural disasters enhance societal trust? Kyklos 67:255–279. https://doi.org/10.1111/kykl.12053

Trentelman C (2009) Place attachment and community attachment: A primer grounded in the lived experience of a community sociologist. Soc Nat Resour 22(3):191–210. https://doi.org/10.1080/08941920802191712

United Nations (2018) World Water Assessment Programme, The United Nations World Water Development Report 2018 (United Nations Educational, Scientific and Cultural Organization, New York, United States). www.unwater.org/publications/world-water-development-report-2018/. Accessed on 15 Sep 2022

Vanderzalm J, Naumann B, Higginson S et al (2020) Australian exemplars of sustainable, economic managed aquifer recharge. Water e-Journal 5(4):1–19. https://doi.org/10.21139/wej.2020.024

Wu C (2021) Did the pandemic shake Chinese citizens' trust in their government? We surveyed nearly 20,000 people to find out. The Washington Post, May 5, 2021, at 6:00 a.m. EDT. https://www.washingtonpost.com/politics/2021/05/05/did-pandemic-shake-chinese-citizens-trust-their-government/

Wu Z, Hemphill E, McKay J et al (2013) Investigating psychological factors of behavioural intention of urban residents in South Australia to use recycled stormwater for non-potable purposes. J Water Reuse Desal 3:16–25. https://doi.org/10.2166/wrd.2013.048

Wu Z, McKay J, Keremane G (2012) Issues affecting community attitudes and intended behaviours in stormwater reuse: a case study of Salisbury. South Australia. Water (Switzerland) 4(4):835–847. https://doi.org/10.3390/w4040835

Xinhuanet (2018) China approves three demonstration zones for sustainable development. Available online http://www.xinhuanet.com/english/2018-02/24/c_136997027.htm

Zhang J, Yu Z, Tian H (2018) An integrated model of emotion and ration on pro-environmental behavior: The role of ecological emotion involvement. Psychol Tech Appl 6:484–492. https://doi.org/10.16842/j.cnki.issn2095-5588.2018.08.005

Part IV
NBS—Governance and Policies

Chapter 14
Policy, Frameworks and Tools for Risk-Informed Uptake of NbS

Ambika Dabral, Krishnakali Ghosh, Ranit Chatterjee, and Rajib Shaw

Abstract The Global Risks Report, 2021 puts extreme weather, climate action failure, human environmental damage, biodiversity loss, and natural resource crises among the top ten risks by likelihood and impact (WEF, 2021). Over the recent years, Nature-based Solutions (NbS) have been increasingly recognized for their role in addressing varied socio-economic and ecological challenges, including the risks identified above. Despite the increasing recognition of NbS, the policy and decision-makers at the local level are not adequately equipped with analytical frameworks, tools and instruments to effectively understand and implement NbS which is most appropriate to their risk landscape and context. Further, they also face the challenge of a lack of adequate knowledge and tools for evaluating and scaling up the NbS. Against this backdrop, the chapter through an extensive literature review will study the existing policy landscape and associated institutional mechanisms including the frameworks and standards for better understanding, selection, implementation and integration of NbS. It will analyse how these instruments are helping the decision-makers and practitioners in undertaking risk-informed and evidence-based selection and uptake of NbS for their local contexts. Further, the chapter will bring forth cases highlighting the key barriers and enablers for mainstreaming NbS in DRM policies and practices.

Keywords Nature-based solutions · Policy · Decision-making · Urban and peri-urban · Disaster risk management

A. Dabral (✉) · K. Ghosh · R. Chatterjee · R. Shaw
Resilience Innovation Knowledge Academy (RIKA) India, New Delhi, India
e-mail: ambika@rikaindia.com

R. Chatterjee · R. Shaw
Keio University, Tokyo, Japan

M. Firoz C et al. (eds.), *Nature-Based Solutions for Urban and Peri-Urban Areas*, Disaster Risk Reduction, https://doi.org/10.1007/978-981-96-5933-3_14

14.1 Introduction

The changing landscape of disaster risks amidst evolving climate and associated uncertainties has put the conventional mitigation and adaptation approaches to the test. This has led to an increasing need and recognition of Nature-based Solutions (NbS) as an appropriate umbrella approach to building socio-economic resilience while holding ecological resilience at its core. Through their underlying principles and functioning that embrace the interconnectedness of socio-economic and ecological systems, NbS extends a multi-dimensional, multi-stakeholder and participatory approach to Disaster Risk Management (DRM). Over the recent years, there has been a growing body of scientific research, global and national standards, guidelines and tools along with a plethora of grassroots initiatives on NbS, underscoring their effectiveness; thus, building momentum on the need to mainstream them into urban development and DRM policies and practices.

With ecosystem-based approaches getting introduced and gaining momentum, nature-based solutions have been increasingly discussed and recognized in research initiatives, disaster risk reduction plans, and international environmental policy agendas as shown in Fig. 14.1. Despite the increasing popularity and efforts that bolster the need for NbS, planning and implementing NbS on the ground remains a complex exercise because of various factors including the lack of guiding resources and tools to better understand and address the inherent functional, temporal, spatial, and jurisdictional trade-offs, among others (Bush and Doyon 2019). The assessment of intangible advantages of NbS (such as biodiversity gains, enhanced well-being, etc.) faces a bottleneck due to a lack of a common framing (Viti et al. 2022). Also, there is a perceived disjoint relationship between science, enabling policies and NbS uptake in the prevailing studies (Lemos et al. 2024) which needs to be explored further to make the findings usable for the practitioners. Bottlenecks and gaps such as these, often make the NbS-related decision-making challenging, particularly, in a complex urban and peri-urban context. Against this backdrop, the current chapter reviews the existing studies, policies, frameworks, standards and associated institutional mechanisms and how these instruments help the urban and peri-urban decision-makers and practitioners in overcoming prevalent challenges in undertaking risk-informed and evidence-based selection and uptake of NbS for their local contexts.

14.2 Materials and Methods

The method employed primarily includes desk-based review and synthesis of existing literature including research studies, techno-legal documents such as policies, plans, guidelines, frameworks, standards and tools available at global, national and local scales along with initiatives showcasing the uptake of NbS by leveraging prevalent policies.

Fig. 14.1 Key events leading to recognition of NbS (Authors)

The literature review is undertaken using keyword searches of academic databases and web portals of inter-governmental agencies, national and local government bodies, relevant United Nations (UN) bodies, and think tanks. Keywords used include thematic terms like nature-based solutions, ecosystem approaches, disaster risk management, climate change adaptation, urban planning, and risk assessment combined with different types of instruments such as policy, frameworks, standards, tools, guidelines, and cases. This was followed by policy analysis wherein the identified instruments including local, national, regional, and international policies, plans and strategies on DRM, urban planning, CCA, environment management, and biodiversity conservation were studied to understand the current levels of integration of NbS in these resources. A case study approach is used to bring forth some of the findings on policy integration (Fig. 14.2).

14.3 Conceptual Framework

This directs one to the conceptual framework for the current chapter which strives to understand the underlying nexus between NbS, DRM and other associated disciplines and sectors. The framework proposes that the effective mainstreaming of NbS in DRM (which itself for its successful implementation requires its mainstreaming in developmental and sectoral policies and planning) is possibly dependent and routed in how DRM, NbS and associated concepts are placed in the existing development and sectoral policies and instruments.

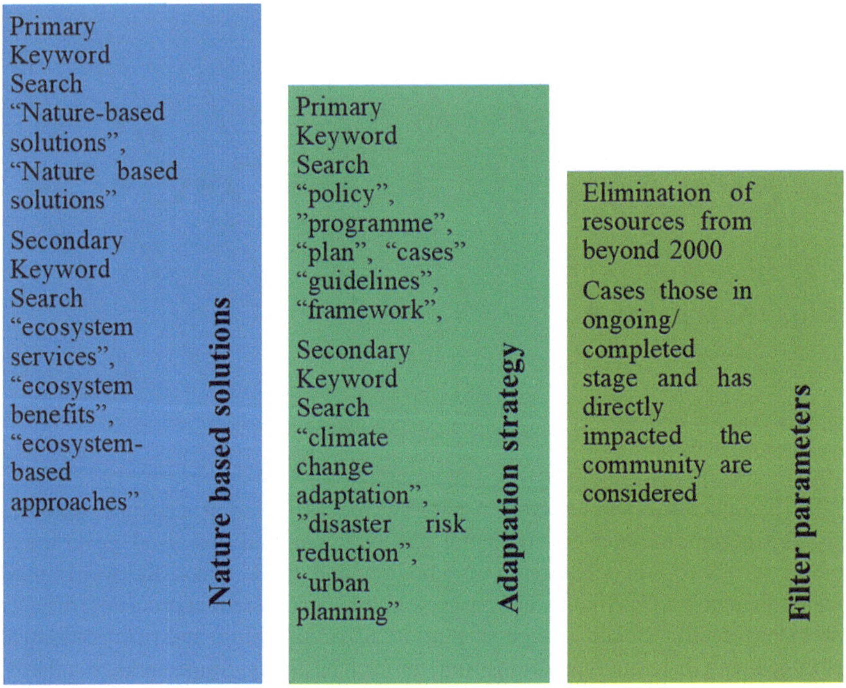

Fig. 14.2 Methodological flow of data collection

14.3.1 Literature Review

(a) Challenges in mainstreaming of NbS

Studies such as Aghaloo et al. (2024) suggest that despite the groundswell recognition of NbS, there are certain ambiguities and vagueness at the local level of governance pertaining to the gains involved, effectiveness and shortcomings which impact the science-based selection of the most appropriate NbS and the associated implementation design in a given context, even when NbS are taken up. The systematic study of the context in terms of biophysical characteristics of involved ecosystems, associated disaster and climate risks (temporal understanding), potential NbS, stakeholder analysis, and inter-alia, is required to address the prevailing ambiguities and vagueness. In this regard, Raymond et al. (2017) recommends using a framework for the design, implementation and evaluation of NbS and its co-benefits. The seven stages of the framework (ibid) are (i) Problem identification, (ii) Selection of appropriate solution and related actions, (iii) Design of implementation plan and process, (iv) Implementation of the solution (v) Regular stakeholder engagement and communication of co-benefits, (vi) Transfer and upscaling of solution, and (vii) Monitoring and assessment of co-benefits. The stages of selection, designing and implementation of NbS necessitate inter- and trans-disciplinarity (Dorst et al. 2019; Hansen and Pauleit

2014) backed by the creation of an evidence base for assisting urban professionals in comparing the effectiveness and complementarities of suitable nature-based and conventional solutions for a given context (Frantzeskaki et al. 2019).

Studies identify another mainstreaming challenge in terms of siloed governance. Under this, various interrelated roles and responsibilities are scattered across different government departments which in the absence of effective coordination and collaborative mechanisms among them leads to "fragmented governance" of NbS (Dorst et al. 2019; Wamsler 2015). Informed by multi-national cases, van der Jagt et al. (2023) put forth pertinent barriers in the mainstreaming pathway of NbS in an urban context. These barriers are (i) inadequate policy support, (ii) limited knowledge (iii) inadequate financial resources, (iv) land use-related competition, (v) low participation of the private sector, (vi) participation of citizens, and (vii) collaborative governance (ibid).

Another apposite challenge often highlighted by local authorities and practitioners is scaling up pilot or demonstrative NbS initiatives. The literature suggests that scaling up is a complex process and not a mere replication of the idea or initiative. It requires in-depth and trans-disciplinary comprehension of multiple jurisdictions, spatial and temporal changes, ecological systems, socio-economic characteristics and institutional and governance mechanisms (Cortinovis et al. 2022; Fastenrath et al. 2020) apart from land availability depending on the nature of NbS (Cortinovis et al. 2022).

(b) Potential solutions

To overcome these barriers, researchers have suggested different approaches including that of mixing different policy instruments, namely regulatory (e.g. master plans, standards), financial (e.g. grants, soft loans, tradable permits) and soft instruments (e.g. knowledge exchange and innovation platforms, tools, competitions) (van der Jagt et al. 2023) and applying differential horizontal and vertical strategies (Wamsler 2015).

Mainstreaming NbS, like its planning and implementation, involves multiple stakeholders including the ones involved in the decision-making and those who are likely to be affected by the interventions, directly or indirectly (and positively or negatively). Based on how these stakeholders are placed with respect to each other in the dynamics of power and authority, Wamsler (2015) recommends differential mainstreaming strategies. As per this framework, horizontal mainstreaming is recommended among actors with lesser power for command and control but who rather can support functions of facilitation and coordination. The proposed strategies for horizontal mainstreaming are (Modified from Wamsler 2015):

- Add-on mainstreaming: Undertaking additional initiatives beyond the core mandate of the organization but directly contributing to NbS;
- Programmatic mainstreaming: Amending the core activities of the organization to integrate NbS;

- Inter-and intra-organizational mainstreaming: Fostering coordination and collaboration within and across the organizations leading to undertaking joint initiatives on NbS.

Similarly, for vertical mainstreaming, which involves stakeholders with powers to dictate and direct, the following mainstreaming strategies are proposed (Modified from Wamsler 2015):

- Managerial mainstreaming: Changing the internal managerial structures of the organization to institutionalize NbS;
- Regulatory mainstreaming: Amending the regulatory institutional mechanism (such as policy, plan and legislation) of the organization to facilitate the integration of NbS;
- Directed mainstreaming: Bringing renewed focus of the organization on NbS through facilitation in funding, and internal capacity building, among others.

(c) Caution for Mainstreaming

Some studies also highlight the criticism associated with NbS uptake and mainstreaming while others recommend taking note of both the best practices and cases which also resulted in some ill-judged consequences or disservices. Some of these criticisms include gentrification (Almenar et al. 2021; Toxopeus et al. 2020; Dorst et al. 2019; Frantzeskaki et al. 2019; Raymond et al. 2017), production of greenhouse gases such as methane (Frantzeskaki et al. 2019), creating sites suitable for nurturing vectors (Frantzeskaki et al. 2019), forced displacement (Toxopeus et al. 2020). So, policy instruments designed for mainstreaming should include preemptive measures to prevent and/or mitigate the potential negative fallout of the NbS intervention. Suitable strategies in this regard include the participatory and collaborative design of NbS along with the provision for effective monitoring and adaptive management for catering to dynamic and evolving characteristics (Raymond et al. 2017).

14.3.2 Policy Analysis

Lemos et al. (2024) use a systematic literature review to identify and study the policies which further NbS for addressing climate change. Having studied around 55 pieces of literature from across the globe, they underscored the absence of any "predominant NbS public policy" (Lemos et al. 2024). Thus, there is an absence of dedicated policies that define and put in place dedicated institutional and financial mechanisms for understanding, planning and implementing NbS in varied socio-economic and ecological contexts. Nevertheless, NbS has been becoming increasingly popular and does find mentions (explicit and implicit) in different techno-legal instruments aimed at disaster risk reduction, ecological restoration, mitigation and adaptation to climate risks, sustainable development and urban sustainability. Van der Jagt et al. (2023) underscores the crucial role of supranational and national governance in pioneering the laying down of the required policy instruments and thereby influencing the local level of governance which are often burdened with routine roles and responsibilities. The same is also visible in the current context of NbS where a varied range of enabling techno-legal instruments have been developed at the global and regional levels compared to those prevailing at the national and sub-national levels. Against this backdrop, it is worth exploring a few existing policy and techno-legal instruments supporting NbS integration or mainstreaming.

At the global level, the Sendai Framework for Disaster Risk Reduction (SFDRR) 2015–2030 makes no explicit mention of NbS. It upholds "ecosystem-based approaches" to DRR, particularly for building resilience and navigating issues on aspects such as transboundary governance of common resources such as river basins, epidemics and displacements. At the national and local levels, the framework presses for ensuring the preservation of critical ecosystem services that support risk reduction while promoting the mainstreaming of DRM into development planning and management of varied ecosystems (UNDRR 2015). The framework further insists on leveraging environmental and natural resource management approaches that embrace the principles of DRR (ibid). Thus, underscoring the interconnectedness among disciplines and sectors such as DRM, development planning, environment management, nature resource management, ecology conservation, and biodiversity conservation, among others through the common goal of building socio-economic and ecological resilience.

In 2015, the Agenda 2030 for Sustainable Development was adopted laying down goals and targets for guiding development at national and sub-national levels. Heads of State pledged to eliminate poverty and hunger and achieve sustainable development with a multi-dimensional socio-economic and environmental lens. The Sustainable Development Goals (SDGs) prioritize ecological protection by promoting principles of sustainability in the management of natural resources, consumption and production processes, and in addressing climate change. The SDGs 2, 6, and 11 emphasize the interconnectedness of ecosystems, Disaster Risk Reduction (DRR), and Climate Change Adaptation (CCA). This includes promoting good practices in these domains.

In December 2015, the Paris Agreement on Climate Change was adopted, resulting in worldwide commitments to reduce carbon emissions and establish guidelines for CCA. The three worldwide policy agreements described above emphasize the importance of healthy ecosystems and sustainable resource management in protecting development gains and increasing resilience to catastrophes and climate change. However, little emphasis is there on the NbS integration into sub-national and local urban policy and planning. Unlike NbS in non-urban environments, those in cities mostly have limited focus on climate change mitigation but are more targeted at CCA and upholding different types of ecosystem services that are relevant to urban policies and practices. Some policy-level interventions are discussed below:

(d) Nationally Determined Contributions (NDC) and National Adaptation Plan (NAP)

National commitments, policies, and action plans are increasingly referencing NbS, including policies and measures that may employ various languages. National commitments via NDCs and NAPs are respectively aimed at climate change mitigation and CCA. WWF-UK reviewed updated NDCs in 2021 and found that 92% (105 out of 114) explicitly reference nature and nature-based methods, with several specifically mentioning "nature-based solutions". A wide range of ecosystems including forests, agricultural lands, wetlands, and coastal and marine ecosystems are part of the NDCs. Countries face challenges with operating NDCs due to monitoring and target setting. One good practice is Thailand's cabinet-approved NDC Roadmap (2021–2030) which outlines mitigation and adaptation actions and targets, including increasing national forest cover to 40% through community participation, with a focus on headwater and mangrove forests to improve adaptive capacity. A similar initiative in India is present where NDCs point to constructing an extra carbon sink of CO2 in the run of 2.5 to 3 billion tonnes by 2030 through forest and tree cover.

(e) European Green Deal

The European Green Deal outlines actions, laws, strategies, and objectives for several policy sectors. The Green Deal prioritizes NbS in climate adaptation and mitigation efforts with a focus on maintaining healthy and resilient seas and oceans also (European Commission 2023). The European Union (EU) aspires to protect its natural capital and citizens' health from environmental hazards and consequences. This highlights the importance of ecosystems and their role in providing critical services such as disaster mitigation and climate regulation. The Green Deal outlines an aspiring EU plan for climate change adaptation, emphasizing the importance of strengthening efforts on climate-proofing, resilience building, prevention, and readiness, as well as ensuring public and private investment in NbS. To achieve these goals, a critical role is being played by instruments such as the EU's biodiversity plan for 2030, farm-to-fork food system strategy, and climate change adaptation strategy.

The 2030 biodiversity policy of the EU aims to guarantee ecosystems are healthy, robust, and rich in biodiversity, while also providing important services for citizens' prosperity and well-being. Recognizing the vital contribution of NbS in climate change mitigation and CCA, the EU has adopted the proposal for Nature Restoration Law as of 2022 with targets that will be legally binding to restore degraded ecosystems, including those with high potential for carbon storage and natural hazard mitigation. It mandates nations to lay down respective restoration plans for meeting the targets. Sustainable management is a key component that highlights the necessity of managing forests, nutrients, water resources, and soil. The goal is to increase spending on nature by at least EUR 20 billion annually through initiatives such as Invest EU (NbS for a Green Recovery), allocation of one-fourth of the EU budget to climate action (mostly through ecosystem restoration), and green public procurement. The strategy prioritizes investing in natural capital to attain these goals and recover from the COVID-19 disaster.

The farm-to-fork plan is a critical component of the Green Deal and focuses on making food systems reasonable, healthy, and environment-friendly. In this context, NbS are specifically acknowledged for their ability to assist in delivering better. To achieve climate and environmental outcomes and increase climate resilience total funding of EUR 10 billion is suggested. This funds activities such as research and innovation in food, bio-economy, agriculture, fisheries, aquaculture, usage of modern technologies and NbS for agri-food systems. The strategy also encourages sustainable management of coastal areas and resources for fish and seafood. The level of support to NbS is classified as medium and the focus on links to climate change and disaster risk management is limited.

(f) Payment for Environmental Services Program

Leading the way in the globe in the use of payments for ecological services is Costa Rica. The Payment for Environmental Services Programme (PES) in Costa Rica pays landowners for preserving their forest cover, which lowers the risk of disasters by preventing flooding and landslides (OECD 2023). Under this, landowners and forest and forest plantation holders receive recognition and support from the National Forestry Financing Fund (FONAFIFO) for the environmental services provided for environment preservation and improvement. Various environmental services on the farms of the owners envisaged for preservation and maintenance under PES include natural forests, forest plantations, agroforestry systems, natural regeneration and sustainable forest management. PES has made it easier and encouraging for stakeholders interested in the growth of Costa Rica's forestry industry to be included. It is linked to several international legal agreements that the nation has ratified including the Regional Agreement for the Development of Forest Plantations and the Management and Conservation of Natural Forest Ecosystems.

Between the years 2010 and 2021, over 5 million hectares of agroforestry systems, twenty thousand hectares of reforestation, twenty-five thousand hectares of natural regeneration and four thousand hectares of forest management have been conserved (Interlace-hub 2024). Some significant enablers of the program are financial incentives and orientation and awareness of landowners that enable them to undertake informed decision-making on aspects of conservation. The institutions associated with these enabling activities have high credibility which further helps in mobilizing community participation for differentiated farm management. In addition, the nation's eco-friendly reputation and conservation initiatives provide legitimacy outside, encouraging foreign players to take part in PES.

(g) All London Green Grid

For the benefit of both humans and wildlife, the All London Green Grid (ALGG) lays down a policy framework arrangement that enables and promotes the creation and implementation of "green infrastructure" throughout the city. The network of green spaces—such as street trees and green roofs—and blue spaces—such as rivers, canals and ponds—that make London's green infrastructure is planned, designed, and maintained to offer several advantages, such as recreation and amenity, healthy living, lowering flood risk, improving air quality, cooling the urban environment, promoting walking and bicycling, and boosting biodiversity and ecological resilience. Three primary components make up the framework: The London Plan, the master plan of the city, contains planning policy on green infrastructure. The ALGG Supplementary Planning Guidance, on the other hand, offers nuanced guidance for policy implementation as per the London Plan and functions as the city's de facto green infrastructure strategy. Lastly, 11 ALGG Area Frameworks offer a thorough evaluation of the opportunities for delivering green infrastructure at the sub-local level, in particular areas of London (Greater London Authority 2012). It is being provided by a range of collaborations prevailing within the city and boroughs. Examples of these include government partners, subregional and cross-border organizations and partnerships, local community groups, and managers of open spaces.

One of the enablers for the uptake of ALGG was the establishment of the mandate for London boroughs to develop open space initiatives that were connected to local biodiversity action plans (CPRE 2014). In their policy documents, boroughs demonstrated a strong comprehension and adherence to the ALGG principles, as evidenced by their specific policy commitment to the initiative. For instance, they acknowledged the significance of green connectivity and emphasized various advantages of strategic management of green spaces. The ALGG also aided in raising awareness among various stakeholders about green infrastructure, which includes areas outside of conventional parks. In addition, the local community (resident organizations, housing associations, schools) was encouraged to collaborate.

(h) Joint Forest Management

According to the Government of India's 1988 Forest Policy, community involvement in and around forested regions is essential to the preservation and growth. On June 1st, 1990, the Indian Government released a comprehensive guideline on how to develop and manage degraded forest land under the care of the staff of the Forest Department with the assistance of voluntary organizations and the local community. States created their resolutions following this directive, which made it easier for Joint Forest Management to spread throughout the states. To safeguard and oversee neighbouring forests, communities band together to form a JFM Committee, which was then in charge of enforcing locally created micro plans and bylaws. The fundamental component of JFM is the ability of communities to control how their members utilize their woods and to keep non-members out (MoEF 2013). They gain by having direct access to and control over the use, and other intangible benefits from the local ecosystem services, like pollination, water recharge, and wildlife habitat. Later, it was also interlinked with the Mahatma Gandhi National Rural Employment Guarantee programme that provides training in animal husbandry, poultry farming, dairy development, and small-business management to individuals living in peripheral forest areas. Therefore, communities must get involved in the preservation of forests and wildlife.

Even though every Indian state has a unique institutional structure, most of them have embraced JFM and passed resolutions approving such partnerships in compliance with the stipulated guidelines. Successful JFM initiatives have been implemented in places like West Bengal, Gujarat, and Maharashtra. These programmes involve people in afforestation, forest preservation and sustainable resource harvesting, improving the health of the forests and generating income from them (Patra 2015). Examples include the *Gaddi* and *Gujjar* tribes in the Himalayan nations, who regulate livestock grazing to limit the growth of wild grass and so contribute to biodiversity preservation. The *Mundas* and *Santhals* in the Chotanagpur area revere the *Mahua* and *Kadamba* trees, and the tribes in Orissa and Bihar revere tamarind and many other trees. However, the JFM Committees meet challenges such as dissolving committees is a discretionary power held by the Forest Department. JFM also faces significant challenges from inefficient management and micro-plan design, low participation of women representatives formally induced in management committees and local conflicts both within and across communities.

14.4 Results and Discussion

The cases studied above illustrate varied instruments and pathways for supporting the planning, implementation and monitoring of NbS. As highlighted by the studies such as (Van der Jagt et al. 2023) and (Wamsler 2015), actors with adequate directing powers or those operating at national, regional and global levels usually lay down instruments that are overarching, multi-sectoral and prescriptive. As suggested by the literature reviewed, cases illustrate the use of varied policy instruments including environmental regulations such as Joint Forest Management, and financial instruments such as Payment for Environmental Services Program, among others. However, literature and cases do bring out some pertinent barriers and opportunities towards integrating and mainstreaming NbS in policy instruments of not just DRM but associated disciplines and sectors. Some of these are briefly listed below.

14.4.1 Barriers and Opportunities for Mainstreaming NbS

(a) Availability and Accessibility to Robust Datasets and Information

There is a lack of availability or limited accessibility to datasets and information that can help urban and peri-urban practitioners in effectively navigating the underlying trade-offs while making NbS-related decisions. For example, (Bush and Doyon 2019) underscores that the lack of datasets on biophysical and spatial characteristics of ecosystems may undermine the full comprehension of both the positive outcomes and any negative unexpected outcome of a given NbS; thereby acting as an impediment to effective decision-making and successful implementation. Accessibility issues also arise because of the presence of such authentic datasets in varied and often non-compatible formats with different government ministries, departments and scientific bodies which usually do not interact frequently and effectively with each other.

(b) Risk-Informed/ Evidence-Based Decision-Making Toolkits

Addressing the concerns of gauging the benefits and co-benefits an NbS could offer or its effective role in DRM and CCA, studies such as (Frantzeskaki et al. 2019) have showcased how tools like TEEB (The Economics of Ecosystems and Biodiversity) tool (TEEB 2010) were helpful in some cities in Tanzania. The tools helped in demonstrating to the stakeholders the relevant benefits of NbS which include greening against climate change and acted as an effective evidence-based advocacy tool. Similarly, a user-friendly risk-informed toolkit that could support the visualization of risk that the NbS will address while also capturing the potential co-benefits of the NbS will be very handy for supporting local authorities in decision making (Ürge-Vorsatz et al. 2014).

(c) Use of Technology for Aiding Decision-Making

A few studies discussed earlier mentioned a criticism of NbS wherein the NbS could result in some unintended negative impacts for the given context. A possible solution in this regard is leveraging the existing technology and technological tools such as digital twin-city modelling. Such a model is an "in-silico experimentation" tool that digitally imitates the characteristics of the study area and could simulate i.e. help determine the pathway of the proposed intervention (NbS, in this case) under a variable set of conditions that are fed as input for simulation (Barat et al. 2021). Thus, such tools could be utilized for modelling the potential outcomes of the proposed NbS and devising appropriate supporting strategies.

(d) Governance via a Mix of Policy Instruments

The transition from governance relying on a single policy instrument (say a regulation) to one leveraging multiple (conventional and new) policy instruments could be a better pathway for mainstreaming NbS (van der Jagt et al. 2023). This is specifically so for an urban context- a complex system of systems involving multiple players operating at different levels and often catering to/ governed by multiple jurisdictions; thus, limiting the efficacy of a single instrument in bringing the envisaged changes. A comprehensive mix of different policy instruments that are in congruence with the desired goal (ibid) could also possibly address the prevailing gap of fragmented governance as highlighted by Dorst et al. (2019) and Wamsler (2015).

14.5 Conclusion

The cases reviewed along with the solutions put forth by different researchers and practitioners provide a way towards navigating the concerns and challenges surrounding the design, implementation and monitoring of NbS discussed above. In this regard, the key aspects that could facilitate the successful integration and mainstreaming of NbS in policy instruments across the world are listed below.

- Involvement of multiple stakeholders: It is to be ensured that local government, non-government, private, community and all other actors are involved in designing and implementing the intervention. Local stakeholders need to be involved at every phase of the process, from identifying the problem to maintaining the solution. The stakeholders must all have equal access to NbS' availability, accessibility, and other features.
- Enabling inclusion of marginalized segments by making youth, women, people with disabilities, and other marginalized segments of the community the primary targets of NbS. It should ideally guarantee inclusive, representative, and responsive decision-making with all parties involved at various levels.
- Scientific concepts are necessary to achieve successful and sustained NbS but the support of indigenous and traditional knowledge will frequently play a key role in providing long-term solutions.
- Integration of multiple sectors: Along with the intention of conservation and enhancing urban resilience, NbS policies and programmes that target economic benefits are more successful. In this aspect, a reflection of ecosystem services on enabling NbS interventions can better inform project evaluations too.
- NbS should form an integral part of climate change plans or local climate action plans. Specifically in the context of cities, the adaptations should be such that they allow for strengthening urban resilience and address specific local needs.
- Robust strategies for the evaluation and monitoring of NbS programs are necessary. This is supported by a fundamental step of defining the key performance indicators that will give pertinent information about how the project will affect people and the environment.

The study has limitations in its comprehensive coverage of the cases of relevant policies and strategies from different geo-political regions to have a comparative understanding of the uptake of NbS integration in these regions and associated enablers.

References

Aghaloo K, Sharifi A, Habibzadeh N et al (2024) How nature-based solutions can enhance urban resilience to flooding and climate change and provide other co-benefits: A systematic review and taxonomy. Urban for & Urban Green. https://doi.org/10.1016/j.ufug.2024.128320

Almenar J, Elliot T, Rugani B (2021) Nexus between nature-based solutions, ecosystem services and urban challenges. Land Use Policy. https://doi.org/10.1016/j.landusepol.2020.104898

Barat S, Parchure R, Darak S et al (2021) An Agent-Based digital twin for exploring localized non-pharmaceutical interventions to control COVID-19 pandemic. Indian Natl Acad Eng. https://doi.org/10.1007/s41403-020-00197-5

Bush J, Doyon A (2019) Building urban resilience with nature-based solutions: How can urban planning contribute? Cities. https://doi.org/10.1016/j.cities.2019.102483

Cortinovis C, Olsson P, Boke-Ol´en N et al (2022) Scaling up nature-based solutions for climate-change adaptation: Potential and benefits in three European cities. Urban For & Urban Green. https://doi.org/10.1016/j.ufug.2021.127450

CPRE London & Neighbourhoods Green (2014) All London green grid: a review of implementation. Available via Google. https://www.google.com/url?sa=t&rct=j&q=&esrc=s&source=web&cd=&cad=rja&uact=8&ved=2ahUKEwjx0_iSwa35AhUT8rsIHYNdDmoQFnoECA4QAQ&url=https%3A%2F%2Flondon.cprelocalgroups.org.uk%2Fresources%2Fitem%2Fdownload%2F750&usg=AOvVaw05MWDsYOyOKuY7_iI6PzlB

Dorst H, van der Jagt A, Ravena R et al (2019) Urban greening through nature-based solutions – key characteristics of an emerging concept. Sustain Cities Soc. https://doi.org/10.1016/j.scs.2019.101620

European Commission (2023) Delivering the European green deal. Available via EU. https://commission.europa.eu/strategy-and-policy/priorities-2019-2024/european-green-deal/delivering-european-green-deal_en#documents

Fastenrath S, Bush J, Coenen L et al (2020) Scaling-up nature-based solutions. Lessons from the Living Melbourne strategy. Geoforum. https://doi.org/10.1016/j.geoforum.2020.07.011

Frantzeskaki N, Mcphearson T, Collier M (2019) Nature-Based solutions for urban climate change adaptation: linking science, policy, and practice communities for Evidence-Based Decision-Making. Biosci. https://doi.org/10.1093/biosci/biz042

Greater London Authority (2012) Green Infrastructure and open environments: the all London green grid supplementary planning guidance. Available via Interlace-HUb. https://interlace-hub.com/sites/default/files/algg_spg_mar2012.pdf

Hansen R, Pauleit S (2014) From multifunctionality to multiple ecosystem services? a conceptual framework for multifunctionality in green infrastructure planning for urban areas. Ambio. https://doi.org/10.1007/s13280-014-0510-2

Harari N, Gavilano A, Liniger, HP (2017) Where people and their land are safer: A Compendium of Good Practices in Disaster Risk Reduction. Bern and Lucerne, Switzerland: Centre for Development and Environment (CDE), University of Bern, and Swiss NGO Disaster Risk Reduction (DRR) Platform, with Bern Open Publishing

Lemos G, Eslabao A, Santos J et al (2024) Nature-based solutions experiences: A systematic literature review for public policies. Nat-Based Solut. https://doi.org/10.1016/j.nbsj.2024.100121

MoEF (2013) Handbook of joint forest management. Publ Minist Environ & For, New Delhi. Available via https://ifs.nic.in/Dynamic/pdf/JFM%20handbook.pdf

OECD (2023) Environmental performance reviews: Costa Rica 2023. Available via OECD iLibrary. https://www.oecd-ilibrary.org/sites/84864bde-en/index.html?itemId=/content/component/84864bde-en

Patra P (2015). Jt for Manag India. https://doi.org/10.1007/978-94-017-9771-9_24

Raymond C, Frantzeskaki N, Kabisch N et al (2017) A framework for assessing and implementing the co-benefits of nature-based solutions in urban areas. Environ Sci Policy. https://doi.org/10.1016/j.envsci.2017.07.008

Toxopeus H, Kotsila P, Conde M et al (2020) How 'just' is hybrid governance of urban nature-based solutions? Cities. https://doi.org/10.1016/j.cities.2020.102839

United Nations Office for Disaster Risk Reduction (2015) Sendai framework for disaster risk reduction. Switzerland. Available via PreventionWeb. https://www.preventionweb.net/sendai-framework/sendai-framework-for-disaster-risk-reduction. Accessed 15 Feb 2024

Urge-Vorsatz D, Herrero S, Dubash N et al (2014) Measuring the Co-Benefits of climate change mitigation. Annu Rev Environ Resour. https://doi.org/10.1146/annurev-environ-031312-125456

van der Jagt A, Tozer L, Toxopeus H et al (2023) Policy mixes for mainstreaming urban nature-based solutions: an analysis of six European countries and the European Union. Environ Sci Policy. https://doi.org/10.1016/j.envsci.2022.10.011

Viti M, Löwe R, JD, Sørup H et al (2022) Knowledge gaps and future research needs for assessing the non-market benefits of Nature-Based Solutions and Nature-Based Solution-like strategies. Sci Total Environ, 841. https://doi.org/10.1016/j.scitotenv.2022.156636

Wamsler C (2015) Mainstreaming ecosystem-based adaptation: transformation toward sustainability in urban governance and planning. Ecol Soc. https://doi.org/10.5751/ES-07489-200230

Chapter 15
Using the Key Player Approach to Design and Manage Adaptive Urban Governance for Nature-Based Solutions Implementation: Study of Vienna and Tucson

Antonija Bogadi

Abstract Nature-based solutions (NbS) are garnering increasing recognition as a viable approach to urban challenges such as climate change, biodiversity loss, and social inequality. Nevertheless, the intricacies of NbS implementation demand an adaptive governance model that fosters sustained collaboration, learning, and innovation across a spectrum of stakeholders. This chapter delineates a method for assessing and bolstering the adaptive capacity of governance, drawing on empirical research from two case studies in Vienna, Austria, and Tucson, USA. The method is anchored in a thorough understanding of the attributes and interrelationships of key actors within governance stakeholder networks. We identify those actors who are most adept at disseminating NbS knowledge (KPP-Pos actors) and assess the diversity of essential actors for upholding network performance (KPP-Neg actors). Our proposed approach is designed to recognize and harness the strengths of key stakeholders to effectively address common issues in NbS implementation. This includes engaging key players to enhance coordination and cooperation at various organizational levels, maintaining focus on NbS to sustain public interest, discerning the dynamics that influence the implementation process, and optimizing the use of stakeholders' knowledge of NbS. Through this empirical investigation, we validate the effectiveness of our method in practical settings, providing insights that are applicable to other scenarios confronting similar urban environmental challenges.

Keywords Nature-based solutions implementation · Adaptive governance · Key player problem · Stakeholder's networks analysis

A. Bogadi (✉)
SYNYO GmbH, Vienna, Austria
e-mail: antonija.bogadi@gmail.com

M. Firoz C et al. (eds.), *Nature-Based Solutions for Urban and Peri-Urban Areas*, Disaster Risk Reduction, https://doi.org/10.1007/978-981-96-5933-3_15

15.1 Introduction

The implementation of Nature-based Solutions (NbS) is too complex to be confined within the problem-solving frameworks of traditional governance. A governance model that integrates both formal and informal institutions, stakeholders with environmental expertise across different urban scales, and a variety of interest groups—and that possesses the resilience to manage crises and change—is one that has the potential to rise to this challenge. This study employs a governance framework as an analytical lens to examine NbS implementation processes. It investigates the capacity of social network analysis to identify and mobilize pertinent knowledge in the implementation of urban green infrastructure.

15.1.1 Literature Review

Addressing climate uncertainties through the implementation of Nature-based Solutions (NbS) is a strategic and spatial planning approach. It involves creating networks of patches and corridors of land dedicated to biodiversity conservation, environmental protection, water resource provision, recreation, cultural uses, urban development control, and contributions to climate change mitigation and adaptation. NbS encompasses greenways, open spaces, greenbelts, urban greening initiatives, cultural landscapes, ecological networks, agricultural lands, and all unbuilt lands that support vegetation. The understanding, planning, and management of NbS require the involvement of stakeholders, planners, and policymakers (Benedict and McMahon 2002, 2012; Ahern 2007; Mell 2008). Given its multidisciplinary, multi-objective, and multi-scalar nature, NbS implementation is a complex and demanding process that necessitates integrative planning approaches.

Employing a network approach to steer governance processes has gained traction. Social networks are pivotal in analysing and evaluating policymaking, public administration, and governance processes (Adam and Kriesi 2007; Kickert et al. 1997; Metz and Brandenberger 2023). Establishing collaborative networks that extend beyond formal hierarchies is increasingly viewed as essential for achieving policy and implementation goals (Hall and O'Toole 2000; Borgatti and Foster 2003; Bressers et al. 1998). This shift from traditional government to a broader governance model reflects a transition from hierarchical structures to multi-actor, networked arrangements (Bache and Flinders 2016; Peters and Pierre 2000; Rhodes 1996; Stoker 1998; Sørensen and Torfing 2005a, b; Sørensen and Torfing 2005a). While governments traditionally create order and facilitate collective action through a political-administrative hierarchy, governance achieves these aims through the involvement of multiple actors across different levels of organization (Stoker 1998). Networks are often characterized as entities tailored to specific problems, capable of fostering collective action, self-organization, and coordination across various scales. This conceptualization of governance underpins the analysis presented in this study

(Berardo and Scholz 2010; Metz, 2017; Provan and Kenis 2008; Schaub and Metz 2020).

Social network analysis (SNA) serves as a tool to unravel the complexities of our interconnected world. This chapter presents the opportunity to shed light on the network structure of the governance under study and to correlate these structures with the desired outcomes through SNA. Originating as a sociological method, SNA is predicated on the notion that all aspects of life—political, social, and economic—can be viewed as networks of relational ties composed of interconnected units (Wasserman 1994). Rooted in matrix algebra and graph theory, SNA facilitates the quantification of these relationships, enabling the analysis of networks through quantitative metrics. Contemporary research offers compelling examples of how network structures—the patterns of connections among actors within a network—correlate with various outcomes (Borgatti and Foster 2003; Burt 2004; Völker and Flap 2004; Friedkin 1981; Borgatti et al. 1998; Burt 1992; Burt 2017; Sandström and Carlson, 2008). When applied to governance, the implication is that the attributes of the actors and the patterns of their interrelationships significantly impact the results of their collective actions.

The adaptive capacity of a stakeholder network is crucial. Holling (1986) identified four phases of change in a system's structure and function: growth or exploitation, conservation or increasing rigidity, collapse or creative destruction, and reorganization or renewal. The ability to adapt and possess sufficient adaptive capacity to progress to the next phase is a key aspect of resilience in social-ecological systems (Folke et al. 2005). In highly adaptable social-ecological systems, actors can reorganize the system within desired states in response to changing conditions and disturbances. This is why adaptive governance is considered a more promising approach to managing complexity than traditional models.

Adaptive governance issues in urban spaces can be reframed in network terms, recognizing how network structures impact governance performance. Network structures evolve through the different phases of the adaptive cycle, influencing network processes. For instance, stagnation in governance (the conservation phase) can be attributed to a lack of diversity among actors. While traditional governance structures with like-minded actors may achieve goals efficiently, they often lack innovation. A different network type is expected to foster greater innovation. Additionally, the failure to mobilize necessary resources for governance processes can be due to the absence of relevant actors. When environmental (NbS) knowledge is essential, as suggested in this study, implementation failures may stem from not involving actors with the necessary knowledge (Wagner et al. 2021).

A high-performing adaptive governance network is characterized by its heterogeneity and a high degree of network closure. Empirical evidence suggests that such networks are dense, centralized, and include a diverse array of actors engaged in cross-boundary interactions (Angst and Brandenberger 2022). Performance variations may correspond to differences in these network characteristics.

First indicator of an adaptive network is network closure. From the perspective of social capital, network linkages serve as resources, namely information providers. As such, establishing links between actors constitutes a resource exchange—a series of interactions between two or more actors where a transaction of material or immaterial resources occurs (Lin 2001). The level of network closure, indicated by the degree of density and centralization, is a key characteristic of network structures and can predict network performance (Ahuja 2000; Blau 1977; Freeman 1978).

High network centralization implies numerous connections to potentially knowledgeable actors and positively influences the mobilization of capital for collective action (Coleman 1994; Reagans and McEvily 2003; Rulke and Galaskiewicz 2000). Conversely, a high degree of centrality may lead to centralized management, resulting in fewer experiments and experimental activities (Leavitt 1951; Bonacich 1972; Everett and Borgatti 1999; Rodan and Galunic 2004).

The second indicator of an adaptive network is network segmentation. This characteristic, crucial for adaptive capacity and analysed in this study, refers to the number of bridges that span structural holes. A certain level of separation between groups within the network is necessary to preserve heterogeneity. Excessive segmentation may hinder the dissemination of knowledge; however, it can also facilitate the spread of diverse information and innovative ideas, often through weak ties among dissimilar entities.

Researchers have explored the relationship between structural holes and heterogeneity. Sandström (2004), for instance, associates a higher degree of heterogeneity with an increased number of structural holes.

Network heterogeneity is evidenced by the presence of bridges and the diversity of actors. Actors from different societal sectors are likely to possess and utilize varied resources. The premise is that a network comprising a more diverse set of actors is expected to have more bridges. Heterogeneity has been linked to enhanced performance (Reagans and Zuckerman 2001; Oh et al. 2004) and is considered a potent attribute. The notion is that actors with varied backgrounds or organizational identities are likely to contribute a wealth of new and diverse resources (Louadi 2008; Gerber et al. 2013; Gronow et al. 2020). In this study, the heterogeneity of actors is identified as a trait that facilitates the crucial process of resource mobilization in governance (Ostrom 2005, 2009).

The Key Player Problem approach is utilized in this study to diagnose the adaptive capacities of stakeholders' networks. The Key Player approach, as outlined by Borgatti (2003), aids in understanding network linkages and segmentation, providing a clearer view of the network's context. It reveals how key players are interconnected and embedded within the network, and the degree to which the network's performance is reliant on them.

The Key Player approach examines network cohesion, such as the network's ability to facilitate knowledge access to many individuals, centralization, which includes numerous links to potentially knowledgeable actors and may lead to centralized management with fewer experimental solutions, and heterogeneity in smaller networks.

Two key player issues warrant consideration: the "Key Player Problem/Negative" (KPP-Neg) and the "Key Player Problem/Positive" (KPP-Pos). KPP-Neg is determined by the network's reliance on its key players for cohesiveness. The need for a KPP-Neg set of actors emerges when selecting individuals to remove from the network to disrupt its performance. For instance, identifying a KPP-Neg set could inform decisions on whom to immunize or quarantine to contain an epidemic or whom to disconnect from a criminal network to hinder coordinated actions.

KPP-Neg involves fragmenting a network into components or increasing path lengths between nodes to the point of disconnection. Removing the KPP-Neg set results in a network with minimal cohesion. The distance-weighted fragmentation criterion F is employed to calculate the KPP-Neg set, considering the internal structure of components and determining which sets most effectively disconnect the network.

KPP-Pos is defined by the extent to which key players are connected to and integrated within their surrounding network. The need for a KPP-Pos set arises when seeking network nodes optimally positioned for rapid information diffusion, behaviour adoption, or goods exchange. For example, a KPP-Pos set could serve as "seeds" for disseminating health-promoting practices or assist organizational management in implementing change by first aligning with a small group of informal leaders.

The KPP-Pos set is highly connected to all other nodes. The underlying principle of KPP-Pos is to identify nodes that can reach the maximum number of other nodes via direct links or the shortest paths. The distance-weighted reach measure, defined as the sum of the reciprocals of distances from the KPP-Pos set S to all nodes, is used for this purpose, where the distance from the set to a node is the shortest possible.

In essence, actors with high KPP-Pos measures can maximize resource utilization—in this case, utilized knowledge about NbS—within the network, while those with high KPP-Neg measures can optimize the advantages of brokerage.

15.1.2 Objectives and Aim

This study explores the ramifications of the interplay between actors' knowledge of Nature-based Solutions (NbS), their affiliations, and their roles within governance networks. Gaining insights into these dynamics can inform strategic network interventions aimed at identifying, targeting, and empowering knowledgeable and/or central actors, thereby fostering more adaptive governance. Consequently, the research question that emerges is: How can a network-centric approach facilitate the mobilization of essential resources, specifically knowledgeable key actors, to enhance the network's adaptive capacity?

15.2 Conceptual Framework

This research adopts the Key Player Problem approach to study the methods by which we can diagnose and develop strategies that enhance the network's adaptive capacity concerning NbS knowledge utilization, as depicted in Fig. 15.1

To estimate the adaptive capacity of the network, we:

a. Use kpp-pos and attribute data (level of utilized knowledge)

The KPP-Pos metric is adapted to interpret group centrality as a tool for strategically diffusing environmental knowledge throughout the network. In this study, we identified the smallest set of nodes that can reach all other actors within one step (link), using the m-reach criterion (m = 1) (Borgatti 2003, 2006). The attribute data regarding the level of utilized knowledge about Nature-Based Solutions of the KPP-Pos set actors will be used as an indicator of the network's adaptive capacity.

b. Use kpp-neg approach and attribute data (affiliation)

Given the limited ability to map the global structure of studied networks, attribute data regarding the positions of involved actors and their affiliations will be used as indicators of heterogeneity. We posit that the greater the heterogeneity among the key actors, the more capable the network is of bridging structural holes. The nature of the interactions will be analysed using the Key Player Neg set approach, which involves calculating the set of players whose isolation from the network would cause maximum disruption (Borgatti 2003, 2006). We will further examine the number and types of institutions comprising the Key Players Neg sets. As a result of this methodological choice, the level of heterogeneity will be gauged by the affiliation diversity of the KPP-Neg actors. The rationale behind this choice is that a more diversified set of actors within a network suggests the presence of more bridges and greater modularity. This characteristic is empirically reflected in the number of different organizational and administrative units involved in the governance process. The actors in this study are categorized by sector: civil society organizations, government departments and

	Adaptive Capacity of the Stakeholders' Network	
Indicators	**Closure**	**Segmentation**
Measurement	Group centrality	Number of bridges / Heterogeneity
Adaptive capacity estimation	Key Player-Pos & level of NbS knowledge	Key Player-Neg & affiliation (sector, institution, department)

Fig. 15.1 Conceptual framework to study the ways for enhancing the network's adaptive capacity

agencies, research institutions, and businesses, as well as by the type of organizations within these sectors: nonprofits, grassroots community organizations, homeowners or site owners in the civil sector, and entities at the national, city, and neighbourhood levels, including universities and businesses.

15.3 Methodology

Our empirical research utilizes a case study design. The case studies presented are two multi-actor networks at the moment of collaboration and collective action for NBS implementation in Vienna, Austria, and Tucson, Arizona, USA. Both are engaged in NBS implementation projects that require the use of scientific and expert knowledge about NBS. At the time of data collection, both projects were in the implementation phase, stable over time, and the decision-making processes involved actors from the civil sector, government agencies, research, and business. When reconstructing their actions, there was a risk that respondents might forget or rationalize their behaviour. This potential weakness in the interview study was minimized by selecting ongoing projects at the time of data collection. We chose two cases to illustrate the variety and intricacy of NBS implementation in diverse settings and to assess the methodology in different governance contexts.

The setting for the first case is Neulerchenfelderstraße in Vienna, focusing on the implementation of green facades. The project was initiated by three Viennese universities and supported by the City of Vienna Department of Environmental Protection and the district-level authority (Gebietsbetreuung) to coordinate the project's execution among homeowners, neighbourhood organizations, related enterprises, and other municipal departments of the city of Vienna. The second case is the El Rio Wash project, which involves desert habitat restoration, rainwater harvesting, and water conservation across an 8 km^2 residential area in Tucson, Arizona, USA. This project was initiated by the NGO Trees Please and supported by the Low Impact Development (LID) working group, a coalition of government, private, and non-profit stakeholders formed to guide NBS implementation in Tucson.

15.3.1 Data Collection

Primary data collection was conducted using the qualitative snowballing interview technique to identify the actors in the studied networks, while the quantitative relational data were obtained from the questionnaires (Table 15.1).

Through the questionnaire, we also collected non-network data that will be combined in the analysis with the network data. The boundary is managed with a combination of 'nominalist criteria,' meaning the interviews began with an open-ended question format for key actors (actors in urban offices and/or official project

Table 15.1 Type of relations, indicators and measurements in the scope of this research

Variable	Description	Indicator	Measurement	References
Level of actor's utilised NbS knowledge	Scale assessing six stages of knowledge utilization	To which extent do you receive/ understand/ mention in references/favour the use/use knowledge about NbS to make professional decisions?	Five-point Likert scale. Score derived from knowledge utilisation scale	Knott and Wildavsky (1980); Landry et al. (2003)
Exposure to formal information exchange	Position in the formal information network	With whom do you talk about work on Project X?	Degree centrality	Fritsch et al. (2010); Brass (1984)

leaders), and 'realist criteria,' which involves using the key actors' lists to ask non-key respondents to nominate actors based on their perceived involvement in the case study project. Eventually, the names most frequently mentioned by the key informants were used to define the network's boundaries.

The interviews were followed by questionnaires designed to collect relational, quantitative data and to gather attribute data about the level of the actors' utilized knowledge and their affiliations—the sector, institution, department, or working unit in which they are engaged.

Respondents in the research are queried about their knowledge related to NbS implementation and are introduced to a scale developed by Knott and Wildavsky (1980) and modified by Landry et al. (2003). This scale is designed to assess a progression of six stages of knowledge utilization: reception, cognition, discussion, reference, effort, and influence, where 'influence' denotes the use of research by actors in their work. Each stage is assigned a score from 0 to 5, and then the first stage is multiplied by 1, the second by 2, and so on, according to their significance.

15.3.2 Data Analysis

Our hypothesis suggests that actors with high NbD knowledge are more likely to be Key Players Pos (actors who can reach all other actors in the fewest number of steps) in NbS implementation. This would indicate a highly adaptive network for NBS implementation.

The second hypothesis posits that there is significant heterogeneity among the Key Actors Neg set, or the group of actors whose removal would most disrupt the network. A high degree of heterogeneity among the Key Actors Neg set would suggest an adaptive network because, in the event of disruptions such as political or administrative changes, funding alterations, etc., networks with high heterogeneity are more likely to disseminate diverse information and devise new solutions.

15.3.3 Estimating Correlation Between the Level of Knowledge Utilization of Actors and Their Belonging to "Key Players Pos" Group

To test the correlation between the independent variable (level of utilized NbS knowledge) and inclusion in the Key Player Pos group within the formal information exchange network for the same set of actors, we constructed a dependent variable. This was done by running the KEY PLAYER software (Borgatti et al. 1999, 2014) Key Player Pos procedure (Analyse > Kpp1 Select Nodes to Utilize > Distance Weighted Reach criterion) and creating the UCINET attribute file. The Key Players Pos set is chosen as groups of actors which, when utilized, reach all nodes in the formal information exchange network in one step, i.e., with reciprocal distance indexes being 100. The Key Player Pos group that reaches 100% of the actors in the examined networks consists of five actors in the Tucson network and four actors in the Viennese network (Table 15.2). The independent variable was calculated as the actor's level of environmental knowledge utilization. The actor's score for their level of utilized knowledge is a normalized score for their level of utilized knowledge. This score is the total score of utilized knowledge, evaluated by the actors themselves, divided by the maximum possible score of utilized knowledge that the node can have, expressed as a percentage. The final step involved regressing the position (dependent variable, belonging to the KPP-Pos set) on the attribute (independent variable, level of knowledge), based on the dataset presented in Table 15.2. UCINET's node-level regression procedure (Tools > Testing Hypotheses > Node Level > Regression) was run to compute basic linear multiple regression statistics by OLS and estimate significance using the random permutations method for constructing sampling distributions of R-squared and slope coefficients.

The next hypothesis proposes that there is a high level of heterogeneity among KPP-Neg actors. The number of sectors, institutions, departments, and working units involved in the Tucson and Vienna cases is presented in Table 15.3.

It is beneficial to consider the internal variations among the studied units. Some departments are heterogeneous, comprising various working units, while others are more homogeneously structured. Thus, it is essential to account for the number of divisions or working units within institutions. As shown in Table 15.3, both networks appear similarly diverse in terms of the number of working units, institutions, and sectors involved in the governance process. This is evident when comparing the maximum potential diversity of each network, that is, if every actor were affiliated with a different department or working unit. However, while attribute data that express the organizational identity of the actors are taken into account, it could be argued that the involvement of actors with diverse attributes does not necessarily guarantee

Table 15.2 The sets of "Key Player Pos" actors and attribute data set—level of NBS knowledge for actors in both Tucson and Vienna networks

Tucson			Vienna		
ID	Belonging to KPP-Pos set	Knowledge level	ID	Belonging to KPP-Pos set	Knowledge level
TP1	0	81	M22	0	96
TP2	1	95	GSG	1	100
ADA	0	67	HO1	0	11
TDT	1	81	M28	0	64
HH	0	12	UNI1	0	79
TCB1	0	100	NG2	1	52
TCB2	0	92	M191	1	79
PAG	0	80	M192	1	100
DS	0	73	BOKU	0	80
TP3	0	52	NG1	0	45
UA1	0	79	NG3	0	61
UA2	0	93	HO2	0	11
TP4	0	57	HO3	0	25
HO	1	29	M46	0	47
TWD	0	53	UNI2	0	45
TCB3	0	100			
RFC	1	20			
SI	1	100			

Table 15.3 The number of different sectors, institutions and departments or working units for the Tucson and Vienna case

Network	Number of actors	Number of represented sectors	Number of represented institutions	Number of represented departments or work units
Tucson	18	3	6	11/18
Vienna	16	4	5	12/16

high heterogeneity. The subsequent step of calculating heterogeneity in KPP-Neg sets will provide further insight into the diversity of the KPP-Neg actors who 'hold' the network together.

By running the KEY PLAYER software (Borgatti et al. 1999, 2014), the Key Player Neg procedure (Analyse > Kpp2 Select Nodes to Remove > Distance Weighted Fragmentation criterion) calculates the set of Key Players Neg for both networks. The Key Players Neg set is chosen as a group of four actors who, when disconnected, most effectively fragment the network. The sets of actors and their non-cohesion measures for both networks are presented in Table 15.4. The non-cohesion measure has a value of 1 when the graph consists entirely of isolates, representing the lowest possible cohesion. The closer the value is to 1, the less cohesive the network is. Comparing the two networks, we observe that the Tucson network has less cohesion and requires fewer actors to be removed for the network to become fragmented. For instance, a set containing four Key Players Neg fragments the Viennese network to just 51%, while in Tucson, it fragments to 80%. Additionally, Vienna has a higher density (0.475) than the Tucson network (0.382). For this study, a set of four Key Player Pos actors is chosen to examine for heterogeneity. Table 15.5 indicates that Tucson's Key Players Neg set is more diverse, with two different sectors represented, along with three different working units, nonprofits, and government offices at the national and city levels. The Viennese network relies on one sector (government), with three entities at the city level and one at the neighbourhood level.

Table 15.4 Key Players Neg sets calculated for both networks

TUSCON		VIENNA	
KPP-Neg ID	Non-cohesion measure	KPP-Neg ID	Non-cohesion measure
TP2	0,228	M22	0,308
TP2 TP1	0,497	M22 M192	0,383
TP2 ADA TP1	0,586	M22 M192 NG1	0,419
TP2 ADA TP1 TWD	0,801	M22 NG2 M192 NG1	0,461

Table 15.5 Affiliation of Key Players Neg sets for both networks

Studied network	KPP-Neg set ID/affiliation	Number of sectors	Number of institutions	Number of departments or work units
Tucson	TP2, nonprofit ADA, govern. national level TP1, nonprofit TWD, govern. city level	2/3	2	3/4
Vienna	M22, govern. city level NG2, govern. neigh. level M192, govern. city level NG1, govern. city level	1/4	2	4/4

The degree of heterogeneity in the Key Players Neg set is determined by contrasting the highest possible and measured heterogeneity of each Key Player Neg set. The highest possible heterogeneity of the network is considered as if every Key Neg actor is affiliated with a different sector, institution, or working unit. In the Tucson Key Player Neg set, two sectors are present, while in Vienna, there is only one. Heterogeneity among all the Key Players Neg sets is lower in Vienna, dependent on the governmental sector, as shown in Table 15.5.

15.4 Results and Discussion

15.4.1 Correlation Between Actors' Knowledge About NBS and Their Belonging to "KPP-Pos" Set

The gathered results are presented here in numerical values and comments, as well as in the form of social network diagrams.

Table 15.6 displays the output from the software, including the calculated correlation coefficients. The dependent attribute is the inclusion in the Key Players group, and the independent variable is the level of actors' utilized NBS knowledge. The vectors for both cases are depicted in Fig. 6.9. The coefficients are generated using standard OLS linear modelling techniques and are based on comparing scores on the independent and dependent attributes of individual actors. Since actors are not independent in this dataset, the estimation of standard errors is conducted through simulation, employing permutation tests rather than standard formulas. These results suggest that the Key Player Pos role does not correlate with the actors' level of utilized NBS knowledge in the formal information exchange network. The visualizations of the network diagrams are performed using the program Net-draw (Borgatti 2002). The layout of the network diagrams is determined by the geodesic distance between the nodes. Points corresponding to more central nodes are typically located at the centre of the diagram. The links in the examined formal information networks are unvalued and undirected.

Figure 15.2 illustrates the composition of actors in both networks. In the network diagrams, an actor's level of utilized knowledge is represented by the size of a node. The greater the level of utilized knowledge an actor possesses, the larger the node appears. An actor's inclusion in a KPP-Pos set is indicated by the node's red colour.

Table 15.6 Outputs for regressing the position (dependent vector, belonging to the KPP-Pos set) on the attribute (independent vector, level of NBS knowledge)

Correlations The Key Player Pos role and actors' level of utilitised environmental knowledge			
	R-sq	p	r
Tucson	0.014	0.635	-0.120
Vienna	0.231	0.057	0.481

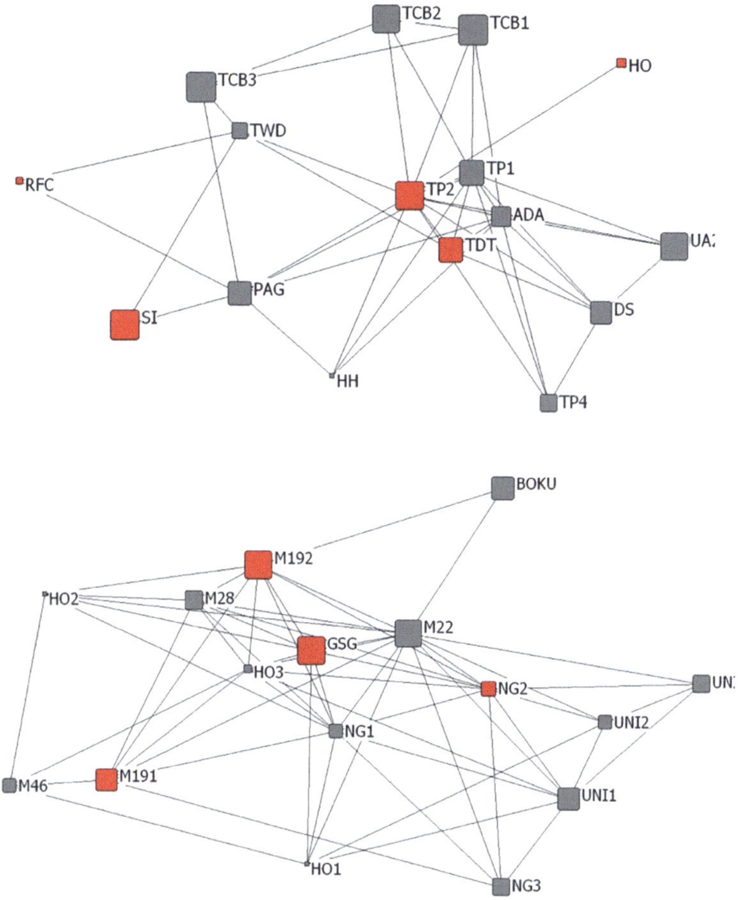

Fig. 15.2 KPP-Pos sets (in red) and actors' level of NbS knowledge (size) in Tucson (above) and Vienna networks

15.4.2 Heterogeneity and Cohesion in Studied Networks

The findings indicate that the Viennese network includes actors from all four sectors, although they are not represented equally; the majority are from the city government. The Tucson network encompasses only three sectors, but these are more heterogeneous, meaning they comprise a greater number of institutions. When assessing network heterogeneity through the diversity of the Key Player Neg set, it becomes apparent that the Viennese network relies solely on one sector (government) for the role of the Key Player Neg actor, which suggests that this set has low heterogeneity (Table 15.7). Figure 15.3 illustrates the composition of Key Player Neg actors in both networks. The colour of each node signifies the actor's affiliation: yellow for civil society organizations and private homeowners, green for government departments and agencies, blue for research institutions, and purple for neighbourhood associations. Actors included in a Key Player Neg group are distinguished by a black border around the node and a label in the network diagram.

Table 15.7 Heterogeneity and cohesion estimations in Tucson and Viennese networks

Governance Network	Heterogeneity whole network	Key Players Neg Set Heterogeneity	Cohesion
Tucson	High	High	Low
Vienna	High	Low	High

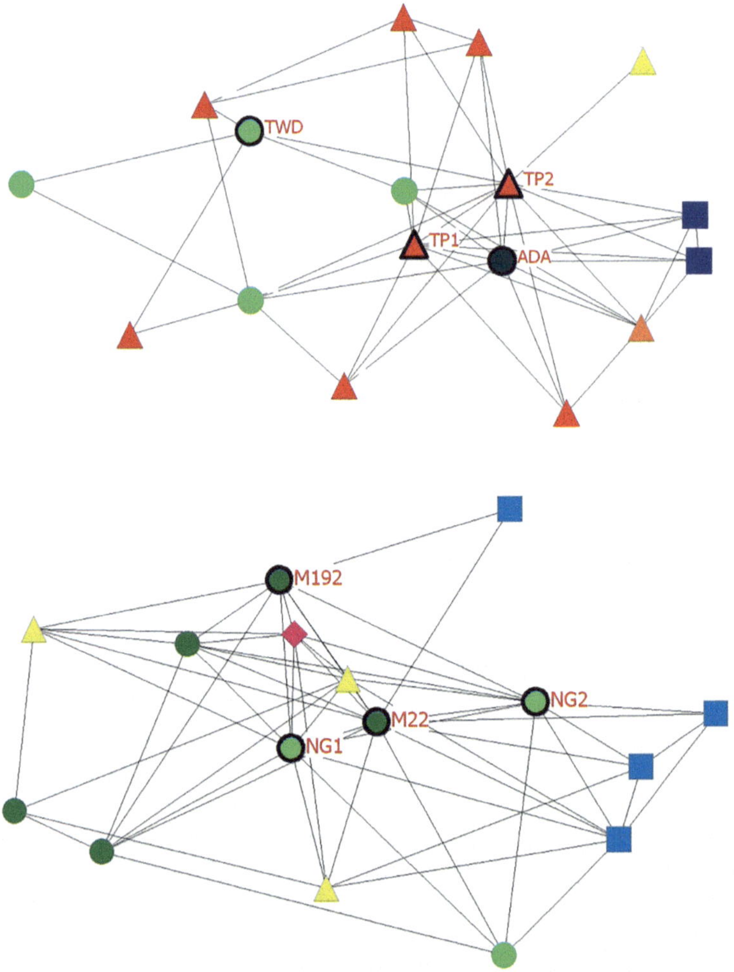

Fig. 15.3 Sets of Key Players Neg (with rim) and their affiliations in Tucson (above) and Vienna networks

At first glance, the network layouts of both cases appear diverse in terms of affiliations to different sectors, institutions, and units. However, when compared to a more 'ideal' model, where all Key Player Neg actors belong to different sectors, it becomes clear that this diversity is less pronounced in the Viennese network. It is accurate to say that the government sector predominates, overshadowing the other three sectors.

15.4.3 Discussion

The empirical findings suggest that a significant weakness in both networks studied is the insufficient recognition of the most knowledgeable actors. This observation underscores the need to identify, empower, and acknowledge these individuals to enable governance networks to implement Nature-based Solutions (NbS) effectively.

The data indicate that actors with a higher level of NbS knowledge are not necessarily more likely to be Key Players in networks focused on NbS implementation. Such insights are valuable for strategic interventions, such as investing in the education of existing Key Player Positive (KPP-Pos) actors or strategically positioning knowledgeable actors in Key Player Pos roles.

By juxtaposing the affiliations of actors with their roles as Key Players, we find that the Viennese network relies solely on the government sector for the role of Key Player Negative (KPP-Neg) actors. This dependency suggests a lower adaptive capacity within the governance network. Analyzing this aspect can guide strategic network management in enhancing institutional heterogeneity—identifying which sectors or institutions are underrepresented, which are favoured, and how to integrate and connect them most effectively.

Social Network Analysis's (SNA) quantitative data and graphical representations provide a clear understanding of network structures and the relationships between individuals and their specific attributes. These features establish SNA as an invaluable tool for planners. The study demonstrates that the Key Player approach can pinpoint how resources crucial for increasing a network's adaptive capacity are distributed. It also offers a solid foundation for developing evidence-based strategies to mobilize these resources. This information is instrumental in addressing common challenges in NbS implementation, such as improving coordination and cooperation across organizational levels, fostering sustained interest in NbS, understanding the relationships that shape planning processes and outcomes, adapting to ongoing changes in governance, and leveraging knowledge about NbS.

One limitation of this study is the quality of the data. Network data is known to deviate from reality to some extent—for instance, 10% of observed ties may not exist, and 10% of non-adjacent pairs may actually be adjacent. Therefore, if the Key Player approach is to serve as a practical tool, it must be robust enough to handle data inaccuracies. This involves not relying solely on the observed dataset but adjusting the data to some degree and optimizing results based on "adjacent" datasets derived from these variations. The outcome would be a Key Player set that, while not a perfect match for the observed dataset, represents the best solution for the graph's overall neighbourhood.

Moreover, the networks examined in this study are confined to organizational units identified through snowball sampling, potentially missing connections beyond the established boundaries. The networks presented also possess a "snapshot character," not accounting for the underlying causes and effects.

Another limitation is the method's focus on a single interaction dimension—formal information exchange—while overlooking other factors like cultural context, established behavioural norms, and past experiences. Investigating these elements is crucial as they shape actors' mental models and influence their decision-making, objectives, and strategies in governance processes.

The results demonstrate that the methodology is effective across different contexts, enabling urban networks to identify and address their unique NbS implementation challenges. The method accounts for the varied and complex factors affecting performance and behaviour in diverse governance scenarios. Although the approach is effective, it demands significant time for data collection and analysis. Consequently, we recommend its application in larger-scale NbS implementations where the benefits justify the investment of time and resources.

15.5 Conclusions

Our proposed method for assessing and managing the adaptive capacity of governance is deeply rooted in understanding the key players within stakeholder networks. By identifying those who can most effectively disseminate NbS knowledge and those critical for maintaining network performance, we have applied our approach to real-world scenarios in Vienna and Tucson.

This research is based on the theoretical concept that the adaptive capacity of a stakeholders' network can be estimated by its closure and heterogeneity. This chapter presents a method for determining the adaptive capacity of stakeholder networks in implementing NbS by analysing the structural positions of the networks' resources: (1) Key Players Pos, i.e., actors able to maximize the spread of knowledge, (2) Key Players Neg, namely, the actors bridging the network, (3) actors' NbS knowledge, and (4) the diversity of institutions and working units involved.

Such analysis is a valuable tool for enhancing adaptive capacity in governance implementing NbS, as it can provide the necessary data for strategic network interventions, e.g., strategies to engage, educate, or empower knowledgeable, central or marginalized actors. The use of the method proposed here could improve NbS knowledge utilization in NbS implementation processes. This approach empowers stakeholders to deeply understand and address common challenges in NbS implementation, as well as paves the way for improving governance coordination and cooperation, and for sustaining public interest in NbS.

The SNA approach is crucial for informing us about the network's adaptive capacity, and future research should also incorporate into this approach the nature of values, motives, and the specific purposes that guide the actors, as well as the quality and diversity of information and knowledge, the power and influence that different actors have in the network, while empirically examining feedback and interactions between the network and its broader social-ecological context.

References

Adam, Silke, Hanspeter Kriesi (2007) The network approach. Theories of the Policy Process, ed. Paul A. Sabatier. Cambridge MA: Westview Press, 189–220

Angst M, Brandenberger L (2022) Information exchange in governance networks—Who brokers across political divides? Governance 35(2):585–608

Ahern J (2007) Green infrastructure for cities: the spatial dimension. In: Novotny V, Brown P (eds) Cities of the future: towards integrated sustainable water and landscape management. IWA Publishing, London, pp 267–283

Ahuja G (2000) Collaboration networks, structural holes, and innovation: A longitudinal study. Adm Sci Q 45(3):425–455

Bache I, Bartle I, Flinders M (2016) Multi-level governance. In Handbook on theories of governance (pp. 486–498). Edward Elgar Publishing

Blau PM (1977) Inequality and heterogeneity: A primitive theory of social structure, vol 7. Free Press, New York

Benedict MA, McMahon ET (2002) Green infrastructure: smart conservation for the 21st century. Renew Resour J 20(3):12–17

Benedict MA, McMahon ET (2012) Green infrastructure: linking landscapes and communities. Island Press

Berardo R, Scholz JT (2010) Self-organizing policy networks: Risk, partner selection, and cooperation in estuaries. Am J Polit Sci 54(3):632–649

Bonacich P (1972) Factoring and weighting approaches to status scores and clique identification. J Math Sociol 2(1):113–120

Borgatti SP (2002) NetDraw: Graph visualization software. Harv: Anal Technol

Borgatti SP (2003) The key player problem (pp. 241–52). na

Borgatti SP (2006) Identifying sets of key players in a social network. Comput Math Organ Theory 12(1):21–34

Borgatti SP, Foster PC (2003) The network paradigm in organizational research: A review and typology. J Manag 29(6):991–1013

Borgatti SP, Everett MG, Freeman LC (1999) UCINET 6.0 Version 1.00. Natick: Anal Technol

Borgatti SP, Everett MG, Freeman LC (2014) Ucinet. Encyclopedia of social network analysis and mining, 2261–2267

Borgatti SP, Jones C, Everett MG (1998) Network Measures of Social Capital. Connections 21(2):27–36

Brass DJ (1984) Being in the right place: A structural analysis of individual influence in an organization. Adm Sci Q, 518–539

Bressers HT, A., and Laurence J. O'Toole. (1998) The selection of policy instruments: a network-based perspective. J Publ Policy 18(03):213–239

Burt RS (2004) Structural holes and good ideas. Am J Sociol 110(2):349–399

Burt RS (2017) Structural holes versus network closure as social capital. In Soc Cap (pp. 31–56). Routledges

Burt R (1992) Structural holes: the social structure of competition. Cambridge, MA: Harvard University Press

Coleman JS (1994) Foundations of social theory. Harvard University Press

Everett MG, Borgatti SP (1999) The centrality of groups and classes. J Math Sociol 23(3):181–201

Folke C, Hahn T, Olsson P, Norberg J (2005) Adaptive governance of social-ecological systems. Annu Rev Environ Resour 30:441–473
Freeman LC (1978) Centrality in social networks conceptual clarification. Soc Netw 1(3):215–239
Friedkin NE (1981) The development of structure in random networks: an analysis of the effects of increasing network density on five measures of structure. Soc Netw 3(1):41–52
Fritsch M, Kauffeld-Monz M (2010) The impact of network structure on knowledge transfer: an application of social network analysis in the context of regional innovation networks. Ann Reg Sci 44:21–38
Gerber ER, Henry AD, Lubell M (2013) Political homophily and collaboration in regional planning networks. Am J Polit Sci 57(3):598–610
Gronow A, Wagner P, Ylä-Anttila T (2020) Explaining collaboration in consensual and conflictual governance networks. Public Adm 98(3):730–745
Hall TE, O'Toole LJ Jr (2000) Structures for policy implementation: An analysis of national legislation, 1965–1966 and 1993–1994. Adm & Soc 31(6):667–686
Holling CS, Dantzig GB, Winkler C (1986) Determining optimal policies for ecosystems. Syst Anal for for Ind 21:453–473
Kickert WJM, Koppenjan JFM, Klijn EH (1997) Managing complex networks: Strategies for the public sector. Manag Complex Netw, 1–224
Knott J, Wildavsky A (1980) If dissemination is the solution, what is the problem? Knowledge 1(4):537–578
Landry R, Lamari M, Amara N (2003) The extent and determinants of the utilization of university research in government agencies. Public Adm Rev 63(2):192–205
Leavitt H (1951) Some effects of certain communication patterns on group performance. J Abnorm Soc Psychol 46:38–50
Lin N (2001) Social capital. a theory of social structure and action. Cambridge University Press, Cambridge
Louadi M (2008) Knowledge heterogeneity and social network analysis–Towards conceptual and measurement clarifications
Mell IC (2008) Green infrastructure: concepts and planning. In FORUM ejournal (8(1), pp. 69–80). Newcastle, UK: Newcastle University
Metz F (2017) From network structure to policy design in water protection: A comparative perspective on micropollutants in the Rhine River Riparian countries. Springer
Metz F, Brandenberger L (2023) Policy networks across political systems. Am J Polit Sci 67(3):569–586
Oh H, Chung MH, Labianca G (2004) Group social capital and group effectiveness: The role of informal socializing ties. Acad Manag J 47(6):860–875
Ostrom E (2009) Understanding institutional diversity. Princeton University Press
Ostrom E (2005) Doing institutional analysis digging deeper than markets and hierarchies. Handbook of new institutional economics. Springer, US, Boston, MA, pp 819–848
Peters BG, Pierre J (2000) Citizens versus the new public manager: The problem of mutual empowerment. Adm & Soc 32(1):9–28
Provan KG, Kenis P (2008) Modes of network governance: Structure, management, and effectiveness. J Public Adm Res Theory 18(2):229–252
Reagans R, Zuckerman EW (2001) Networks, diversity, and productivity: The social capital of corporate R&D teams. Organ Sci 12(4):502–517
Reagans R, McEvily B (2003) Network structure and knowledge transfer: the effects of cohesion and range. Adm Sci Q 48(2):240–267
Rodan S, Galunic C (2004) More than network structure: How knowledge heterogeneity influences managerial performance and innovativeness. Strateg Manag J 25(6):541–562
Rhodes RAW (1996) The new governance: governing without government. Polit Stud 44(4):652–667
Rulke DL, Galaskiewicz J (2000) Distribution of knowledge, group network structure, and group performance. Manage Sci 46(5):612–625

Sandström A (2004) Innovative policy networks: the relation between structure and performance (Doctoral dissertation, Luleå tekniska universitet)
Sandström A, Carlsson L (2008) The performance of policy networks: the relation between network structure and network performance. Policy Stud J 36(4):497–524
Schaub S, Metz F (2020) Comparing discourse and policy network approaches: Evidence from water policy on micropollutants. Polit GovAnce 8(2):184–199
Sørensen E, Torfing J (2005a) The democratic anchorage of governance networks. Scand Polit Stud 28(3):195–218
Sørensen E, Torfing J (2005b) Network governance and post-liberal democracy. Adm Theory & Prax 27(2):197–237
Stoker G (1998) Governance as theory: five propositions. Int Soc Sci J 50(155):17–28
Völker B, Flap H (2004) Social networks and performance at work. Creat Returns Soc Cap 9:133
Wagner PM, Ylä-Anttila T, Gronow A, Ocelík P, Schmidt L, Delicado A (2021) Information exchange networks at the climate Science-policy interface: Evidence from the Czech Republic, Finland, Ireland, and Portugal. Govance 34(1):211–228
Wasserman S (1994) Social network analysis: methods and applications. Camb Univ Press Google Sch 2:131–134

Chapter 16
Drivers of Successful Implementation of Nature-Based Solutions Initiatives: Challenges and Policy Recommendations

Charleton Dunn L. Llosa, Edgar M. Reyes Jr., and Casper Boongaling Agaton

Abstract Nature-based solutions (NbS) are promising mitigation and adaptation measures that sustainably address societal and global problems, while simultaneously providing numerous benefits. However, its implementation is challenged by various issues such as financing, social acceptance, and governance concerns. This study aims to analyze how different drivers affect the successful implementation of NBS initiatives applying the Reporting Standards for Systematic Evidence Syntheses (ROSES) in environmental research. Results discuss the factors affecting NBS implementation from the themes based on the International Union for Conservation of Nature (IUCN) Global Standards for NbS including: (1) The NbS effectively addresses societal challenges, (2) The design of the NbS is informed by scale, (3) The NbS initiatives result in a net gain to biodiversity and ecosystem integrity, (4) NbS are economically viable, (5) NbS are based on inclusive, transparent, and empowering governance processes, (6) NbS equitably balance the trade-offs between achievement of primary goal(s) and the continued provision of multiple benefits, (7) NbS are managed adaptively, based on evidence, and (8) NbS are sustainable and mainstreamed within an appropriate jurisdictional context. Moreover, results found that NbS initiatives are challenged by low participation of stakeholders, limited funding opportunities, and supporting laws and policies. These findings provide recommendations including a participatory planning process, coordination between and among stakeholders, transparency and documentation, flexible operation and implementation, and establishment of monitoring and evaluation mechanisms. Overcoming these hurdles is essential to successfully implement NbS initiatives, maximize its benefits for the environment, economy, and society, and create a more resilient and sustainable urbanization.

Keywords Nature-based solutions · Implementation · IUCN Global Standard · ROSES · Sustainable urbanization

---

C. D. L. Llosa · E. M. Reyes Jr. · C. B. Agaton (✉)
Department of Community and Environmental Resource Planning, College of Human Ecology, University of the Philippines Los Baños, Laguna, Philippines
e-mail: cbagaton@up.edu.ph

M. Firoz C et al. (eds.), *Nature-Based Solutions for Urban and Peri-Urban Areas*, Disaster Risk Reduction, https://doi.org/10.1007/978-981-96-5933-3_16

16.1 Introduction

In recent decades, there has been a growing recognition of the crucial role of nature-based solutions (NbS) in addressing various environmental and societal challenges while fostering sustainable development. These challenges include climate change, food security, water resources, and disaster risk management (Pauleit et al. 2017). NbS utilizes the power of nature to enhance sustainable urbanization, restore degraded ecosystems and increase their resilience, develop climate change strategies, and improve environmental risk management (Lafortezza et al. 2018).

Based on the perceived and documented effectiveness, readiness, scalability, and affordability of NbS relative to technological solutions, its popularity has rapidly grown particularly in policy, research, and business agendas (Seddon 2022). Yet, their successful implementation faced several challenges from financial and technical constraints to institutional, governance, and socio-political barriers (Agaton and Guila 2024; Nelson et al. 2020). Hence, understanding the key problems, enablers, and successful cases in the implementation of NbS initiatives is crucial to devising targeted interventions and policy measures to maximize their potential in addressing pressing environmental and societal issues.

This chapter explores the drivers contributing to the successful implementation of NbS initiatives, identifies the challenges encountered in their deployment, and provides policy recommendations to overcome these barriers. This employs a systematic review of the NbS literature applying the RepOrting standards for Systematic Evidence Syntheses (ROSES) framework. Based on the International Union for Conservation of Nature and Natural Resources (IUCN) Global Standards for NbS implementation, the systematic analysis aims to shed light on the intricate dynamics shaping the implementation of NbS and offer actionable insights for policymakers, practitioners, and stakeholders involved in NbS initiatives.

16.2 Literature Review

16.2.1 Nature-Based Solution

As first mentioned in 2008, the concept of NbS emerged as an innovative solution to manage natural systems by working with nature toward a resilient, resource-efficient, and green economy (Sowińska-Świerkosz and García 2022). A more comprehensive definition was provided by the IUCN as "actions to protect, sustainably manage, and restore natural and modified ecosystems in ways that address societal challenges effectively and adaptively, to provide both human well-being and biodiversity benefits" (Cohen-Shacham et al. 2016).

NbS interventions encompass a range of strategies that utilize nature and its ecosystem services to address societal challenges, offering multifaceted benefits ranging from disaster risk reduction, climate change mitigation and adaptation, enhancing biodiversity, and improving human well-being (Kolokotsa et al. 2020; Liu et al. 2021; Nehren et al. 2023; Seddon 2022; Vasiliev and Greenwood 2022). Notable examples of NbS include natural and man-made ecosystems to protect river catchments or coastal areas from flooding and erosion; creation of forests and wetlands in urban areas for cooling and flood risk reduction; reforestation to help mitigate climate change and protect biodiversity; and constructed wetlands to treat wastewater (Agaton 2024; Seddon 2022). These interventions aim to enhance both human well-being and biodiversity conservation simultaneously, promoting the harmonious coexistence of people and nature.

Its wide range of applications resulted in the implementation of NbS initiatives spreading across different regions and sectors globally. Yet, before 2020, there was a lack of specific, detailed criteria for NbS resulting in several actions that can be regarded as complementary or related to NbS but that do not fulfill all the requisite criteria to be called NbS (Sowińska-Świerkosz and García 2022). Hence, the IUCN provided a Global Standard, which frames green/blue interventions as NbS actions.

16.2.2 IUCN Criteria for NbS Implementation

The IUCN (2020) Global Standard presents a systematic learning framework that will help guide and strengthen the effectiveness, adaptability, and sustainability of NbS methods among decision-makers. The framework is composed of eight (8) general criteria as discussed below:

Criterion 1. NbS effectively addresses societal challenges. This criterion ensures that the NbS is created in response to social issues and that the community directly affected by the issues is prioritized. This also guarantees that all stakeholders involved in the process of determining the most important challenges are included in the decision-making process, particularly those who have rights and will benefit from the NbS initiatives.

Criterion 2. The design of NbS is informed by scale. This criterion promotes NbS designs that acknowledge the complexity and unpredictability present in dynamic, land and water environments, including biophysical, geographical, and economic systems, as well as policy frameworks and cultural viewpoints.

Criterion 3. NbS results in a net gain to biodiversity and ecosystem integrity. This criterion requires that the design and implementation of NbS avoid undermining the integrity of the system and proactively attempts to enhance the ecosystem's functioning and connection ensuring the long-term resilience and sustainability of the NbS initiatives.

Criterion 4. NbS are economically viable. This criterion requires a thorough analysis of the NbS initiative's long-term economic viability and sustainability considering important success indicators such as ROI, intervention efficacy and efficiency, and equitable benefit–cost distribution.

Criterion 5. NbS is based on inclusive, transparent, and empowering governance processes. To meet this requirement, NbS initiatives must recognize, involve, and address the concerns of various stakeholders, particularly the rights of the communities.

Criterion 6. NbS equitably balances trade-offs between the achievement of their primary goal(s) and the continued provision of multiple benefits. To meet this requirement, proponents of the net-benefit system must accept the trade-offs between managing land and natural resources while adhering to objective, transparent, and inclusive procedures for balancing and fairly managing NbS initiatives over time and geographic space.

Criterion 7. NbS are managed adaptively, based on evidence. This criterion requires that the implementation of NbS plans must have clauses that allow adaptive management in response to uncertainties and as an option to efficiently utilize ecosystem resilience.

Criterion 8. NbS are sustainable and mainstreamed within an appropriate jurisdictional context. To meet this criterion, NbS initiatives must be planned and implemented considering the long-term sustainability as well as the existing sectoral, national, and other policy frameworks.

16.2.3 Barriers and Drivers on NbS Initiatives

The literature provides several reviews of barriers and drivers to successfully implement NbS initiatives. For instance, Sarabi et al. (2019) systematically reviewed the definition of NbS from the theoretical point of view, its significance in addressing societal challenges, the key actors in NbS planning, implementation, and management, as well as the major challenges and drivers of NbS initiatives. In another study, Mendonça et al. (2021) conducted a systematic literature review evaluating the various policy instruments supporting the adoption of NbS initiatives. In terms of climate solutions, Schulte et al. (2021) presented a systematic review clarifying natural climate solutions, which include NbS and other ecosystem-based adaptation initiatives; presented the state-of-the-art including the trends, opportunities, gaps, and hotspots; and critically identified the factors affecting the implementation of NbS initiatives in different geographic spaces. In terms of green financing, Toxopeus and Polzin (2021) carried out a systematic review to identify the key challenges and corresponding strategies to finance urban NbS. They also identified two overarching barriers to NbS finance including the coordination between private and public financiers and the integration of NbS benefits into valuation and accounting methods. In terms of NbS governance, Morita and Matsumoto (2021) analyzed the development and implementation of NbS initiatives in Asia with a particular focus on climate change mitigation and adaptation, disaster risk reduction, and infrastructure.

To the best of the authors' knowledge, there has been no literature review focusing on the IUCN standard of the implementation of NbS. This standard provides a systematic learning framework for improving and evolving the applications, leading to greater confidence in NbS among decision-makers from policy making to NbS project planning and implementation.

16.3 Methodology

The review protocol in this chapter utilized the ROSES method. This protocol carefully considers systematic mapping, narrative, and quality synthesis methods in developing the highest possible reporting standards in conducting systematic reviews (Haddaway et al. 2018). By employing the ROSES Flow Diagram for systematic reviews, the study ensures that relevant methodological information is reported in the review, including the analysis of its reliability and validity.

The literature search was conducted in January 2023. Two databases were included, namely, Scopus and Web of Science. Furthermore, to include additional relevant articles for the review, manual searching was also utilized from the Google Scholar database.

The three main processes involved in reviewing the literature include identification, screening, and eligibility.

Searching/Identification. The first step pertains to the process of selecting suitable keywords to be utilized in searching for relevant articles in the review. The keywords were determined from the PICo tool employed in the formulation of the main research question, which were *nature-based solutions*, *successful implementation*, and *factors*. Shown in Table 16.1 is the list of extracted keywords from the main research question, along with related terms and synonyms that were utilized in the search process.

For the search string, the Boolean operators "AND" and "OR" were used to indicate that all keywords must appear in any parts of the article while also narrowing down the results. The results of the identification process for the databases acquired a total of 83 articles. Together with the results from the manual searching which resulted in 15 additional articles, the first stage of the systematic review process accumulated a total of 98 articles from the three databases.

Screening. The screening process comes after identification which entails the secondary selection of articles based on set criteria formulated to represent the identified topic and the main research question of the study. This process involves removing duplicates, title screening, abstract screening, and retrieving only full-text articles. For full-text screening, four (4) criteria were set as shown in Table 16.2. After applying the set criteria, 62 were excluded and only 36 articles remained.

Critical appraisal and synthesis/Eligibility. The last process pertains to eligibility, wherein the screened articles were manually assessed by examining their whole content to evaluate whether they have fulfilled the set criteria and to ensure that they

Table 16.1 Extracted keywords and related terms from the main research question

Keywords and related terms/synonyms		
Factors	Nature-based solutions	Implementation
Elements, aspects, component	Nature-based solutions, ecosystem-based adaptation, natural infrastructure, green infrastructure, ecological engineering	Application, operation

Table 16.2 Inclusion and exclusion criteria

Criteria	Inclusion	Exclusion
Access	Open access	Non-open access
Timeline	2017–2022	< 2017
Literature type	Journals (peer-reviewed research articles)	Reports
Language	English	Non-English

fit the objectives of the study. Of the 36 screened articles, 8 were removed after critical appraisal due to their irrelevant content that do not aid in answering the main research question, and another 5 were eliminated after narrative synthesis due to the same reason. Overall, there were only a total of 19 articles selected for analysis. Figure 16.1 summarizes the entire 3-step systematic review process applied in the study using the ROSES Flow Diagram. Moreover, the summary of results for the systematic review is shown in Table 16.3.

16.4 Results

16.4.1 NbS Effectively Addresses Societal Challenges

NbS plays an integral role in addressing contemporary local and global challenges including the worsening impacts of climate change; effects of urban heat; various hazards such as flooding; landslides; air, water, and soil pollution; and most importantly, the improvement of human health (Antuña-Rozado et al. 2019). Approaches such as the redevelopment of degraded urban areas, landslide and flooding risk reduction, and stormwater management are just some of the various functions that NbS can offer. Additionally, societal challenges, such as in the case of a water treatment under dry-weather conditions in a dense neighborhood in the Great Metropolitan Area in Costa Rica, were resolved through NbS by utilizing a co-design process in the formulation of novel plans (Albert et al. 2021).

Approaches utilized for the inclusion of this criterion mainly focused on ensuring a participatory planning process for the stakeholders in the NbS projects to identify and solve the problems they encounter. Community consultations, which enable an inclusive planning process and may allow the co-creation of citizens and other stakeholders throughout the project, are a prevalent practice (Albert et al. 2020; Antuña-Rozado et al. 2019; Mekala and Hatton MacDonald 2018; Solheim et al. 2021). This practice prompts the interests of the stakeholders especially if their views, about the direct challenges faced, are heard. Dushkova and Haase (2020), on the other hand, explored a different lens, wherein emotional needs were provided through a sense of community approach, providing importance to the intangible values that are usually underrepresented (e.g., belonging, protection, safe spaces),

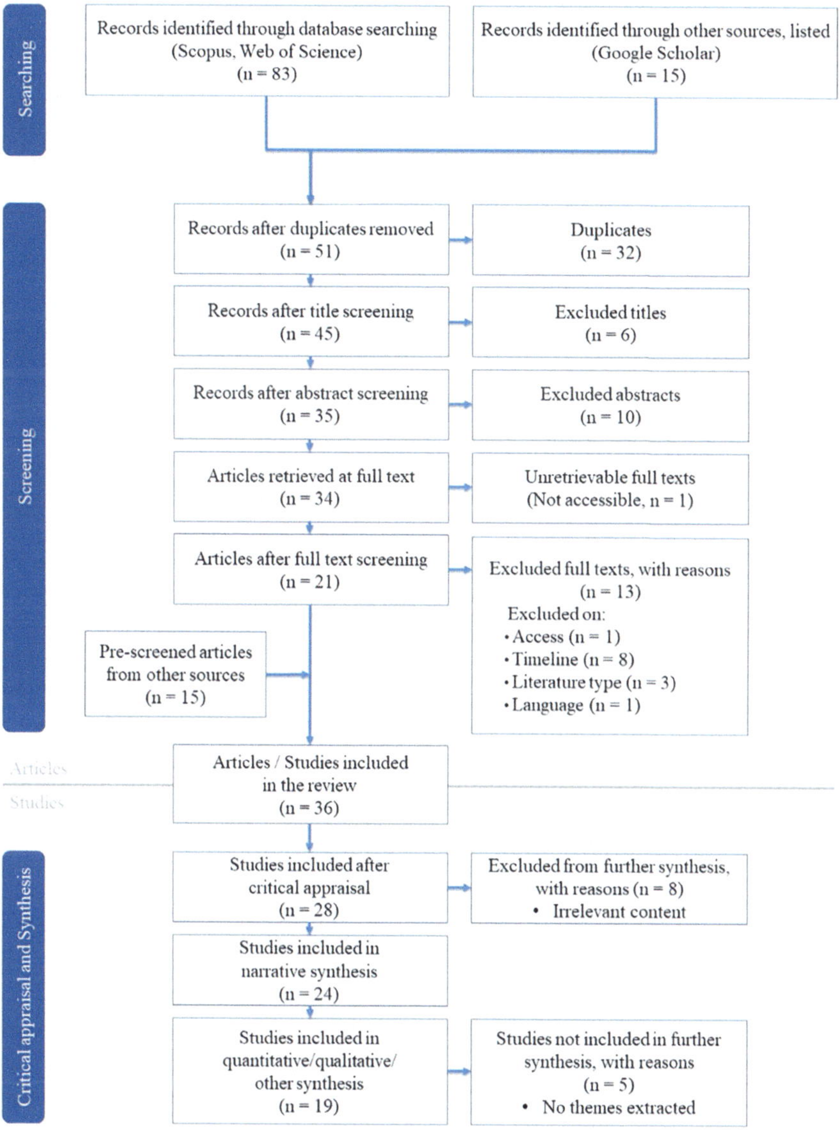

Fig. 16.1 Flow diagram of the systematic review

including place attachment and place-making. In line with this, local acceptance, wherein community needs, benefits, and consistent quality of life are aligned and ensured, may require social restructuring, especially in cases where path-shifting transformations in the social system are needed to reorganize stakeholder networks for community empowerment (Palomo et al. 2021; Schulte et al. 2021). Moreover,

Table 16.3 NbS Systematic review summary of results

Authors (year)	Criteria							
	1	2	3	4	5	6	7	8
Emilsson and Ode Sang (2017)			■		■			■
Wüstemann et al. (2017)			■	■	■	■	■	
Mekala and Hatton MacDonald (2018)	■	■	■	■	■			
Antuña-Rozado et al. (2019)	■	■	■		■			■
Cohen-Shacham et al. (2019)		■					■	■
Frantzeskaki (2019)					■	■	■	
Sarabi et al. (2019)		■		■	■	■	■	
Dushkova and Haase (2020)	■	■	■	■	■	■	■	■
Kumar et al. (2020)			■		■		■	
Zingraff-Hamed et al. (2020)		■	■	■	■		■	
Albert et al. (2020)	■	■	■		■		■	■
Albert et al. (2021)	■	■	■		■			
Cohn et al. (2021)		■	■			■		■
Egusquiza et al. (2021)		■		■	■		■	
Palomo et al. (2021)	■		■	■	■		■	■
Ruangpan et al. (2020)	■	■	■		■	■		
Schulte et al. (2021)	■	■	■	■	■		■	■
Solheim et al. (2021)	■			■	■			■
Wickenberg et al. (2021)		■	■	■	■	■	■	

Ruangpan et al. (2020) stress that a fair and equitable manner of participation can carefully outline the main and co-benefits of NbS while also promoting transparency.

16.4.2 Design of NbS is Informed by Scale

Specific study areas, including their relevant characteristics, are essential in the design of NbS initiatives. One such case is the continuous land conversion in the Victorian Volcanic Plain in Australia for agricultural use (Mekala and Hatton MacDonald 2018). The area's remaining native grasslands were declared critically endangered, with more than 25 threatened flora and fauna species identified, making it a priority for conservation and protection initiatives including NbS. Contrastingly, Antuña-Rozado et al. (2019) focused on Tampere City in Finland, a known center for technology, culture, research, education, and sports. The city aims to turn into a global demonstrator for NbS, transforming itself into an Urban Nature Lab (UNaLab) funded by the EU Horizon 2020 program. Although these large-scale areas were highly suitable for NbS, Zingraff-Hamed et al. (2020) found that successful implementation remains at a pilot area level or a local government scale.

Place-specificity, as a significant aspect of NbS, ensures that local conditions and challenges are acknowledged in planning, and aids in resource efficiency and resilience to change (Albert et al. 2020; Ruangpan et al. 2020). Numerous methods allow the achievement of this mechanism, including inventory mapping to monitor large-scale areas; implementation of urban labs to act as demonstration areas; spatial allocation to define potential specific locations; multi-criteria analysis for integrated evaluation especially to multiple stakeholders; and ecological assessment for the design and development of natural areas (Antuña-Rozado et al. 2019; Cohn et al. 2021; Ruangpan et al. 2020; Sarabi et al. 2019). Thus, NbS benefits only accrue when space availability and the corresponding social-ecological systems within the local context are considered (Albert et al. 2021; Dushkova and Haase 2020). Moreover, it aids in the formulation of informed decision-making through practical and solution-specific data and averts implications that may likewise harm various ecosystems (Egusquiza et al. 2021; Schulte et al. 2021; Wickenberg et al. 2021).

Moreover, Cohen-Shacham et al. (2019) found that landscape-scale planning recognizes the different interactions among ecosystems, which may either enhance or hinder the potential for ecological management success. This was supported by Cohn et al. (2021) as the holistic understanding of the environmental setting aids in conforming to the design of PPAs about local conditions, leading to increased resiliency.

16.4.3 NbS Results in a Net Gain to Biodiversity and Ecosystem Integrity

Numerous methods have been determined, including a hierarchical process to strategically select and prioritize NbS initiatives; the addition of green spaces to existing open areas; modeling approaches for the multi-factor risk analysis of various climate stimuli; NbS integration to management plans in local, regional, and national levels; and biodiversity monitoring and restoration (Dushkova and Haase 2020; Emilsson and Ode Sang 2017; Kumar et al. 2020; Palomo et al. 2021). All these aid in achieving supplementary co-benefits such as additional ecosystem services, environmental justice, and increased resilience (Albert et al. 2020; Cohn et al. 2021; Mekala and Hatton MacDonald 2018).

In addition to the studies of Ruangpan et al. (2020), Cohn et al. (2021), and Wickenberg et al. (2021), which concentrated on biodiversity net gain and ecosystem integrity, other abundant benefits have also been traced by other studies including climate change initiatives, pollutants fixation, and disaster risk reduction such as for flood water retention (Albert et al. 2021; Cohn et al. 2021; Dushkova and Haase 2020; Guila et al., 2024; Schulte et al. 2021; Wüstemann et al. 2017; Zingraff-Hamed et al. 2020). Ultimately, these benefits assist in safeguarding socio-economic and human well-being to also heighten societal resilience (Ruangpan et al. 2020).

16.4.4 NbS Are Economically Viable

For NbS to be economically viable, diverse mechanisms were lifted from the studies, including the availability of funds for ecosystem-based climate policies; the presence of different transaction costs (e.g., information costs, coordination costs, strategic costs); shared opportunity costs and economic risks; and the operational integration of financing, governance, and business models (Dushkova and Haase 2020; Egusquiza et al. 2021; Mekala and Hatton MacDonald 2018; Sarabi et al. 2019; Wüstemann et al. 2017). These techniques were shown to be effective, especially when the real value of the NbS co-benefits were quantified over time, and the financing capacities of the economic models utilized were both innovative and inclusive (Solheim et al. 2021; Wickenberg et al. 2021).

The benefits of the economic variable are primarily reflected in the social aspect, wherein the development of innovative financing models drove stronger citizen action (Zingraff-Hamed et al. 2020). Furthermore, NbS creates employment opportunities for the locals, land value creation, and the inception of a community conservation area for sustainable livelihoods (Palomo et al. 2021; Samantela and Maquiling 2024). Concerning this, labor availability, market competitiveness, and accessibility were some of the co-benefits provided by NbS, supporting the local livelihoods (Schulte et al. 2021).

16.4.5 NbS is Based on Inclusive, Transparent, and Empowering Governance Processes

Participatory planning serves as an essential facet for the effective implementation of NbS. Several approaches were observed from the studies such as the active involvement of the community enabling co-creation, co-design, and co-management of NbS initiatives; stakeholder analysis and capacity building; joint reflection and regular community consultations; multi-stakeholder engagement for social inclusion and cohesion; and the vertical and horizontal cooperation between and among stakeholders (Antuña-Rozado et al. 2019; Dushkova and Haase 2020; Emilsson and Ode Sang 2017; Kumar et al. 2020; Mekala and Hatton MacDonald 2018; Palomo et al. 2021; Sarabi et al. 2019; Schulte et al. 2021; Zingraff-Hamed et al. 2020).

On the other hand, in terms of policy coherence for NbS, actions involved aimed at better alignment and coordination of planning laws and regulatory frameworks; taking into consideration the pre-existing political, social, and economic conditions; recognition of stakeholder preferences for better decision-making, and for a systematic and transparent planning process; selection of appropriate governance models based on context; and the exploration of governance innovation, about governance that go beyond the government to include state and non-state actors for social inclusion (Albert et al. 2020, 2021; Egusquiza et al. 2021; Solheim et al. 2021; Wickenberg et al. 2021; Wüstemann et al. 2017).

Inclusive approaches for NbS were observed to drive innovative ideas that are better customized according to the needs and demands of the community, improve the longer-term commitment for further development, and shift the perceptions of stakeholders in terms of new ways of collaboration, especially with the locals (Antuña-Rozado et al. 2019; Frantzeskaki 2019; Ruangpan et al. 2020). About this, Emilsson and Ode Sang (2017) recognized the significance of collaborative processes to enhance the viability of NbS particularly in terms of site adaptation.

16.4.6 NbS Equitably Balances Trade-Offs Between the Achievement of Their Primary Goal(S) and the Continued Provision of Multiple Benefits

Trade-offs can serve as one of the main hindrances in successful NbS implementation. Some of these trade-offs were drawn on public perceptions, questionable initial funding investments, community buy-in, and the lack and/or absence of multi-stakeholder cooperation for planning (Cohn et al. 2021). These unexpected and undesired issues are just some of the unfortunate realities of NbS interventions that were poorly planned (Dushkova and Haase 2020).

To aid in mitigating the presence and effects of trade-offs, some of the approaches utilized in the studies include the use of a quantitative (optimization) model to enable a broader understanding of synergies between climate policy and nature conservation; employment of scientific reports to provide scientific bases for management decisions; planning for NbS through a landscape approach; the acknowledgment of risk reduction and social and ecological outcomes for both the communities and ecosystems (Frantzeskaki 2019; Ruangpan et al. 2020; Sarabi et al. 2019; Wüstemann et al. 2017). Nevertheless, a collaborative setting is essential co-creation of knowledge and building capacities needed to value and weigh the possible solutions, especially if the net gain of the trade-offs is involved (Wickenberg et al. 2021).

16.4.7 NbS Are Managed Adaptively, Based on Evidence

Adaptive management is practiced in various forms and mechanisms. These can be through a local and national assessment of ecosystems including their services; institutionalization of monitoring programs within organizations and stakeholder groups; utilizing the sense of place and belongingness of the communities; and the prioritization of local grown/traditional practices, which were known to work (Cohen-Shacham et al. 2019; Dushkova and Haase 2020; Frantzeskaki 2019; Wüstemann et al. 2017). In terms of knowledge types, it was observed that the inclusion of various knowledge types, or even their combination is essential in transforming adaptive management practices in the changing times (Egusquiza et al. 2021; Palomo et al. 2021; Wickenberg et al. 2021).

The recognition of local actors and norms in the planning and management of NbS presents vast benefits, primarily on the encouragement of trust and social learning; enhancement of wider uptake by the end users; and community empowerment for ecosystem stewardship, all contributing to the likelihood of successful NbS outcomes (Kumar et al. 2020; Sarabi et al. 2019; Schulte et al. 2021). Moreover, evidence-based practices can lead to reliable actions and recommendations, particularly on empirically grounded interventions (Albert et al. 2020).

16.4.8 NbS Are Sustainable and Mainstreamed Within an Appropriate Jurisdictional Context.

The sustainability of NbS can be fulfilled through various measures, such as through the recognition of characteristics of the urban and/or rural morphology; knowledge generation, and sharing from multiple actors (e.g., citizens, businesses, NGOs, scientists, social innovation networks); the use of frameworks to evaluate the processes and outcomes of transformative change, such as in the case of community engagement and management practices (Dushkova and Haase 2020; Emilsson and Ode Sang 2017; Palomo et al. 2021). Additionally, the importance of policy coordination between sectors, and the alignment of standards and laws among different levels (Schulte et al. 2021).

Employing sustainability approaches for NbS in appropriate contexts leads to multifold benefits, some of which include the increased potential to scale up solutions for local/international partnerships and financial support; the promotion of broader programmatic principles and policy support system; and the long-term protection of vulnerable environments and improved resiliency (Antuña-Rozado et al. 2019; Cohen-Shacham et al. 2019; Cohn et al. 2021). Furthermore, the continuous monitoring of NbS initiatives can also strengthen the co-benefits felt over time, particularly for the affected stakeholders (Solheim et al. 2021).

16.5 Discussion

The result of the systematic review can be summarized into two parts: the issues faced during NbS implementation and subsequent recommendations.

16.5.1 Issues and Challenges in the Implementation of NbS Initiatives

Participation. The low level of participation among different stakeholders, especially among the implementers and beneficiaries, along with the lack of constructive collaboration, can hinder the success of NbS implementation (Mekala and Hatton MacDonald 2018). In some cases, the raw data collected to develop ecological indicators are often not shared publicly, adding to the difficulties faced by the stakeholders in informing the development of NbS. Moreover, institutional fragmentation, also known as sectoral silos, is another barrier faced. This pertains to the split among responsibilities of various departments as they usually work in terms of their vision, procedures, and legal frameworks, and utilize their sectoral language, limiting the incorporation of novelty for NbS planning and management, and the confusion on the distribution of roles and responsibilities (Llosa et al. 2024; Sarabi et al. 2019).

In small, rural municipalities, limited personnel results in individuals handling multiple responsibilities (Solheim et al. 2021). This can lead to lesser co-creation and co-design approaches with stakeholders, leading to skepticism. Changes in the administration also matter in this context, as NbS projects should stand independently from political platforms.

Funding. The lack of financing mechanisms is a significant obstacle to NbS implementation. While the co-benefits of NbS are usually realized in the long term, available funding sources are limited and usually in the short term (Agaton et al. 2024; Antuña-Rozado et al. 2019; Cohen-Shacham et al. 2019). In addition, reliance on local government resources alone puts pressure on the allocation of expenditures, as these are usually limited (Guila et al. 2024; Sarabi et al. 2019). Thus, exploration of private investments is encouraged for additional fiscal resources. Also, the need for long-term coordination, collaboration, and knowledge integration adds to the challenge of procuring funding schemes; therefore, understanding the different stakeholders' benefits from NbS including the business models behind it can contribute to the viability of funding opportunities (Wickenberg et al. 2021).

Laws and Policies. Pertinent laws and policies can also stand as a barrier in the NbS planning process. For instance, ineffective enforcement can lead to the slow progress of implementing PPAs, as well as the inability to impose relevant policies which are usually contingent on technical capacity and labor availability (Schulte et al. 2021). Eventually, this may result in overly complex regulations, which tend to increase the costs of the proposals and may render them to be technically infeasible.

16.5.2 Policy Recommendations on the Implementation of NbS Initiatives

Participatory Planning. Enabling a participatory and collaborative governance approach is a central component of the success of NbS initiatives (Palomo et al. 2021). Co-creation and co-design with various stakeholders, particularly with implementers and community members, motivate the formulation of appealing and socially acceptable NbS, wherein citizens are most likely to appreciate and contribute to their management (Frantzeskaki 2019). These collaborative efforts are advised to be seen as a continuum, starting from the design to the implementation stages, up until the operation and monitoring and evaluation stages. Moreover, when local citizens are involved or even lead in the process, it tends to be more sustainable as changes in the political sphere become disregarded (Antuña-Rozado et al. 2019; Devanadera et al. 2024).

Coordination. Direct coordination between and among stakeholders, such as project managers and community beneficiaries, including public and private sectors, is crucial for the creation of integrated approaches between existing frameworks to widen the range of the social and environmental benefits offered by NbS (Cohen-Shacham et al. 2019). Also, the value of public–private partnerships is essential, especially in terms of the development of co-financing strategies for implementation (Wickenberg et al. 2021).

Documentation. NbS should be designed in a manner that the lessons for its effectiveness can easily be gathered and replicated in other locations (Frantzeskaki 2019). Although NbS is usually contextual, generally, it can contribute to more resilient ecosystems and livable communities, especially when critical factors for its viability are highlighted. Thus, it is also essential for the communities to learn from others and to build their capacities, especially in terms of effective governance practices.

Monitoring and Evaluation. It was observed that NbS still lacks a fully developed monitoring and evaluation mechanism, particularly for the social benefits it offers (Sarabi et al. 2019). This is vital as effective standardized evaluation systems act as an enabler to aid in facilitating increased NbS uptake. Nonetheless, at the local level, simple monitoring and evaluation approaches can be utilized that fit the context and technical capacities of stakeholders.

Experimentation. Through a controlled environment, experimentation can offer stakeholders the chance to collaborate and discuss possible innovative measures shifting the passive experience to a participatory one (Sarabi et al. 2019). This can also aid in the careful transition to formal planning especially when the NbS initiatives are relatively new and unique (Wickenberg et al. 2021). Furthermore, the presence of trade-offs might be detected earlier; thus, relevant actions and strategies can be initially planned (Cohn et al. 2021).

16.6 Conclusion

Nature-based Solutions are projected to provide innovative and sustainable solutions to the challenges faced by society and the environment. Despite their benefits, its implementation is challenged by several factors. While the literature provided several overviews of these barriers and drivers, this chapter presented a systematic review of NbS implementation using the IUCN Global Standard.

Results from the systematic review revealed that the 5th criterion (inclusive, transparent, and empowering governance process) was the most prevalent theme, followed by the 3rd criterion (net benefit to biodiversity and ecosystem integrity) and the 2nd criterion (NbS design informed by scale). Meanwhile, criterion 1 (addressing societal challenges), criterion 8 (sustainable and mainstreamed in appropriate contexts), and criterion 7 (equitably balanced trade-offs) were the least prevalent. Among the main issues and challenges identified include community participation, funding sources, supporting laws, and implementation policies. These findings provide recommendations for the successful implementation of NbS initiatives including a participatory planning process, coordination between and among stakeholders, transparency and documentation, flexible operation and implementation, and establishment of monitoring and evaluation mechanisms.

This chapter was limited to the utilized databases including Scopus, Web of Science, and manual search using Google Scholar. Future studies might consider other publications not included in the study, beyond the relevant keywords employed from the search string, gray literature, and recent studies (2022–2024). Furthermore, the reviewed literature has limited studies covering all the IUCN criteria. Future studies and NbS initiatives should consider these criteria to promote consistency, transparency, and accountability in the planning and implementation of NbS initiatives at local, national, and global scales and maximize their potential benefits for both humans and the environment while minimizing potential risks and trade-offs.

References

Agaton CB (2024) Real options analysis of constructed wetlands as nature-based solutions to wastewater treatment under multiple uncertainties: A case study in the Philippines. Sustainability 12(22):9797. https://doi.org/10.3390/su16229797

Agaton CB, Guila PMC (2024) Success factors and challenges: implications of real options valuation of constructed wetlands as nature-based solutions for wastewater treatment. Resources 13(1):11. https://doi.org/10.3390/resources13010011

Agaton CB, Guila PMC, Rodriguez ADH (2024) Economic analysis of NbS for wastewater treatment under uncertainties. In: Jegatheesan V, Velasco P, Pachova N (eds) Water treatment in urban environments: A guide for the implementation and scaling of nature-based solutions. Springer Nature. https://doi.org/10.1007/978-3-031-49282-2_4

Albert C, Brillinger M, Guerrero P, Gottwald S, Henze J, Schmidt S, Ott E, Schröter B (2020) Planning nature-based solutions: Principles, steps, and insights. Ambio 50(8):1446–1461. https://doi.org/10.1007/s13280-020-01365-1

Albert C, Hack J, Schmidt S, Schröter B (2021) Planning and governing nature-based solutions in river landscapes: Concepts, cases, and insights. Ambio 50(8):1405–1413. https://doi.org/10.1007/s13280-021-01569-z

Antuña-Rozado C, Herzog CP, Freitas T, Cagnin C, Wiedman G (2019) Nature Based Solutions (NBS) for sustainable and resilient cities: experiences from Europe and Brazil. IOP Conf Ser: Earth Environ Sci, 297(1). https://doi.org/10.1088/1755-1315/297/1/012001

Cohen-Shacham E, Andrade A, Dalton J, Dudley N, Jones M, Kumar C, Maginnis S, Maynard S, Nelson CR, Renaud FG, Welling R, Walters G (2019) Core principles for successfully implementing and upscaling Nature-based Solutions. Environ Sci Policy 98:20–29. https://doi.org/10.1016/j.envsci.2019.04.014

Cohen-Shacham E, Walters G, Janzen C, Maginnis S (2016) Nature-based solutions to address global societal challenges. IUCN. https://doi.org/10.2305/IUCN.CH.2016.13.en

Cohn JL, Copp Franz S, Mandel RH, Nack CC, Brainard AS, Eallonardo A, Magar V (2021) Strategies to work towards long-term sustainability and resiliency of nature-based solutions in coastal environments: A review and case studies. Integr Environ Assess Manag 18(1):123–134. https://doi.org/10.1002/ieam.4484

Devanadera MCE, Estorba DS, Guerrero KM, Lecciones A (2024) Social acceptability assessment for nature-based solution for wastewater treatment. In: Jegatheesan V, Velasco P, Pachova N (eds) Water treatment in urban environments: A guide for the implementation and scaling of nature-based solutions. Springer Nature. https://doi.org/10.1007/978-3-031-49282-2_5

Dushkova D, Haase D (2020) Not simply green: nature-based solutions as a concept and practical approach for sustainability studies and planning agendas in cities. Land, 9(1). https://doi.org/10.3390/land9010019

Egusquiza A, Arana-Bollar M, Sopelana A, Babí Almenar J (2021) Conceptual and operational integration of governance, financing, and business models for urban nature-based solutions. Sustain, 13(21). https://doi.org/10.3390/su132111931

Emilsson T, Ode Sang Å (2017) Impacts of climate change on urban areas and nature-based solutions for adaptation. In: Nature-based solutions to climate change adaptation in urban areas, pp 15–27. https://doi.org/10.1007/978-3-319-56091-5_2

Frantzeskaki N (2019) Seven lessons for planning nature-based solutions in cities. Environ Sci Policy 93:101–111. https://doi.org/10.1016/j.envsci.2018.12.033

Guila PMC, Agaton CB, Rivera RRB, Abucay ER (2024) Household willingness to pay for constructed wetlands as nature-based solutions for wastewater treatment in Bayawan City, Philippines. J Hum Ecol Sustain, 2 (1), 5. https://doi.org/10.56237/jhes23018

Haddaway NR, Macura B, Whaley P, Pullin AS (2018) ROSES reporting standards for systematic evidence syntheses: pro forma, flow-diagram and descriptive summary of the plan and conduct of environmental systematic reviews and systematic maps. Environ Evid, 7(1). https://doi.org/10.1186/s13750-018-0121-7

IUCN (2020) Global Standard for nature-based solutions: a user-friendly framework for the verification, design and scaling up of NbS (First ed.). International Union for Conservation of Nature and Natural Resources. https://doi.org/10.2305/IUCN.CH.2020.08.en

Kolokotsa D, Lilli AA, Lilli MA, Nikolaidis NP (2020) On the impact of nature-based solutions on citizens' health & well being. Energy Build, 229. https://doi.org/10.1016/j.enbuild.2020.110527

Kumar P, Debele SE, Sahani J, Aragão L, Barisani F, Basu B, Bucchignani E, Charizopoulos N, Di Sabatino S, Domeneghetti A, Edo AS, Finér L, Gallotti G, Juch S, Leo LS, Loupis M, Mickovski SB, Panga D, Pavlova I, Zieher T (2020) Towards an operationalisation of nature-based solutions for natural hazards. Sci Total Environ, 731. https://doi.org/10.1016/j.scitotenv.2020.138855

Lafortezza R, Chen J, van den Bosch CK, Randrup TB (2018) Nature-based solutions for resilient landscapes and cities. Environ Res 165:431–441. https://doi.org/10.1016/j.envres.2017.11.038

Liu H-Y, Jay M, Chen X (2021) The role of nature-based solutions for improving environmental quality. Health Well-Being Sustain, 13(19). https://doi.org/10.3390/su131910950

Llosa CDL, Reyes Jr EM, Agaton CB, Cordero-Bailey KSA (2024) Thematic and multicriteria analysis of the readiness, factors, and strategies for successful implementation of nature-based solutions initiatives in Victoria, Laguna. J Hum Ecol Sustain, 2(1),9. https://doi.org/10.56237/jhes23017

Mekala GD, Hatton MacDonald D (2018) Lost in transactions: Analysing the institutional arrangements underpinning urban green infrastructure. Ecol Econ 147:399–409. https://doi.org/10.1016/j.ecolecon.2018.01.028

Mendonça R, Roebeling P, Fidélis T, Saraiva M (2021) Policy instruments to encourage the adoption of nature-based solutions in urban landscapes. Resources 10(8):81. https://doi.org/10.3390/resources10080081

Morita K, Matsumoto KI (2021) Governance challenges for implementing nature-based solutions in the Asian Region. Polit Gov, 9(4):102–113. https://doi.org/10.17645/pag.v9i4.4420

Nehren U, Arce-Mojica T, Barrett AC, Cueto J, Doswald N, Janzen S, Lange W, Vargas AO, Pirazan-Palomar L, Renaud FG, Sandholz S, Sebesvari Z, Sudmeier-Rieux K, Walz Y (2023) Towards a typology of nature-based solutions for disaster risk reduction. Nat-Based Solut, 3. https://doi.org/10.1016/j.nbsj.2023.100057

Nelson DR, Bledsoe BP, Ferreira S, Nibbelink NP (2020) Challenges to realizing the potential of nature-based solutions. Curr Opin Environ Sustain 45:49–55. https://doi.org/10.1016/j.cosust.2020.09.001

Palomo I, Locatelli B, Otero I, Colloff M, Crouzat E, Cuni-Sanchez A, Gómez-Baggethun E, González-García A, Grêt-Regamey A, Jiménez-Aceituno A, Martín-López B, Pascual U, Zafra-Calvo N, Bruley E, Fischborn M, Metz R, Lavorel S (2021) Assessing nature-based solutions for transformative change. One Earth 4(5):730–741. https://doi.org/10.1016/j.oneear.2021.04.013

Pauleit S, Zölch T, Hansen R, Randrup TB, Konijnendijk van den Bosch C (2017) Nature-based solutions and climate change—four shades of green. In: Nature-based solutions to climate change adaptation in urban areas, pp 29–49. https://doi.org/10.1007/978-3-319-56091-5_3

Ruangpan L, Vojinovic Z, Plavšić J, Doong D-J, Bahlmann T, Alves A, Tseng L-H, Randelović A, Todorović A, Kocic Z, Beljinac V, Wu M-H, Lo W-C, Perez-Lapeña B, Franca MJ (2020) Incorporating stakeholders' preferences into a multi-criteria framework for planning large-scale Nature-Based Solutions. Ambio 50(8):1514–1531. https://doi.org/10.1007/s13280-020-01419-4

Samantela SS, Maquiling KSM (2024) Examining institutional challenges of land value capture: The case of implementing land-based Taxes in the Philippines. J Hum Ecol Sustain, 2(1). https://doi.org/10.56237/jhes23021

Sarabi, Han, Romme, Vries, Wendling (2019) Key enablers of and barriers to the uptake and implementation of nature-based solutions in urban settings: A review. Resources, 8(3). https://doi.org/10.3390/resources8030121

Schulte I, Eggers J, Nielsen JØ, Fuss S (2021) What influences the implementation of natural climate solutions? A systematic map and review of the evidence. Environ Res Lett, 17(1). https://doi.org/10.1088/1748-9326/ac4071

Seddon N (2022) Harnessing the potential of nature-based solutions for mitigating and adapting to climate change. Science 376(6600):1410–1416. https://doi.org/10.1126/science.abn9668

Solheim A, Capobianco V, Oen A, Kalsnes B, Wullf-Knutsen T, Olsen M, Del Seppia N, Arauzo I, Garcia Balaguer E, Strout JM (2021) Implementing nature-based solutions in rural landscapes: Barriers experienced in the PHUSICOS project. Sustain, 13(3). https://doi.org/10.3390/su13031461

Sowińska-Świerkosz B, García J (2022) What are nature-based solutions (NBS)? Setting core ideas for concept clarification. Nat-Based Solut 2. https://doi.org/10.1016/j.nbsj.2022.100009

Toxopeus H, Polzin F (2021) Reviewing financing barriers and strategies for urban nature-based solutions. J Environ Manag, 289. https://doi.org/10.1016/j.jenvman.2021.112371

Vasiliev D, Greenwood S (2022) Making green pledges support biodiversity: Nature-based solution design can be informed by landscape ecology principles. Land Use Policy, 117. https://doi.org/10.1016/j.landusepol.2022.106129

Wickenberg B, McCormick K, Olsson JA (2021) Advancing the implementation of nature-based solutions in cities: A review of frameworks. Environ Sci Policy 125:44–53. https://doi.org/10.1016/j.envsci.2021.08.016

Wüstemann H, Bonn A, Albert C, Bertram C, Biber-Freudenberger L, Dehnhardt A, Döring R, Elsasser P, Hartje V, Mehl D, Kantelhardt J, Rehdanz K, Schaller L, Scholz M, Thrän D, Witing F, Hansjürgens B (2017) Synergies and trade-offs between nature conservation and climate policy: Insights from the "Natural Capital Germany—TEEB DE" study. Ecosyst Serv 24:187–199. https://doi.org/10.1016/j.ecoser.2017.02.008

Zingraff-Hamed A, Hüesker F, Albert C, Brillinger M, Huang J, Lupp G, Scheuer S, Schlätel M, Schröter B (2020) Governance models for nature-based solutions: Seventeen cases from Germany. Ambio 50(8):1610–1627. https://doi.org/10.1007/s13280-020-01412-x

The manufacturer's authorised representative in the EU is Springer Nature Customer Service Centre GmbH, Europaplatz 3, 69115 Heidelberg, Germany. If you have any concerns regarding our products, please contact ProductSafety@springernature.com

Printed and bound by CPI Group (UK) Ltd, Croydon, CR0 4YY
01/10/2026
02225913-0015